Blockchain for Smart Cities

Blockchain for Smart Cities

Edited by

Saravanan Krishnan
Assistant Professor, Department of Computer Science and Engineering, Anna University Regional Campus, Tirunelveli, India

Valentina Emilia Balas
Full Professor, Department of Automatics and Applied Software, Faculty of Engineering, "Aurel Vlaicu" University of Arad, Arad, Romania

E. Golden Julie
Department of CSE, Regional campus, Anna University, Tirunelveli, India

Y. Harold Robinson
Post Doctoral Fellow, School of Information Technology and Engineering, Vellore Institute of Technology, Vellore, India

Raghvendra Kumar
Associate Professor, Department of Computer Science and Engineering, GIET University, Gunupur, India

ELSEVIER

Elsevier
Radarweg 29, PO Box 211, 1000 AE Amsterdam, Netherlands
The Boulevard, Langford Lane, Kidlington, Oxford OX5 1GB, United Kingdom
50 Hampshire Street, 5th Floor, Cambridge, MA 02139, United States

Library of Congress Cataloging-in-Publication Data
A catalog record for this book is available from the Library of Congress

British Library Cataloguing-in-Publication Data
A catalogue record for this book is available from the British Library

ISBN: 978-0-12-824446-3

Publisher: Joe Hayton
Acquisitions Editor: Brian Romer
Editorial Project Manager: Sara Valentino
Production Project Manager: Maria Bernard
Cover Designer: Christian J. Bilbow

Typeset by TNQ Technologies

Contents

5. Blockchain-based energy-efficient smart green city in IoT environments

K. Suresh Kumar, A.S. Radha Mani, S. Sundaresan, T. Ananth Kumar and Y. Harold Robinson

6. Applications of blockchain in smart cities: detecting fake documents from land records using blockchain technology

R. Dinesh Kumar and V.N.S. Manaswini

Contributors

A. Amuthan, Department of CSE, Pondicherry Engineering College, Puducherry, India

T. Ananth Kumar, Department of CSE, IFET College of Engineering, Villupuram, Tamil Nadu, India

V. Ashok, Department of CSE, Rajalakshmi Institute of Technology, Chennai, Tamil Nadu, India

J.V. Bibal Benifa, Department of Computer Science and Engineering, Indian Institute of Information Technology, Kottayam, Kerala, India

Usharani Chelladurai, University College of Engineering BIT Campus, Anna University Tiruchirappalli, Tiruchirappalli, Tamil Nadu, India

R. Dinesh Kumar, Siddhartha Institute of Technology and Sciences, Hyderabad, Andhra Pradesh, India

Lakshmi Prabha Ganesan, Department of Computer Applications, Sri Sarada College for Women, Tirunelveli, Tamil Nadu, India

Amol Goje, Society for Data Science, Pune, India

K. Gokulakrishnan, Anna University-Regional Campus, Tirunelveli, Tamil Nadu, India

E. Golden Julie, Anna University-Regional Campus, Tirunelveli, Tamil Nadu, India

Y. Harold Robinson, School of Information Technology and Engineering, Vellore Institute of Technology, Vellore, Tamil Nadu, India

S. Jeba Prasanna Idas, Department of CSE, College of Engineering and Technology, International University of Technology Twintech, Sanaa, Yemen

A. Jaya Kumar, Department of CSE, IFET College of Engineering, Villupuram, Tamil Nadu, India

J. Jesu Vedha Nayahi, Anna University-Regional Campus, Tirunelveli, Tamil Nadu, India

D. Jeyabharathi, Department of Information Technology, Sri Krishna College of Technology, Coimbatore, Tamil Nadu, India

D. Kesavaraja, Department of CSE, Dr. Sivanthi Aditanar College of Engineering, Tiruchendur, Tamil Nadu, India

Saravanan Krishnan, Department of Computer Science and Engineering, Anna University Regional Campus, Tirunelveli, Tamil Nadu, India

Sakthi Kumaresh, M.O. P. Vaishnav College for Women, Chennai, Tamil Nadu, India

V.N.S. Manaswini, Siddhartha Institute of Technology and Sciences, Hyderabad, Andhra Pradesh, India

A.S. Radha Mani, V.V. College of Engineering, Tirunelveli, Tamil Nadu, India

J. Nulyn Punitha Markavathi, Department of CSE, PSN College of Engineering, Tirunelveli, Tamil Nadu, India

C.M. Naga Sudha, Anna University-MIT Campus, Chrompet, Tamil Nadu, India

R. Nishanth, Cochin University College of Engineering Kuttanad, CUSAT, Kuttanad, Kerala

V. Padmapriya, Innovation Cluster Lab, Department of Computer Applications, B. M. S. College of Engineering, Bengaluru, Karnataka, India

Sneha Pandey, Department of Information Technology, Netaji Subhas University of Technology, New Delhi, India

Seethalakshmi Pandian, University College of Engineering BIT Campus, Anna University Tiruchirappalli, Tiruchirappalli, Tamil Nadu, India

S. Porkodi, Dr. Sivanthi Aditanar College of Engineering, Tiruchendur, Tamil Nadu, India

N.R. Rajalakshmi, Computer Science and Engineering, Vel Tech Rangarajan Dr. Sagunthala R & D Institute of Science and Technology, Chennai, Tamil Nadu, India

T. Sangeetha, Department of Information Technology, Sri Krishna College of Technology, Coimbatore, Tamil Nadu, India

V.V. Satyanarayana Tallapragada, Tallapragada Department of ECE, Sree Vidyanikethan Engineering College, Tirupati, Andhra Pradesh, India

Rishabh Setiya, Department of Information Technology, Netaji Subhas University of Technology, New Delhi, India

Madhavi Shamkuwar, Zeal Institute of Business Administration, Computer Application and Research, Savitribai Phule Pune University, Ganesh Khind, Pune, Maharashtra, India

Deepak Kumar Sharma, Department of Information Technology, Netaji Subhas University of Technology, New Delhi, India

Neha Sharma, Analytics and Insights, Tata Consultancy Services, Pune, India

Ayush Kumar Singh, Department of Information Technology, Netaji Subhas University of Technology, New Delhi, India

Inderjit Singh, Vara Technology Pvt Ltd, Pune, India

D.N. Sujatha, Innovation Cluster Lab, Department of Computer Applications, B. M. S. College of Engineering, Bengaluru, Karnataka, India

S. Sundaresan, Department of ECE, SRM TRP Engineering College, Trichy, Tamil Nadu, India

K. Suresh Kumar, Department of CSE, IFET College of Engineering, Villupuram, Tamil Nadu, India

Angelin Merling Thava, Department of Information Technology, Jayaraj Annapackiam CSI College of Engineering, Nazareth, Tamil Nadu, India

V. Usha, Computer Science and Engineering, Vel Tech Rangarajan Dr. Sagunthala R & D Institute of Science and Technology, Chennai, Tamil Nadu, India

G. Venifa Mini, Department of Computer Science and Engineering, Noorul Islam Centre for Higher Education, Nagercoil, Tamil Nadu, India

Biographies

Dr. Saravanan Krishnan is a Senior Assistant Professor in the Department of Computer Science and Engineering at Anna University, Regional Campus, Tirunelveli, Tamilnadu. He did his ME in Software Engineering and PhD in Computer Science Engineering. His research interests include cloud computing, software engineering, Internet of Things, and smart cities. He has published research papers in 14 international conferences and 27 international journals. He has also written 9 book chapters and 5 books with international publishers. He has done consultancy works for Municipal Corporation and Smart City schemes. He is an active researcher and academician. Also, he is a reviewer for many reputed journals in Elsevier, IEEE, etc.

Valentina E. Balas is a Professor in the Department of Automatics and Applied Software at the Faculty of Engineering, "Aurel Vlaicu" University of Arad, Romania. She holds a PhD Cum Laude in Applied Electronics and Telecommunications from Polytechnic University of Timisoara. Dr. Balas is author of more than 350 research papers. Her research interests are in intelligent systems, fuzzy control, soft computing, smart sensors, information fusion, modeling, and simulation. She is the Editor-in-Chief of IJAIP and IJCSysE journals in Inderscience, member of Editorial Board of several national and international journals, and evaluator expert for national and international projects and PhD theses. Dr. Balas is the Director of Intelligent Systems Research Centre and Director of the Department of International Relations, Programs, and Projects in Aurel Vlaicu University of Arad. She is recipient of the "Tudor Tanasescu" Prize from the Romanian Academy for contributions in the field of soft computing methods (2019).

E. Golden Julie is a Senior Assistant Professor in the Department of Computer Science and Engineering, Anna University, Regional Campus, Tirunelveli. She has received PhD in Information and Communication Engineering from Anna University, Chennai, in 2017. She has more than 13 years of experience in teaching. She has published more than 35 papers in various international journals and presented more than 20 papers in both national and international conferences. She has written 10 book chapters in books published by Springer and IGI Global. Her research areas include wireless sensor ad-hoc networks, soft computing, block chain, fuzzy logic, neural network, soft computing techniques, clustering, and IoT. She is also an active life-time member of the Indian Society of Technical Education.

Dr. Y. Harold Robinson is currently working in the School of Information Technology and Engineering, Vellore Institute of Technology, Vellore. He received his PhD in Information and Communication Engineering from Anna University, Chennai, in 2016. He has more than 15 years of experience in teaching. He has published more than 100 papers in various international journals and presented more than 45 papers in both national and international conferences. He is acting as an editor for many research books. His research areas include wireless sensor networks, ad-hoc networks, soft computing, blockchain, IoT, and image processing.

Dr. Raghvendra Kumar is an Associate Professor in Computer Science and Engineering Department at GIET University, India. He received BTech, MTech, and PhD in Computer Science and Engineering, India, and Postdoc Fellow from Institute of Information Technology, Virtual Reality, and Multimedia, Vietnam. He has published number of research papers in international journal and conferences. He also published 13 chapters in edited book published by IGI Global, Springer, and Elsevier. His research areas are computer networks, data mining, cloud computing and secure multiparty computations, theory of computer science, and design of algorithms. He has authored and edited 23 computer science books on IoT, data mining, and biomedical engineering.

Chapter 1

Smart cities with blockchain technology: a comprehensive survey

Saravanan Krishnan[1], **Lakshmi Prabha Ganesan**[2]
[1]*Department of Computer Science and Engineering, Anna University Regional Campus, Tirunelveli, Tamil Nadu, India;* [2]*Department of Computer Applications, Sri Sarada College for Women, Tirunelveli, Tamil Nadu, India*

Chapter outline

1. Introduction

Because blockchain has a disruptive and decentralized nature, it has attracted research communities and has many applications, such as smart contracts, cryptocurrency, secure payments, securing healthcare systems, and so forth. Due to its decentralized nature blockchain systems have no third-party authority to be trusted on the transactions; instead, every node in the system should trust each other for successful implementation of the systems. To attain consensus among different types of nodes in a system there are several common consensus mechanisms proposed.

Blockchain for Smart Cities. https://doi.org/10.1016/B978-0-12-824446-3.00016-8

To maintain equality and fairness among the nodes, the consensus models have specific objectives, and each block added to a transaction must satisfy a set of rules. The common objectives include coming to an agreement, collaboration among the nodes, cooperation, equal rights for all the nodes in the system, and active participation of every person in the system. Fig. 1.1 shows the workflow model of blockchain.

Consensus mechanisms will only work if all the elements in the system work in harmony. If any of the nodes malfunction, then the whole system will fail. This will lead to the Byzantine Generals' Problem. The main problem with Byzantine is to reach an agreement. If even a single fault occurs, nodes cannot come to an agreement. In this paper the different consensus mechanisms used in different applications in a smart environment have been discussed, particularly proof of work (PoW), proof of stake (PoS), and their variations, different Byzantine fault tolerance (BFT) mechanisms and their applications in smart environments. The consensus algorithms are reviewed based on three features, namely their real-time implementation (Castro and Liskov, 1999), security, and fault tolerance. In the PoW systems, there are mainly two features that have contributed to the wide popularity of this consensus protocol: (1) it is hard to find a solution for the mathematical problem; and (2) it is easy to verify the correctness of that solution. Perhaps the features there are also some shortcomings of this protocols they are 51% of risk, resource consumption, time consuming, and the time elapsed during the confirmation of the transaction. BFT is the property of a system that is able to continue operating even if some of the nodes fail or act maliciously. BFT is a crucial part of an effective blockchain and there are multiple ways in which it can be implemented. Deciding which approach to take requires weighing the nature and priorities of the community associated with the blockchain an

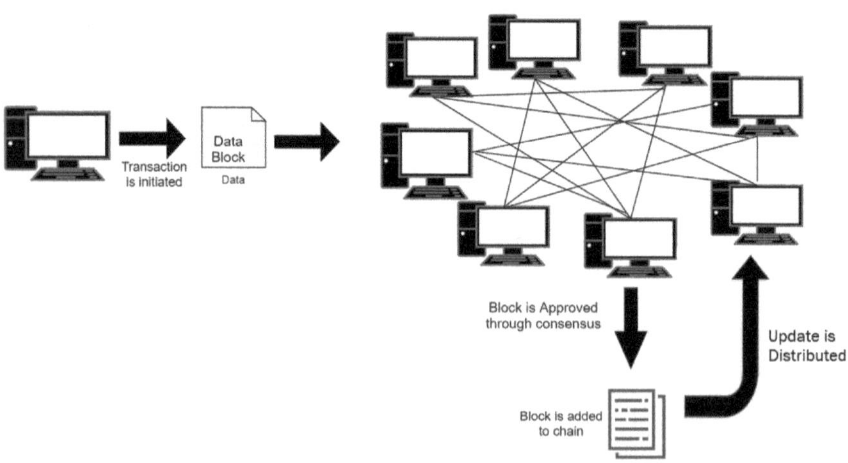

FIGURE 1.1 Shows the process of updating blockchain.

organization wants to build (Sun and Duan, 2014). The variants of BFTs have higher efficient when combined with other consensus methods. To compare the consensus mechanisms factors such as preserving node identity, vulnerability and mining processes are taken into account.

2. Related work

A Smart city environment completely relies on data to meet the needs of a growing urban population. Secured wireless connectivity and the Internet of Things (IoT) technology are the factors mandatory for real-time deployment of smart cities. To achieve this, blockchain is integrated with the IoT in smart city environment (Botello et al., 2020). The key elements of blockchain technology (BCT) include distributed ledger, immutable transactional records, and smart contracts, which makes the smart environment more reliable and secured. Blockchain is a sequence of blocks where blocks have a complete record of all the transactions. Fig. 1.2 shows the structure of a blockchain. A block is the basic fundamental unit of a blockchain, which has two divisions, namely, (1) block header and (2) block body (Chandel et al., 2019). The header part contains the information about the block version, timestamp, nonce, hash of its parent block, Merkle tree root hash, where the information about all the other blocks in the blockchain has been located, and nBits. The block body contains the actual data and the hash value. The workflow of a blockchain framework has been shown in Fig. 1.1 where a user initiates the transaction request and the transaction data is packaged into a block and sent through the network to all its members. Then the transaction is validated through consensus mechanism and approved by the network. Finally the block is added to the chain through its hash value in the Merkle tree.

The correctness, latency, and security of a blockchain integrated model is achieved through consensus mechanisms and there are several algorithms for

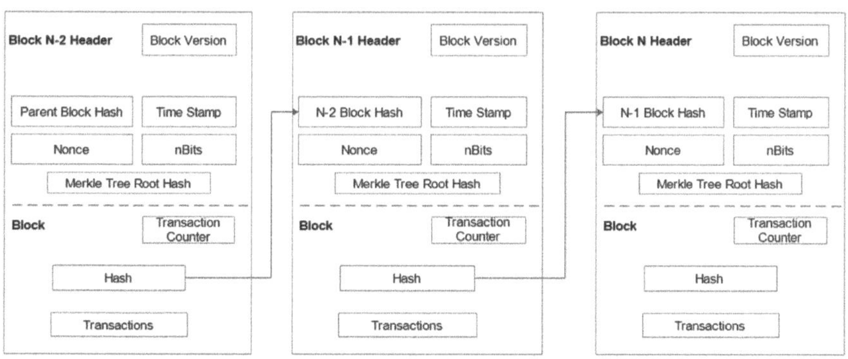

FIGURE 1.2 The structure and how the blocks are organized.

consensus have been proposed in the literature. A tamper-resistant memory-hardened PoW model has been proposed in Sharma and Park (2018) to ensure the security and privacy in the blockchain framework. Alfandi et al. (2020) has proposed an integrated IoT model with fog computing and a BFT consensus algorithm, namely, pBFT, where all the devices in the system will participate in updating blockchain and it can tolerate up to one-third of the total devices being faulty, hence the system is fault-tolerant and minimizes the computational overhead and latency. The proposed system has two types of vehicles: (1) critical vehicles for which secured data transfer is required; and (2) normal vehicles. To reduce the computational overhead, consensus is applied only for critical infrastructures, and the data is finally updated in the fog. Sharma and Park (2018) has proposed a hybrid distributed framework for meeting the architectural challenges in a smart city environment such as scalability, network bandwidth, single point of failure, privacy and security, and high latency. To reduce the single point of failure, edge nodes preprocess the raw data from IoT devices and send the data to the core network if necessary; the core network, which has miners, update the blockchain by processing and verifying the PoW, so they had stated that their work is well-suited for a smart city environment.

3. Permissioned and permissionless blockchain in the smart city environment

There are several public as well as private sector applications where the BCT can be applicable such as financial services (Bitcoin, Ethereum, etc.,), healthcare systems, public security (Cash & Bassiouni, 2018), digital identity, and so on. BCT has a decentralized nature over the data so it is not applicable for sharing the confidential data in which data security and privacy is a major concern. So a variation of blockchain known as permissioned BCT found many applications where the data confidentiality is at the top. Some degree of decentralization is gained in permissioned BCT, and data privacy, integrity, and security can be maintained for organizational purposes, enterprises, several public sectors, and healthcare systems.

3.1 Permissionless BCT

Permissionless blockchains are the first generation of blockchain. These are the distributed ledgers where the data is transparent to all the participating nodes and the governing rules can be on-chain or off-chain. It is completely decentralized. Bitcoin and Ethereum are the well-known permissionless blockchain applications. The anonymity of the nodes participating in the blockchain network is a major challenge in the permissionless BCT. However, the data once batched into block and added to the blockchain by attaining consensus among the participants in the network cannot be modified by any node in the network. Hence the data integrity is achieved.

3.2 Permissioned BCT

Permissioned BCT is run by the members of a consortium. Only preapproved entities can run the nodes that validate transaction blocks and execute smart contracts on the blockchain. It is applicable where the interaction between the participants must be secure and they do not even have to trust each other, so the consensus mechanism should be fault tolerant and have fast finality of transactions. Hence BFT consensus can be used in such systems. The Hyperledger project is one of the distributed ledgers hosted by the Linux foundation (Sukhwani et al., 2017). The nuances between the permissioned and permissionless block chain systems are

(1) **Decentralization:** Permissionless BCT must be decentralized and distributed so that no one entity can bring the network down, whereas the measure of decentralization is less in permissioned BCT when compared to the former. It is based on the structure of business relationships between the consortium members.
(2) **Transparency:** In permissionless BCT the transaction information such as how the transaction is ordered and batched into blocks is transparent to all the miners in the network, since they have cryptoeconomic incentives (Androulaki, Barger et al., 2018). But this is not the case in permissioned BCT because they do not have cryptoeconomic incentives built into them; instead they focus on how to minimize the cost, time of the information, and security in sharing the information.
(3) **Privacy or Anonymity:** In permissionless BCT, the miners and other participants in the network can stay anonymous; for example, Bitcoin and Ethereum can be considered as pseudo-anonymous. A permissioned BCT offer fine-grained access to the transactions and assets so degree of data privacy is high when compared to the former.
(4) **Governance:** There is no preexisting level of trust is needed between the participating nodes in a permissionless BCT network. All the peers participate in validating and processing of transactions through consensus. It offers both off-chain (e.g.,: Etherum, Bitcoin) and on-chain (e.g., Tezos) governance. In permissioned BCT it needs some degree of preexisting trust and the transactions are validated and processed only by those who are already recognized by the ledger.
(5) **Security:** Due to complete decentralization and distributed ledger technology (DLT) permissionless blockchain is resistant to hacking and DDoS attacks. The security of a public blockchain relies completely on its proof of work. The data privacy and scalability are the major challenges in a public blockchain, so this limits the adaptation of security token in a public blockchain. A private blockchain, on the other hand, can offer higher scalability and performance than the public blockchain but is more vulnerable to hacking and DDoS attacks.

4. Applications of blockchain technology in smart cities

4.1 Sharing economy services

Sun et al. (2016) proposed a conceptual framework for smart cities that is based on human need, organization, and technology. The service relationships among these three factors make the framework trust-free, decentralized, secured, and democratized. Decentralization is achieved by replacing central authorities with a community of peers to form a peer-to-peer network. This work focuses on how a BCT can be adopted in a smart city environment through sharing economy services and service relationships. Their work elicited the researches toward how a blockchain can contribute to sharing economy services in a smart city environment.

Gori et al., (2015) narrated the interdependence between the smart city concept and sharing economy. They have stated that the cities become even smarter if they have implemented the sharing economy services. Sharing platforms can contribute to the economic success of a city through meeting the needs of the citizen by sharing personal assets, microbusiness needs, and sharing transportation services. However a framework for sharing economy services has not been discussed in the paper. This paper narrates the difficulties in sharing economy services in terms of law and regulations and how start-up and microeconomy services can be benefited through smart sharing.

4.2 Smart vehicles

The work of Sharma, Moon, and Park (2017) proposed a vehicular network architecture using blockchain for a smart city environment. They have framed a vehicular network named Block-VN, which primarily aims at high scalability. There are two types of nodes in the proposed system, namely, controller/miner nodes and the ordinary nodes. The system is highly scalable, and the use of BCT by miner nodes makes the system reliable and the availability of networks can be enhanced through interaction between the miner nodes. A detailed schema of their proposed system is not provided so the challenges that may occur during the implementation of such a system may not be addressed.

A permissioned BCT is implemented by Michelin et al. (2016) where the vehicles, road side infrastructures (RSI) and service providers (SP) are the entities. Their proposed system is resistant to Sybil attacks as the ordinary vehicles are identified using location-based trust model. The RSI and SP are only authorized to manage the blockchain where the ordinary vehicles only store their public key hash value in the Merkle tree. This can reduce storage of unnecessary data where a large number of vehicles can participate in the proposed system of smart city environment.

4.3 Smart Healthcare Services (SHS)

The work of Solanas et al. (2014) stated smart health services (SHS) as a provision of ICT infrastructures and technology in the smart cities. They have stated that SHS is the augmentation of smart city infrastructures, mobile health, and e-health services. This paper reviewed the opportunities of their term smart health in different healthcare apps and medical services. They have also discussed about various challenges in the implementation of SHS in real-time scenarios. The following works are the extents of Solanas et al. (2014). A detailed schema of SHS ecosystem has been discussed in (Capossele et al., 2018). A descriptive model for interaction between the stakeholders in the development of SHS applications has been presented in the paper. Here, blockchain is proposed to form the core of middleware, which is placed at the top of the diverse set of assets and SHS applications. The legal, technical, and commercial challenges in implementing such a model have been discussed. The privacy, access control of the assets, trust, and data integrity are major factors to be considered while implementing such SHS apps. They have paved the way for future research directions in terms of compliance and sustainable SHS apps.

The sensitivity of medical data of the patients and healthcare systems, data privacy, and security challenges are addressed in Tripathi, Ahad, and Paiva (2020) using BCT in smart healthcare services. They have proposed a secured SHS named S2HS to provide data transparency to all the actors in the system while maintaining the privacy of the patients. The middleware consists of permissioned Hyperledger BCT. To preserve integrity of the data, encryption and digital signatures schemes have been used. The system is resistant to eavesdropping and data theft. The patient's EHR is stored in the blockchain using the hash value despite the patient's identity so the privacy of the patient is preserved while achieving the data transparency and also the mismatching of EHR is avoided.

An integration of IoT, blockchain, and machine learning is done in (Chakraborty et al., 2019) to monitor the patient's health condition continually and so the quality of living and healthcare systems in a smart city can be improved. The IoT module collects the patient's health data through sensors and wearable devices which are stored externally using the block chain technology and the machine learning module monitors the data collected and find out the anomalies if any. A permissioned BCT has been used in their proposed model to provide accurate data. A consortium model of hospital, doctors and patients using permissioned BCT and cryptographic security has been proposed (Ramani et al., 2018). They have proposed a secured model of healthcare systems in which updating blockchain requires the authentication of doctor's identity and patient's approval. The authentication mechanism is done by using elliptic curve cryptography (ECC).

5. Case studies

5.1 Smart Dubai using Blockchain

Dubai is a fast-growing city that has developed as a global hub for business, trade, and tourism over the last four decades. The Dubai smart city project aims to encourage collaboration between the public and private sectors to achieve targets in six focus areas: smart life, smart transportation, smart society, smart economy, smart governance, and smart environment (Kerr, 2016).

As smart phones become more ubiquitous and technology rapidly changes our lives, Dubai is marching toward a new era of improvement and embracing a higher quality of life. To provide a better communication, exchange of data and information and seamless connectivity for public and private sectors "Dubai Data Law" has been proposed and it consists of four data principles open data, shared data, big data, and rich data (Khan et al., 2017).

- Open data: Data published by the government or private sector to be used or exchanged with individuals or third parities openly.
- Shared data: Data to be available for sharing and reuse among entities.
- Big data: Large amounts of data to be analyzed to reveal patterns, trends, and associations.
- Rich data: Big data to be further utilized to deeper, often qualitative analyses, and extracting insights.

The Dubai Government has set up the Global blockchain Council, which is a public private initiative that brings government entities with local and international business/start-ups to foster blockchain development with test cases. Fig. 1.3 describes the blockchain strategy of Dubai Government. Dubai Multi Commodities Center is engaged in a test case related to the authentication and the transfer of Kimberley certificates. The Emirates Integrated Telecommunication Corporation (Du) is piloting a use case for health records to share data records between SP (Michelin et al., 2016).

5.1.1 Use cases of Blockchain strategy in Dubai Government Services

(1) Identity Management and Record Keeping: Emirates ID Authority can manage electronic identity of UAE citizens, for record keeping and the use of this identity for any other government services.

(2) Value Registry and Assets Management: Dubai Land Department (DLD) has achieved another technical milestone by becoming the world's first government entity to adopt BCT. Following the launch of the Dubai blockchain Strategy in October 2016, which aims to make Dubai the first government in the world to apply all transactions through this network by 2020, DLD created a smart and secure database that records all real estate contracts. This strategy is in the stage of infancy and it is likely to be implemented in the sectors of FinTech, banking, and real estate to increase the efficiency of government's operations.

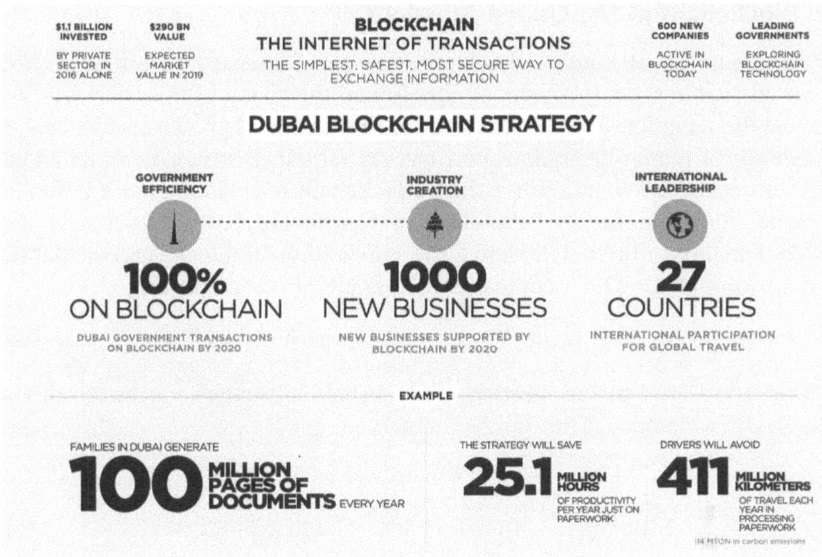

FIGURE 1.3 Future directions of smart Dubai through the implementation blockchain technology. *From https://schoolofdisruption.com/exponential-technologies/blockchain/.*

(3) Voting System: Voting system with BCT can be used by the government for Federal National Council elections.

(4) Healthcare Record keeping: Dubai Health Authority may use BCT to securely maintain health record data and enable the sharing of data.

(5) Smart Cities and IoT: The Dubai Government has recently lunched Dubai Internet of Things strategy which aims at building the world's most advanced IoT ecosystem for Dubai smart city to improve people's lives. The integration of blockchain with IoT has many use cases as it allows peer-to-peer communication between IoT devices, and hence it enables the peer-to-peer market. Moreover, the blockchain enables the tracking of assets throughout the supply chain using IoT devices without the manual intervention from users (Christidis & Devetsikiotis, 2016).

(6) Smart Tourism ObjectTech is working with the Dubai Immigration and Visas Department to develop digital passports that can potentially eliminate manual checks at Dubai International airport (Nam, Dutt, Chathoth, & Khan, 2021). Their proposed system combines biometric verification and BCT, and will use a "preapproved and entirely digitized passport" to authorize passengers' entrance into the country. The system will further verify individuals through a three-dimensional scan via a short tunnel as they walk from the aircraft to claim their baggage. Using BCT, the digitized passport will incorporate a feature called "self-sovereign identity" for privacy protection, which it claims allows passengers to control which parties can view their passport information.

5.2 Blockchain in the United Kingdom

The combination of continued urbanization and increased requirements for enhanced citizen participation continues to fuel the vision of the smart city. To promote the adoption of BCT across the government and private sectors across the country a nonprofit organization namely British Blockchain Association has been set up (DKV, nd). The British Blockchain Association has a vision to accelerate progress and collaboration of blockchain based projects in the United Kingdom. Table 1.1 shows the application of blockchain in public services through GovTech companies of the UK government (Fig. 1.4).

TABLE 1.1 **Shows the application of blockchain in public services through GovTech companies of UK government.**

Application	Partners/ platform	Type of block chain/technology used	Description
Welfare spending	GovCoin	Public	Claimants are paid in cryptocurrency rather than Sterling. This is an important way of gathering more data about both how the funds are spent but also about the individual doing the spending. The hope is that this data will allow for the creation of better targeted goods and services.
UK energy sector	Centrica electron EDF energy	Private, developed by LO3 energy (US) ethereum and IPFS decentralized web projects collaborates with Electron's platform. Blockchain-led peer-to-peer technology	Testing energy trading transactions including multiparty peer-to-peer unlike traditional trading platforms that match buyers and sellers on a one-to-one basis, Electron's platform leverages blockchain technology to allow multiple parties to coordinate and share the value of a single consumer's action. Electron's platform has also been awarded 2017 tech pioneer status by the world economic forum (WEF). One of the first energy companies to be involved in a government project implementing blockchain technologies.

TABLE 1.1 Shows the application of blockchain in public services through GovTech companies of UK government.—cont'd

Application	Partners/ platform	Type of block chain/technology used	Description
Electoral systems	ChainVote	Open vote network, along with DRE-i and DRE-ip encryptions.	A transparent voting system, which is based in the UK and the US.
Charity	English Heritage and GiftCoin platform	Ethereum, BitCoin	Used to track donor payments from the time they have been gifted to when the funds are actually been spent.

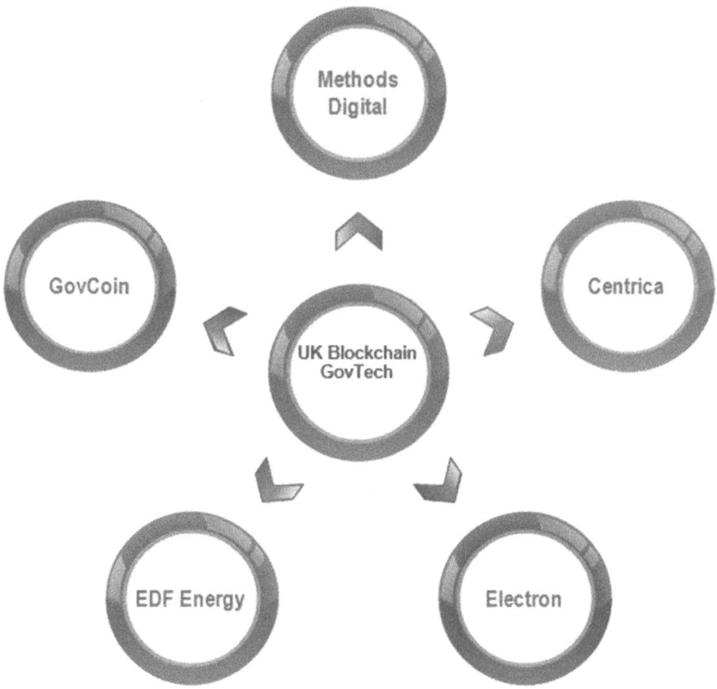

FIGURE 1.4 GovTech companies that are identified for public services.

5.2.1 UK projects related to Blockchain

1. **Blockchain technology for algorithmic regulation and compliance (BARAC):** BARAC investigates the feasibility of using BCT for automating regulation and compliance producing a proof-of-concept platform and facilitating knowledge transfer by means of a bottom-up cross-disciplinary approach developed together with industry and regulators. This project aims to have impact in several sectors including finance, food industry, education, business models and managerial practices.
2. **ARCHANGEL - Trusted Archives of Digital Public Records:** The aim of ARCHANGEL is to ensure the long-term sustainability of digital archives though the design, development and trialing of transformational new DLT to promote accessibility and ensure integrity of content, while maximizing its impact through novel business models for commodification and open access. ARCHANGEL will leverage cutting-edge machine learning to collect robust digital signatures derived from digitized physical and born-digital content, within a permissioned DLT. Both signatures and programmatic code to render content and verify its provenance and integrity will be encoded within the DLT.
3. **Smart Money: Precision Data Management for Distributed Ledger enabled Central Bank issued Digital Currencies:** Investigates the data management, governance and social utility challenges concerned with how central banks, clearing banks and other stakeholders. Central to this research is a governance framework which uses smart contracts to manage and police access rights to payment transaction data sets held on the DL for all certified and warranted stakeholders. Access to the content on the DL will be facilitated through a public key system held by individuals, government agencies, and the Ministry of Justice.
4. **Trusted and Transparent Voting Systems** aims to explore applications of distributed ledger technologies (Blockchain) in domains involving voting and collective decision making.
5. **Household-Supplier Energy Market (HoSEM):** the project is exploring "distributed ledger" technology to develop a platform to support P2P household-level energy trading EDF Energy is the industrial partner in this project and is currently testing the concept of peer-to-peer trading between households as part of a trial within a block of flats in London. The Universities of Exeter and Leicester are the other partners on HoSEM project. Exeter researchers are looking at what factors will encourage households/groups to join this peer-to-peer market, while Leicester researchers are taking an algorithmic and game-theoretic view on the peer-to-peer trading (University of Bristol, n.d.).
6. **Distributed Ledgers and Decentralized Energy in sub-Saharan Africa:** In collaboration with GSMA's Mobile for Development Utilities program and ElectriCChain, the developers of the SolarCoin Blockchain,

Distributed Ledgers, and Decentralized Energy in sub-Saharan Africa will design and test a platform for peer-to-peer electricity exchanges in East Africa with the aim of accelerating access to off-grid solar energy distribution (EPSRC, n.d.).

7. **Co-operative Models for Evidence-based Healthcare Redistribution (CoMEHeRe):** CoMEHeRe aims to transform personal healthcare for the benefit of individuals through the use and management of biometric information created by wearable devices. To do this it will combine data from an individual's wearables with DLT (blockchains) and machine learning to securely store and access data to enable the individual to share and benefit from their generated information (UKRI, n.d.).

5.3 UK and India Consortium

According to Professor Muttukrishnan Rajarajan, this consortium will be exploring the use of a privacy-preserving blockchain architecture for IoT applications in healthcare data sharing (Healthcare, nd), using attribute-based encryption (ABE) to provide greater security for the devices. Following a recent outbreak of cyber attacks exposing the inadequacies of IoT security, a healthcare-focused consortium, led by network security engineers from City University London and the Indian Institute of Technology Kharagpur, is looking to blockchain for answers. The consortium, drawn from academia and industry, was set up after the WannaCry ransomware attack in May, which hit organizations across the globe, including the UK's National Health Service. There, the attack forced hospitals to cancel operations and left staff recording vital patient data using pen and paper.

6. Conclusion

In a modern smart city environment the disruptive nature of blockchain can serve the urban citizens by solving the problems of data communications that include privacy, integrity, security, transparency and availability of assets. This paper presents a comprehensive survey of the smart city applications that use BCT. The proposals of different application models has been studied and analyzed in terms of privacy preserving and data integrity. The implementation of BCT in different applications in a smart city such as healthcare, smart transportation, and smart economy services has been discussed in the paper. According to the case studies the application of blockchain in the public and financial services are still in infancy stage and yet to be promoted to improve the quality of life of the urban citizens and to provide a secure data and financial transactions.

References

Alfandi, O., Otoum, S., & Jararweh, Y. (2020a). Blockchain solution for IoT-based critical infrastructures: Byzantine fault tolerance. In *In proceedings of IEEE/IFIP network operations and management symposium 2020: Management in the age of softwarization and artificial intelligence, NOMS 2020*. Institute of Electrical and Electronics Engineers Inc. https://doi.org/10.1109/NOMS47738.2020.9110312.

Alfandi, O., Otoum, S., & Jararweh, Y. (2020b). Blockchain solution for IoT-based critical infrastructures: Byzantine fault tolerance. In *NOMS 2020 - 2020 IEEE/IFIP network operations and management symposium, Budapest, Hungary* (pp. 1–4). https://doi.org/10.1109/NOMS47738.2020.9110312.

Androulaki, E., Barger, A., Bortnikov, V., Muralidharan, S., Cachin, C., Christidis, K., et al. (2018). Hyperledger fabric: A distributed operating system for permissioned blockchains. In *Proceedings of the 13th EuroSys conference, EuroSys 2018*. Association for Computing Machinery, Inc. https://doi.org/10.1145/3190508.3190538.

Capossele, A., Gaglione, A., Nati, M., Conti, M., Lazzeretti, R., & Missier, P. (2018). Leveraging blockchain to enable smart-health applications. In *2018 IEEE 4th international forum on research and technology for society and industry (RTSI), Palermo* (pp. 1–6). https://doi.org/10.1109/RTSI.2018.8548357.

Cash, M., & Bassiouni, M. (2018). Two-tier permission-ed and permission-less blockchain for secure data sharing. *2018 IEEE International Conference on Smart Cloud (SmartCloud)* (pp. 138–144). New York, NY. https://doi.org/10.1109/SmartCloud.2018.00031.

Chakraborty, S., Aich, S., & Kim, H. (2019). A secure healthcare system design framework using blockchain technology. In *2019 21st international conference on advanced communication technology (ICACT), PyeongChang Kwangwoon Do, Korea (south)* (pp. 260–264). https://doi.org/10.23919/ICACT.2019.8701983.

Chandel, S., Cao, W., Sun, Z., Yang, J., Zhang, B., & Ni, T.-Y. (2019). A multi-dimensional adversary analysis of RSA and ECC in blockchain encryption. *Advances in Biochemical Engineering*, 988–1003.

Christidis, K., & Devetsikiotis, M. (2016). Blockchains and smart contracts for the internet of things. *IEEE Access, 4*, 2292–2303. https://doi.org/10.1109/ACCESS.2016.2566339.

EPSRC. (n.d.). http://gow.epsrc.ac.uk/NGBOViewGrant.aspx?GrantRef=EP/P031854/1.

Gori, P., Parcu, P.L. Stasi, M. (2015). Smart Cities and Sharing Economy. Robert Schuman Centre for Advanced Studies Research Paper No. RSCAS 2015/96, Available at SSRN: https://ssrn.com/abstract=2706603 or https://doi.org/10.2139/ssrn.2706603.

Khan, M. S., Woo, M., Nam, K., & Chathoth, P. K. (2017). Smart city and smart tourism: A case of Dubai. *Sustainability, 9*(12). https://doi.org/10.3390/su9122279.

Michelin, R. A., Dorri, A., Steger, M., Lunardi, R. C., Kanhere, S. S., Jurdak, R., et al. (2016). Dubai turns to blockchain for domestic challenges. In *SpeedyChain. Proceedings of the 15th EAI international conference on mobile and ubiquitous systems: Computing, networking and services*. MobiQuitous '18.

Nam, K., Dutt, S. C., Chathoth, P., & Khan, M. S. (2021). Blockchain technology for smart city and smart tourism: latest trends and challenges. *Asia Pacific Journal of Tourism Research, 26*(4), 454–468. https://doi.org/10.1080/10941665.2019.1585376.

Ramani, V., Kumar, T., Bracken, A., Liyanage, M., & Ylianttila, M. (2018). Secure and efficient data accessibility in blockchain based healthcare systems. In *2018 IEEE global communications conference (GLOBECOM), Abu Dhabi, United Arab Emirates* (pp. 206–212). https://doi.org/10.1109/GLOCOM.2018.8647221.

Sharma, P. K., Moon, S. Y., & Park, J. H. (2017). Block-VN: A distributed blockchain based vehicular network architecture in smart city. *Journal of Information Processing Systems, 13*(1), 184–195. https://doi.org/10.3745/JIPS.03.0065.

Sharma, P. K., & Park, J. H. (2018). Blockchain based hybrid network architecture for the smart city. *Future Generation Computer Systems, 86*, 650–655. https://doi.org/10.1016/j.future.2018.04.060.

Solanas, A., Patsakis, C., Conti, M., Vlachos, I., Ramos, V., Falcone, F., et al. (2014). Smart health: A context-aware health paradigm within smart cities. *IEEE Communications Magazine, 52*(8), 74–81. https://doi.org/10.1109/MCOM.2014.6871673.

Sun, J., Yan, J., & Zhang, K. Z. K. (2016). Blockchain-based sharing services: What blockchain technology can contribute to smart cities. *Financial Innovation, 2*(1). https://doi.org/10.1186/s40854-016-0040-y.

Tripathi, G., Ahad, M. A., & Paiva, S. (2020). S2HS- A blockchain based approach for smart healthcare system. *Healthcare, 8*(1). https://doi.org/10.1016/j.hjdsi.2019.100391.

UKRI. (n.d.). https://gtr.ukri.org/projects?ref=EP%2FP03196X%2F1.

University of Bristol. (n.d.). https://www.bristol.ac.uk/cabot/what-we-do/peer2peer-energy/#:~:text=The%20EPSRC%2Dfunded%20'Household%20Supplier,utilities%20or%20other%20third%20parties.

Further reading

Botello, J. V., Mesa, A. P., Rodríguez, F. A., Díaz-López, D., Nespoli, P., & Mármol, F. G. (2020). BlockSIEM: Protecting smart city services through a blockchain-based and distributed SIEM. *Sensors, 20*(16), 4636. https://doi.org/10.3390/s20164636.

Castro, M., & Liskov, B. (February 1999). Practical byzantine fault tolerance. In *Third symposium on operating systems design and implementation.*

DKV. (n.d.). http://analytics.dkv.global/data/pdf/Blockchain-in-UK-Executive-Summary.pdf.

Falazi, G., Hahn, M., Breitenbücher, U., Leymann, F., & Yussupov, V. (2019a). Process-based composition of permissioned and permissionless blockchain smart contracts. In *2019 IEEE 23rd international enterprise distributed Object computing conference (EDOC), Paris, France* (pp. 77–87). https://doi.org/10.1109/EDOC.2019.00019.

Falazi, G., Khinchi, V., Breitenbücher, U., & Leymann, F. (2019b). *Transactional properties of permissioned blockchains.* SICS Software-Intensive Cyber- Physical Systems (To be published).

Healthcare. (n.d.). https://www.gov.uk/search/all?keywords=UK+and+India+healthcare+services&order=relevance.

Kerr, S. (October 7, 2016). *Dubai turns to blockchain for domestic challenges* [Online]. Available https://www.ft.com/content/48ff547c-53e7-11e6-9664-e0bdc13c3bef. (Accessed 28 April 2017).

Sukhwani, H., Martinez, J. M., Chang, X., Trivedi, K. S., & Rindos, A. (2017). Performance modeling of PBFT consensus process for permissioned blockchain network (hyperledger fabric). In *2017 IEEE 36th symposium on reliable distributed systems (SRDS).* https://doi.org/10.1109/srds.2017.36.

Sun, F., & Duan, P. (September 2014). *Solving byzantine problems in synchronized systems using bitcoin.* Self-published Paper [Online]. Available https://allquantor.at/blockchainbib/pdf/sun2014solving.pdf.

Chapter 2

Future of Sustainable Smart Cities: an insight

V. Padmapriya, D.N. Sujatha
Innovation Cluster Lab, Department of Computer Applications, B. M. S. College of Engineering, Bengaluru, Karnataka, India

Chapter outline

1. Introduction

Globally, the human race is experiencing a tremendous change in the sphere of social, economic, demographic, political, and industrial sectors. Concurrently, our planet's natural ecosystem is harmed, and resources are depleting swiftly. As a result, the life and environment are deeply impacted. Numerous environmental organizations and global communities have raised alarm in this direction, but we have a long way to go. Additionally, population explosion and urbanization are the two major factors for environmental hazards. Urban growth has intensively contributed to environmental damage. Several studies

Blockchain for Smart Cities. https://doi.org/10.1016/B978-0-12-824446-3.00015-6

show that the big cities in the world are the first ones to experience the full force of climate change. They relentlessly face extreme weather changes, floods, depletion of green spaces, and dwindling of natural resources. This acute scarcity of natural resources, increasing infrastructure needs, environmental pollution, etc. are making the cities uninhabitable. Therefore, there is a pressure to work toward sustainable living in urban areas. The global urban population is steadily increasing, and World Health Organization (WHO) reports that the global urban population is expected to grow approximately by 1.84% per year between 2015 and 2020, 1.63% per year between 2020 and 2025, and 1.44% per year between 2025 and 2030 (Urbanization Causes and Impacts, 2020; Urban Population Growth Available, 2020). With this rise in the population, the cities are facing issues of insufficient resources and a large infrastructure crunch. On the same lines, it is found that there is a huge gap between environmental realities and the economic cravings of the human race. Today, the lifestyle in the cities is predominantly driven by technology. The cities are working toward improving the quality of life. The cities are considered as the coherent organizational units where people innovate, coordinate, and act with the cultural and spatial connect. Cities are embarking on different paths, and they have different capacities to handle this situation. Studies have found that every year there is a birth of a megacity and the population is touching the 10-million threshold (Global Urbanization Problems, 2017).

This large rise in population brings a plethora of questions to the mind:

- Do we have enough space on earth?
- Can we provide a sustainable living to the next generation?
- Can we accommodate this exponential rise of population and still have a comfortable living?
- Do we have the technological support to handle the situation and arrive at feasible solutions?

The urban areas are further affected by the increased industrialization and are scumbling to the pressure of innovation. The urban population juggles with the problems of an unhealthy lifestyle and reduced physical activity, housing problem, development of slums, sanitation problem, segregation and disposal of trash and increase in crime rate, and many more. Hence, it is comprehended that urbanization is a never-ending phenomenon, and numerous studies have predicted that the future of the planet will fundamentally be urban areas and they will be fountainheads of unsustainability. Thus, there is an urge to build smart and sustainable cities (Components of Smart City, 2018; Global Urbanization Problems, 2017).

Although the urban areas propel growth, they are the hub of various economic and environmental problems. Hence, there is a need for a framework that uses technologies and platforms to improve life in cities. A statistical report by the International Data Corporation (Worldwide Digital Transformation

Spending Guide, 2020; Zettabytes of the Data World - Prediction for 2025, 2018) points that the digital information created, used, and replicated has increased by 62% in 2009, it reached 1.2 million petabytes, further grew to 9 billion in 2018, and by 2020 it is estimated to be 44 times larger than what it was in 2009. As a consequence of these needs, the urban areas have to adapt the information and communication technology to meet the growing urban challenges, and thus came the concept of Smart City. Now let us define Smart City and further look into the ways to incorporate sustainability through technology inclusion.

1.1 What is a sustainable smart city?

A sustainable smart city (SSC) is a city that has developed technological infrastructure to develop, deploy, and nurture sustainable operations and facilitate to meet the growing urbanization needs. It opens the door to integrate technological dynamism into the local governance, education, healthcare, energy, infrastructure and mobility sectors, etc. In the beginning, the back-end systems of these sectors always remain a *black box* to the end-users. But, as technological integration gets more visible to the end-users, there will be a huge transition in urban living in smart cities. Next, we move to the understanding of the core infrastructural elements that are required for the development of a Smart City.

1.2 Components of SSC

Smart cities are made up of components that are self-regulated, well-connected, responsive, and sustainable components. A smart city is visualized as a network of sensors that connects various devices, aggregates the collected data, optimizes, analyzes, and predicts the pattern of usage of the resource. The innovations in technology support the development of SSC. An SSC includes components such as *Smart Governance, Smart HealthCare, Smart Transportation, Smart Water Management. Renewable Energy, Smart Agriculture, Smart Waste Management, Smart Connectivity, and Smart Building*, which are depicted in Fig. 2.1. A detailed study of smart cities by Mohanty, Choppali and Kougianos (2016) highlights the components, attributes, themes, and infrastructure for a smart city. Further, the author categorizes *sustainability, quality of life, urbanization,* and *smartness* as attributes and society, *economy, environment,* and *governance* as the themes with the *Information and Communication Technology* and *services* as the supporting infrastructure. The technology inclusion such as Internet of Things (IoT) and Blockchain makes the networked data collected from various sources and devices, located at different places, available to all the stakeholders through an open access mechanism with the promise of security and privacy of data. In the next subchapter, let us look into the global perspective of Smart Cities.

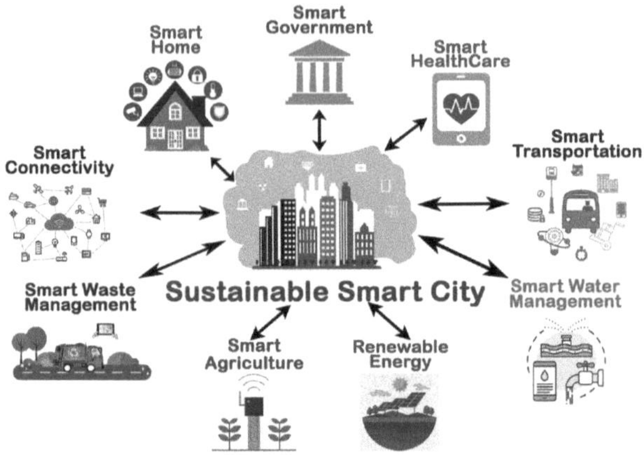

FIGURE 2.1 Components of sustainable smart city.

1.3 The emergence and growth of sustainable smart cities across the globe

The United Nations (UN) has proposed themes for sustainable living. They have set 17 thematic targets for all the countries across the globe. These targets are agreed upon by the international community and the cities are striving to implement the 10 primary actionable themes for developing a sustainable community. The top actionable themes recommended by the UN are *Zero Hunger, Clean Water & Sanitization, Affordable and Clean Energy, Industry, Innovation & Infrastructure, Sustainable Cities and Communities, Responsible Consumption & Production, Climate Action, Life below Water, Life on Land, Peace & Justice, and Strong Institutions* (Urban Solutions Handbook 2017, 2019). Worldwide, the cities are tackling global challenges such as climate change, poverty, and clean air, and are all striving to provide technology-based solutions to handle these challenges. The emergence of SSC shows progress toward these initiatives. Table 2.1 lists the top 10 smart cities identified across the globe for the year 2020 (ranked in February 2020).

This ranking is based on the cities continuous efforts in managing various sectors of their cities for being the benchmark for other cities across the globe (Elements of Smart City, 2018; Top 10 Growing Cities in the World, 2020):

The next chapter provides an overview of the smart city initiatives, technological challenges, and advancements in the implementation of SSC in India.

1.4 The position of SSCs in India

Globally, countries have been striving hard to work toward sustainable living. The Government of India has a mission to promote healthier, wealthier, and

TABLE 2.1 Top 10 smart cities across the globe listed in 2020.

Rank	Name of the city	Characteristics, vision, and goals
1	Singapore	The city tops the list with smart transportation, smart energy, smart water management, a monitoring system for the elderly. By the year 2022, the government plans to implement energy-efficient lighting for public roads, and solar lighting
2	Dubai	This city is ranked the second position for its implementation and continual efforts in areas of smart transportation, communication, infrastructure, electricity, economic services, and urban planning. In Dubai, 90% of government jobs are digitized. This city is monitoring cybercrimes and has plans to introduce autonomous police stations. Further, the city is working on a concrete structure using a 3D printer technology.
3	Oslo	The city of Oslo has implemented smart energy consumption and has embraced sensors for lighting, cooling, and heating. This city has a goal to cut carbon emission by 36% by 2020 and 95% by 2030
4	Copenhagen	Copenhagen Solutions Lab received an award in 2017 for a system that monitors traffic, air quality, waste management, energy use, and other items and compares operations in real time. It connects parking systems, traffic lights, buildings, smart metering, and charging systems for electric vehicles to direct traffic in real time and optimize energy use according to fuel prices, traffic movement, and weather. This is a green city with half of the city uses bikes.
5	Boston	Boston has an "Innovation District" that works on the concept of "participatory urbanism." This participative group collects citizen's issues on potholes, child's school van tracking, parking information, etc. Boston is infamous for its traffic snarls, hence the city plans to install smart sensors with the micro hub to ease the traffic and decongest the roads.
6	Amsterdam	This city has an "IoT living lab" that connects and collects user's data using Bluetooth devices. Amsterdam has adapted the green transportation modes such as driving cycles, carpool, and autonomous vehicles for shuttle services in parks and subway areas. It has plans for a smart grid and smart housing development and uses carbon dioxide to generate electricity.
7	New York	The New York city municipal authorities have placed several low-powered sensors forming a wide-area network spanning its business district. The city boasts of having completely covered by the Internet.

Continued

TABLE 2.1 Top 10 smart cities across the globe listed in 2020.—cont'd

Rank	Name of the city	Characteristics, vision, and goals
8	London	This city has introduced the "Connected London" program that aims to provide 5G connectivity to the entire city. Use of drones to find the vacant spaces where cellular antennas can be installed. The iconic "London Lamppost" is connected with sensors and charging points for electric vehicles
9	Barcelona	The city of Barcelona is identified as the first city to embrace "Smart Systems." It hosted the first smart city expo in 2011 to promote a self-sufficient city of productive neighborhoods at human speed, inside a hyperconnected zero-emissions metropolitan area.
10	Hong Kong	Hong Kong has launched several initiatives to boost "Smart Governance" and "Smart Economy." This city's lamppost will be connected to mobile devices using 5G. The city is also geared up to introduce digital security, biometric, voice, and face recognition systems for banking systems. The city will get a mobile-friendly city digital dashboard to mark all the major places in the city.

happier cities for better urban life using state-of-the-art technology. India has a mission of developing smart cities under the National Sustainable Habitat and Smart cities scheme. In India, the objective of building smart cities is to harness technology, improve mobility, connectivity, infrastructure, and provide healthy living. The mission of arriving at an SSC aims to bridge the gap by providing a means for the interaction of citizens with the local government, social, and volunteer groups. The policies in India are working toward developing smart metering systems for energy, water, intelligent traffic management, e-governance to provide transparency in the system, established sustainable solutions for waste management, waste-to-recycle projects, reduction of waste, etc. (Challenges of Implementing Smart City Initiatives in India, 2017; Indian Smart Cities, 2015).

Further, the governmental authorities have laid out a comprehensive plan by flagging off new schemes such as the "Make in India," "Swachh Bharat," "SmartNet," and "Incredible India" for developing and promoting the smart cities that include the following features:

- Promotion of the mix-land usage, developing "unplanned-areas" in the cities
- Expanding housing opportunities for all strata of society
- Creating walkable localities that aim to reduce congestion, air pollution, depletion of natural resources, boost the economy, and strive to promote healthy interaction in the society with safety

- Preserving and developing open spaces such as parks, playgrounds, and other recreational spaces. The open and green spaces provide ecological balance and help in reducing urban heat.
- Promoting multimodal transport services and promote the use of public transport like metro trains, suburban trains, or buses, and green vehicles.
- Bringing transparency and accountability in governance. Providing user-friendly, mobile app-based services.
- Forming citizen groups for creating a collaborative environment and adapting crowdfunding for community work
- Facilitating a video tour of the worksites to facilitate concerned government authorities and officers to monitor civil and public service work in online mode

The upcoming smart cities in India with a vision for promoting sustainable and technology-enabled cities are Bhuvaneshwar, Pune, Jaipur, Surat, Kochi, New Delhi, Ahmedabad, Jabalpur, Vizag, Solapur, Davanagere, Indore, Coimbatore, Kakinada, Belgaum, Udaipur, Guwahati, Chennai, Ludhiana, and Bhopal (Smart City Features: Govt. of India, 2017; Smart City Profiles, 2018).

Most of these cities have smart solutions in the following areas listed:

- E-governance: electronic service delivery, citizen engagement, public information, and grievance redressal systems, crime monitoring.
- Waste management: waste management system, waste to compost, waste to recycle, waste segregation, etc.
- Urban mobility: smart parking, intelligent traffic management, multimodal transport, etc.
- Energy management: smart meters, renewable energy, green buildings, etc.

One of the biggest challenges faced for the implementation of Smart cities is on the technology front. Smart city implementation includes setting up a network of sensors, cameras, routers, and meters, handling handheld devices, managing mobile app, creating, updating, and managing websites, and handling datacenter issues. The biggest challenge is handling the privacy and security of the citizen's data (Make In India Sectors, 2018; Sethi, 2015; UITP - International Association of Public Transport (India), 2019; Urban Mission in India, 2017). Thus, it is very evident that IoT and Blockchain technologies are favorable to provide robust, convenient, and secure interactions in Smart cities.

2. Characteristics and technologies used in SSC

With the understanding of the popular and emerging smart cities across the globe and their position in India, it is inferred that technology plays a significant role in the emergence of SSC. The SSCs have a pronounced impact on information and communication technology, with the sensors deployed to collect data, capacitate the Internet to push the collected data onto the cloud and the infrastructure managed and by e-governance. Leveraging the technology for building sustainable cities backed with powerful ideas, supported by the public, monitored by the government should be controlled by the reality of its implementation. In the next subchapter, we will discuss in detail the technical support required for an SSC. The essential characteristic and technological support for the development of SSC are depicted in Fig. 2.2. The desired characteristics for developing a sustainable smart city necessitates the cities to promote and adopt sustainable techniques, strive to bring in innovative solutions, be economical, and provide simple solutions, and all these without harming the environment. The technological support for SSC needs the support of sensors and network elements, ICT tools, and embracing the e-governance for better administration of the resources.

Thus, it is inferred that the smart cities can be realized by using the two major disruptive technologies namely the IoT and Blockchain (Smart City - IoT Agenda, 2017).

The next subchapter describes the path to develop a smart city using these two popular disruptive technologies. The IoT has affected our lives visibly and invisibly.

We are living in the world of connected devices that are loaded with sensors, fueled with data, facilitating mobility, and seamlessly integrate with heterogeneous devices. According to IoT Analytics (Smart City Features, 2017), the IoT market research and analytics organization have estimated that by the end of 2019, there would be around 9.5 billion connected IoT devices and would reach 28 billion by 2025. The three major driving factors for this explosion are the following: increase consumer devices specifically the

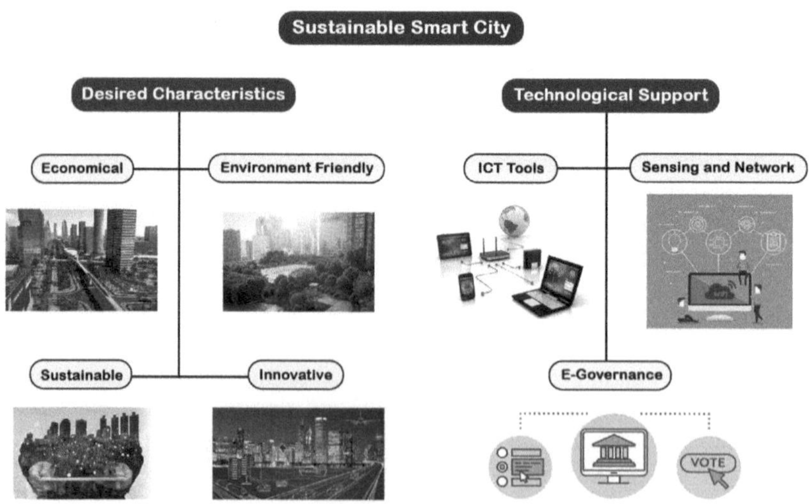

FIGURE 2.2 Components of sustainable smart city.

domestic devices, propelled by the cellular IoT or the M2M connectivity, and massive production of devices across the globe. The author Stephan Haller, Karnouskos, and Schroth (2008) defines IoT as follows:

> *"A world where physical objects are seamlessly integrated into the information network, and where the physical objects can become active participants in business processes. Services are available to interact with these, 'smart object', over the Internet, query their state and any information associated with them, taking into account security and privacy issues."*

Thus, IoT, which is loaded with tiny sensing "smart objects," is considered to be one of the most promising and emerging disruptive technologies of this decade. IoT provides network connectivity across the globe connecting devices, establishing networks, facilitating data exchange, and storage. Although this technology has been around here for several years; today, it is perceived as one of the key players and is predicted as the future of the Internet. This domain of connected devices has opened doors for several new opportunities in the domain of healthcare, manufacturing, transportation, agriculture, supply chain, etc. It spans a wide range of applications ranging from home automation to the wide range of applications catering to different areas fueled by the tiny "smart objects."

With the recent advancement in the field of technology inclusion in developing the urban network, it is found that IoT has the capacity and capability to handle the pressure of urban demands very efficiently. This field of the network of sensors has geared up to facilitate innovative, comfortable, and affordable security for day-to-day living in urban areas. Hence, they have become the hub of digital activities.

A report by Raconteur shows that IoT-enabled smart cities are changing the urban lifestyle and are speeding up the service delivery time. Further, it is predicted that services such as emergency and safety, transportation, healthcare, and environmental issues are highly sought after IoT-enabled services in urban areas. On the same lines, the smart city spending predicted for the year 2017–22 is estimated to be around 62% for office wearables, 49% on the vehicle to everything connectivity, with 25% share to open data, 23% spending on the smart thrash collection, and 23% on smart city platforms, etc. (IOT - Building Smarter Cities, 2020; IOT-Business-2020, 2020). All these statistics point out that the popularity of IoT-enabled urban life. It also forces people to think on an uncorrupted, immutable, distributed yet secured, and digital ledger management system facilitated by Blockchain. Hence, the IoT network blended with the powerful features of blockchain is perceived as an unspoken force that revolutionizes the technology-driven and sustainable smart cities. These two technologies are robust enough for creating efficient, sustainable, and resilient smart cities.

In the next subchapter, we will discuss how the two disruptive technologies are supporting various applications propelling the growth of SSCs and its future directions.

3. Future of IoT and Blockchain and its impact on the various sectors of smart cities

3.1 Smart home and smart society

3.1.1 Background

Intelligent systems are the ones that sense the surroundings and the environment. They provide economical, safe, comfortable living with a touch of automation to the urban population. Smart home and smart society fall in the ambit of these intelligent systems. The two primary concerns of this system are to achieve energy efficiency and enhance the safety of living.

Today, urban homes are predominantly run by various appliances and all of them extensively use energy drawn from natural resources. The modern household appliances include kitchen essentials such as the refrigerator, blenders, hobs, lighting systems, temperature controllers, entertainment systems, utility devices like washing machines, vacuum cleaners, curtains raisers, garage door operating systems, and cleaning utilities. The smart society includes sensible lighting for the roads, parks, and society walkways. All these devices are energy-intensive and usually, the resources are wasted due to human negligence. Therefore, automation comes to the rescue for these problems. The next big thing is the safety of elders and children in urban homes. Smart homes and communities provide tailor-made solutions to handle safety issues.

Until recently, smart homes were considered as luxury and affordable only to the upper class of society. There has always been a debate on the economy versus luxury and it is found that luxury wins with the upper strata of society and middle-class households settle with the economy part. But, experts forecast that this scenario is drastically changing in both developed and developing countries. They find a huge market potential in the home automation sector. This sector that covers smart home management, IoT-enabled home security systems, and the new trend urban space is the IoT-enabled indoor or rooftop home garden management providing smart gardening solutions for urban homes.

3.1.2 Technology adaptation

The first change begins from home with the concept of connected household appliances. A smart home is a popular IoT use case and is influencing people around the world. The refrigerator reminds the lady of the house about the vegetables and dairy products at home and what needs to be ordered, and the calorie value of the food stored in the refrigerator. Further, the refrigerator automatically checks out the cart of products billed in the name of its master. These smart appliances allow homemakers to increase efficiency and reduce carbon footprints but are vulnerable to cyberattacks and criminal activities. This is foreseen as the most desirable technology if it is authenticated and supported by blockchain technology. Blockchain technology assures that the transactions made by an appliance are safe, immutable, and traceable, and it facilitates a faster settlement return process. Similarly, the home security systems can be compromised by a hacker making the house vulnerable to robbery and other criminal activities, and Blockchain is the savior here too (Smart Home Management, 2018).

3.1.3 Description of the use case

The United States of America was the first country to enter into the market of Smart homes and are still the leaders in this area. A study by a leading technology media platform dedicated to IoT and its allied areas shows that presently 32% of American households are automated and are expected to rise to 53.1% by 2020. The households in America have smart lights, smart furnishing, climate control systems, and smart entertainment systems that consume more energy and hence are heavy on utility bills. The economy of America supports this kind of lifestyle. On the contrary, most of the home automation projects in Europe target resource-saving, both financial and natural. Scandinavia was the first country to adopt the automation of natural resources, where energy is produced from thermal springs and water tides. These economic methods are a boon to energy-dependent smart homes.

In the case of Eastern Europe and Central Asia, it was found that technology inclusion is quite challenging. The cities in these countries are majorly old residential buildings with obsolete meters Moreover, it is found that the larger population is middle-class, and the number of people who can afford smart homes is very low. On the contrary, the United Arab Emirates is one of the leaders in the adaptation of smart automation urges for climate control and is worries less about the home security systems. The cities here face extreme weather conditions thus temperature controllers are very popular and the cities have already good laws and order in place defeating the need for home security systems.

The next strange, innovative, and astonishing adaptation of smart systems is by "Japan." This country is famous for its "Japanese Toilet," where toilets greet people songs, the toilet seats are warmed up after every use, and the seats give its customers dietary recommendations.

Finally, India is witnessing a huge shift in the urban sprawl, the economic growth, and the emergence of megacities have brought in a variety of opportunities in the smart home segment. This growth has propelled the realty business across the country. A study by the Park Associates in 2018 points that 75% of urban customers who plan to buy a house want smart buildings and homes integrating appliances and devices for a comfortable, well-informed, and secure living for the future. The most popular one also includes technology for remote monitoring of the elderly and children at home. Thus, home security is another most sought blended technology with a very large market in Indian Smart home segment. Shortly, with the depleting green spaces, smart farming is foreseen as the next big wave in India.

With all these adaptations, it is very clear the IoT and Blockchain help us to build very versatile, robust, flexible, and safe urban spaces. These areas will become the hotbeds of research activities in both disruptive technologies (Farmanbar, Parham, Arild, & Rong, 2019; IoT Solutions, 2016).

3.2 Smart grid

3.2.1 Background

Electricity was a landmark discovery witness by the second industrial revolution. For more than 150 years, electricity has been the driving force for the industry. Industry 4.0 is the new paradigm that includes numerous devices, powering networks, running across oceans, and reaching consumers across the globe. All of them run are driven by "electricity" or "energy." For years, energy is derived from fossil fuels, and with the rapid increase in population and urbanization, fossil fuels are fast depleting. As energy is crucial for modern society, the electric grid plays a major role in balancing the distribution of electricity across the cities. Several studies find that the energy consumption of the cities is around 75%–80% of the world's energy consumption, and it has led to 80% of total greenhouse gas emission (Farmanbar et al., 2019). Further, there is a scarcity of renewable resources that necessitates the usage of smart energy management techniques. The energy can be managed using the "smart grids." A "grid" is a network of electricity transmission lines, substations, transformers, etc. that deliver electricity to homes and offices.

The smart grid (SG) is an updated version of the traditional energy infrastructure that uses technology to cater to the sustainable energy needs of smart cities. This infrastructure commits for the reliable, energy-efficient, and uninterrupted power supply to the cities. Thus, smart grids are imperative for urban space. The smart grids include smart metering, distribution automation, smart homes, public lighting, device management, energy storage and grid management, and electric vehicle charging and grid maintenance systems.

Many electric grids across the globe are old and are running for many years producing several million megawatts of power every day. The smart grids aim to integrate with the existing infrastructure and merge renewable and nonrenewable resources, reduce environmental pollution, and modernize the power systems facilitating real-time monitoring, automation, and controlling. These features of SGs aim to provide smart and sustainable solutions to the energy needs in the cities.

The smart grid provides several benefits that include the following:

- Technology-assisted power distribution and management
- Reduces operations and maintenance cost
- Quick restoration of electricity and power disturbances
- Maintaining the balance between the power supply and demand
- Reducing the power cost on customers
- Checking power leakages and thefts
- Facilitating and promoting renewable energy systems
- Improving security at the power stations

3.2.2 Technology adaptation

Generally, energy distribution is managed by governmental agencies or energy distribution boards. Worldwide, the energy distribution is traditionally unidirectional with fixed tariffs set by the government. The consumer premises is installed with an analog meter that records the consumption of energy. The consumers are billed according to their consumption and it provides no mechanism to interact with the authorities in case of problems. These meters are not capable of two-way data exchange. Hence, there is a need for an infrastructure that resolves this issue. The Advanced Metering Infrastructure with the smart meters provides a solution to this problem. The smart meters are embedded with sensors that provide real-time usage information, billing, and tariff information, alert consumers if they have exceeded the limit of usage, provide information on carbon emission in their premises, and facilitate the usage reports. The transactions of these smart meters can be tracked using Blockchain technology, which secures the transaction with smart contracts. The feature of immutability to the transaction history helps in auditing the payment records or to resolve the disputes (Agung & Handayani, 2020).

3.2.3 Description of the use case

The concept of the smart grid gained momentum in the year 2000. Many countries have embraced this technology with their unique policies, objectives, and approaches.

The United States of America introduced a Federal policy with the passage of the Energy Independence and Security Act of 2007. This law provisioned funding for establishing smart grid capabilities and created a Grid Modernization Commission to assess its benefits. The Federal Energy Regulatory Commission (FERC) laid policy statements and action plans for SGs. The FERC assures that the SGs accommodate renewable energy resources, handle the demand response systems, and assist to promote the usage of energy storage systems and electric vehicles. Brazil's smart grid efforts are aligned to diversify the generation of energy sources and enhance the electric grid infrastructure. This country has rolled out the smart meter project in 2012, and it estimates to install 27 million smart meters by 2030. Brazil illustrates that Biomass is the most efficient alternative fuel and is pro to nature and has initiated the process of electricity generation diversity by using alternate fuels.

Europe has established the European Technology Platform (ETP) for the implementation of SGs in the year 2006. The ETP uses the smart grid to generate innovative products and services and make plans for intelligent monitoring, control, communication, and building self-healing networks. Europe plans to install 200 million smart meters for electricity and 45 million for gas.

South Africa has put tremendous efforts to build a smart grid system with the focus to increase the penetration of renewable energy generation, improving network reliability and availability. India is the fourth-largest energy consumer in the world. With the rise in population and rapid urbanization, there is a surge in the demand for energy in India. The Ministry of Power, Government of India has launched the National Smart Grid Mission with a vision to transform India poser sector. It aims to provide secure, adaptive, sustainable, reliable, digitally enabled, and quality energy to all the stakeholders. Currently, India has 14 smart grid pilot projects that will be implemented by several state-owned distribution utilities (Chen, 2010; Paaso, Kushner, Bahramirad, & Khodaei, 2018; Role of Smart Grids in Smart Cities, 2017; Smart Grid - Indian Power Sector, 2018; Smart Grid Platform - Control and Monitor Smart Devices, 2019; Smart Grids in South Africa, 2018; Smart Grid - United States, 2019).

3.2.4 Future directions

Smart grids are a boon to smart cities. It provides a better quality of life for its citizens. Increasing the intelligence of the infrastructure enables the citizens to better manage their utilities efficiently. Further, the smart grid has the potential to reduce carbon emissions, achieve renewable energy goals, improve economic efficiency, thwart power thefts, and improve power distribution.

3.3 Smart healthcare

3.3.1 Background

Technology has touched all the parts of our lives and has evolved over the years. Technology inclusion and digital transformation have a profound influence on the healthcare sector. Healthcare is a very vast field where innovation is crucial for sustainability. Technology inclusion in healthcare improves the quality and longevity of life. Today, medical aid is facilitated with remote consultations, telemedicine, targeted treatment, and various mobile apps.

3.3.2 Description of the use case

The reports from the WHO point out that the world is rapidly moving toward urbanization as a result of which is going to significantly impact the standard of living, lifestyle, social behavior, and health. Further, urbanization has set in a sedentary lifestyle, people work under pressure, mental stress, competitiveness with unhealthy choices of food, and increased automation leading to several health issues. Many health-related issues in urban areas are related to environmental pollution, injury during the violence, road accidents, fire breakout, building collapse, noncommunicable or lifestyle-based diseases such as diabetes, heart ailments, chronic respiratory diseases, or cancer, etc.

3.3.3 Technology adaptation

Technology has brought in a consequential shift in the healthcare industry. The doctors, paramedical staff, nursing staff, the staff of the medical diagnostic center, etc. are the stakeholders of the healthcare sectors that are technologically assisted to handle the patients. All the stakeholders have access to the diagnostic tools, the benefit of adopting cutting-edge technology, minimally invasive, less painful, and quicker healing methods in surgeries and treatments.

A plethora of healthcare treatments include remote consultations with specialists, targeted treatments, availability of mobile apps, and superior nursing treatment technologies leading to an enhanced quality of treatment.

The technology has changed the healthcare systems, the following are its digitalization path:

- Digitization of health records
- Mobile app-assisted patient monitoring
- Electronic health record (EHR)
- Telemedicine/telehealth
- Remote consultation with video conferencing
- Suggestive and preventive treatment
- Identify risk factors
- Usage of 3D printing and RFID tag-based techniques for dental, prosthetics, hearing aid, and patient care
- Virtual reality (VR)-based assistance for patients with symptoms of depression, anxiety, mental illness, and helping patients with postoperative care
- Augmented reality (AR) to help the surgeon and his team to perform the operation
- AR/VR device to help improve the patient's health
- Robotic surgeries and gene therapies for various ailments

Most healthcare systems adapt various components of IoT to form a wireless network. This network is used to consistently monitor vitals of patients, track elderly patients or infants, alert medical staff about emergency conditions, and even observe the behavioral patterns of patients. The smart healthcare system built using IoT devices helps to collect the patient's data and vital information, use various mechanisms to upload it to cloud storage that can be used by doctors to remotely monitor the health of the patients. The patient's records hold confidential information and need to be secured. The privacy of the data is maintained and all the transactions are secure with the assurance from the Blockchain technology. Both IoT and Blockchain are extensively studied by several researchers. Numerous research works are carried out to analyze the patient data, design, and develop ubiquitous data access techniques to collect and publish the globally available data and facilitate the collection and integration of data from different platforms (Chakraborty, Aich, & Kim, 2019; Smart Cities and Healthcare: A Systematic Review, 2019).

3.3.4 Future directions

Today, worldwide, healthcare is one of the most technologically adaptive sectors. The technology adaptation in this area is a big game changer. IoT and Blockchain provide cutting-edge applications that save a life, provide preventive and supportive care with hassle-free, secure, decentralized transactions, and support for customers' data and payment issues. Both of these disruptive technologies are multifaced to handle the healthcare sector. The Health Internet of Things promises to provide applications that collect data from sensors, gather real-time data, monitor patients, and alert the doctors in case of emergencies.

The inclusion of Blockchain in healthcare assures secure, distributed, and immutable records of the transactions. The digital-ledger technology of the Blockchain creates a sequence of transactions that can be securely shared among the members in the network with the assurance of immutability for records. Thus these two technologies provide timely, secure, and a single source of truth (BlockChain for Smart Cities, 2020; Internet of Things and Blockchain - Oracle White Paper, 2018).

4. Conclusion

Urbanization is a global phenomenon. Many cities around the world are in a dire state to improve the quality of life and provide sustainable solutions to common problems. Cities are the epitome of opportunities where every product, every piece of infrastructure, and every bit of information can absorb technology. Therefore, the cities are the playground for implementing smart, secure, and sustainable solutions. Further, it is very essential to identify the key players in all the areas, recognize the priority sectors for improvement, build a blueprint to bring in change, and apply the relevant technologies to provide long-term sustainable solutions. The IoT and Blockchain are two promising technologies that have the potential to provide smart solutions to build sustainable smart cities.

References

Agung, A. A. G., & Handayani, R. (2020). Blockchain for smart grid. *Journal of King Saud University - Computer and Information Sciences*. https://doi.org/10.1016/j.jksuci.2020.01.002.

BlockChain for smart cities. (2020). https://iiot-world.com/smart-cities-buildings-infrastructure/smart-cities/blockchain-for-smart-cities-simplified/.

Chakraborty, S., Aich, S., & Kim, H. C. (2019). A secure healthcare system design framework using blockchain technology. In *International conference on advanced communication technology, ICACT* (Vol. 2019, pp. 260–264). Institute of Electrical and Electronics Engineers Inc. https://doi.org/10.23919/ICACT.2019.8701983.

Challenges of implementing smart city initiatives in India. (2017). http://smartcities.gov.in/content/innerpage/challenges.php.

Chen, T. (2010). Smart grids, smart cities need better networks. *IEEE Network, 24*(2), 2–3. https://doi.org/10.1109/MNET.2010.5430136.

Components of smart city. (2018). https://inside.battelle.org/blog-details/4-components-to-a-smart-city.

Elements of smart city. (2018). https://www.information-age.com/elements-smart-city-123473906/.

Farmanbar, M., Parham, K., Arild, Ø., & Rong, C. (2019). A widespread review of smart grids towards smart cities. *Energies, 12*(23).

Global urbanization problems. (2017). https://www.conserve-energy-future.com/urbanization-problems.php.

Smart grids in South Africa. (2018). https://en.wikipedia.org/wiki/Smart_grids_by_country#South_Africa.

Haller, S., Karnouskos, S., & Schroth, C. (2008). The internet of things in an enterprise context. In *Future internet symposium* (pp. 14–28).

Indian smart cities. (2015). https://geographyandyou.com/indias-smart-cities/.

Internet of things and blockchain - oracle white paper. (2018). http://www.oracle.com/us/industries/healthcare/iot-blockchain-opp-hc-wp-5187127.pdf.

Iot - building smarter cities. (2020). https://www.visualcapitalist.com/iot-building-smarter-cities/.

IoT-business-2020. (2020). https://res.cloudinary.com/yumyoshojin/image/upload/v1/pdf/iot-business-2020.pdf.

IoT solutions. (2016). https://www.iotforall.com/about-iot-for-all/.

Make in India sectors. (2018). https://www.makeinindia.com/sectors.

Mohanty, S. P., Choppali, U., & Kougianos, E. (2016). Everything you wanted to know about smart cities. *IEEE Consumer Electronics Magazine, 5*(3), 60–70. https://doi.org/10.1109/MCE.2016.2556879.

Paaso, A., Kushner, D., Bahramirad, S., & Khodaei, A. (2018). Grid modernization is paving the way for building smarter cities [technology leaders]. *IEEE Electrification Magazine, 6*(2), 6–108. https://doi.org/10.1109/MELE.2018.2816848.

Role of smart grids in smart cities. (2017). http://bwsmartcities.businessworld.in/article/Role-of-Smart-Grids-In-Smart-Cities/03-02-2017-112409/.

Sethi, M. (2015). Smart cities in India: Challenges and possibilities to attain sustainable urbanisation. *Naga, 47*(3), 20–37.

Smart city features. (2017). Govt. of India. http://smartcities.gov.in/content/innerpage/smart-city-features.php.

Smart city - IoT Agenda. (2017). https://internetofthingsagenda.techtarget.com/definition/smart-city.

Smart grid - United States. (2019). https://en.wikipedia.org/wiki/Smart_grid_in_the_United_States.

Smart city profiles. (2018). http://smartcities.gov.in/content/innerpage/smart-city-features.php.

Smart home management. (2018). https://www.iotforall.com/use-case/smart-home-management/.

Smart grid - Indian power sector. (2018). https://www.ecoideaz.com/expert-corner/smart-grid-systems-in-india.

Smart grid platform - control and monitor smart devices. (2019). https://opensmartgridplatform.org/.

Smart Cities and Healthcare. (2019). A systematic Review. *Technologies, 58.* https://doi.org/10.3390/technologies7030058.

Top 10 growing cities in the world. (2020). https://www.asme.org/topics-resources/content/top-10-growing-smart-cities.

Uitp - international association of public transport (India). (2019). https://www.uitp.org/regions/india/.

Urban mission in India. (2017). https://smartnet.niua.org/missions.

Urban population growth available. (2020). https://www.who.int/gho/urban_health/situation_trends/urban_population_growth_text/en/.

Urban solutions Handbook 2017. (2019). https://www.wwf.se/dokument/urban-solutions-handbook/.

Urbanization causes and impacts. (2020). https://www.nationalgeographic.com/environment/habitats/urban-threats.

Worldwide digital transformation spending Guide. (2020). https://www.idc.com/getdoc.jsp?containerId=IDC_P32575.

Zettabytes of the data world - prediction for 2025. (2018). https://www.networkworld.com/article/3325397/idc-expect-175-zettabytes-of-data-worldwide-by-2025.html.

Chapter 3

Artificial intelligence and machine learning approaches for smart transportation in smart cities using blockchain architecture

S. Sundaresan[1], K. Suresh Kumar[2], R. Nishanth[3], Y. Harold Robinson[4], A. Jaya Kumar[2]

[1]*Department of ECE, SRM TRP Engineering College, Trichy, Tamil Nadu, India;* [2]*Department of CSE, IFET College of Engineering, Villupuram, Tamil Nadu, India;* [3]*Cochin University College of Engineering Kuttanad, CUSAT, Kuttanad, Kerala;* [4]*School of Information Technology and Engineering, Vellore Institute of Technology, Vellore, Tamil Nadu, India*

Chapter outline

1. Introduction

Artificial intelligence (AI) and machine learning (ML) are the wide and major areas of computer science that brings effectiveness with some of the machine functions similar to the human brain. In traditional techniques there are lots of difficulties, and challenges were addressed by various research works in the

Blockchain for Smart Cities. https://doi.org/10.1016/B978-0-12-824446-3.00009-0

field of computation (Rusul, Hussein, Sohani, & Asadi, 2019). AI was initially brought to reality by John McCarthy in 1956, but it was botched to accomplish its main purpose (Artificial Intelligence in Transportation, 2007), and due to the nonexistence of revolutions in computational technology, it has not attained much attention and made it less promising. Artificial Neural Networks (ANNs) are considered to be one of the most esteemed AI systems employed for the analysis of various applications. The most similar kind of ANN is the feed-forward neural network, by which the data travels from the input layer into the hidden layer into the output layer in one direction. The Convolutional Neural Network (CNN) (Jin, McCann, Froustey, & Unser, 2017; Krizhevsky, Sutskever, & Hinton, 2012; Kim, 2017) along with the Recurrent Neural Network (RNN) (Caterini & Chang, 2018; Du & Swamy, 2014; Zhang, Yin, Zhang, Liu, & Bengio, 2018) is considered as the alternate forms of ANN. For image processing tasks, the CNN performs better at the time when the RNN might progresses a sequence for the input data on the way to develop a well-suited system for many applications like the languages, written contents along with text recognition. Because of the many hidden layers built in their architectures they are also called deep learning techniques. There are several inconsistencies and differences in the data that cannot be overcome with conventional techniques. Thus, the use of AI provides certain uncertainties in association with certain models that generates a typical relationship among the cause and effect of many number of real-life circumstances by integrating accessible data with premises and prospects intended for improved investigation (Patterson, 1990). Transport glitches are becoming much higher and provide a challenge for the system; the actions of users are too complex for modeling and forecasting the most renowned travel patterns. To address the various challenges that determine the growing demand of travel along with CO_2 pollution determination, safety and privacy related issues in addition with environmental degradation, AI is therefore considered to be a strong match for transportation systems. These types of challenges might keep track of the steak growth from the steady growth, considering the rural and urban traffic conditions that occur due to the increasing inhabitant's statistics, particularly in most developing countries. In countries like Australia, the population is predicted to rise around 30 million by 2031 and the level of congestion is identified on the way to reach around 53.3 billion (Australian Infrastructure Audit, 2015). Smart transportation is enabled by the usage of number of technologies like Internet of things (IoT), GPS, sensing and wireless technologies, and so forth. The smart transportation-related applications are commonly used applications that are widely spread throughout the work. The related applications are considered to be a tedious or complicated one as the data and the information it can handle is of dynamic, i.e., continuously varying, data. Some of the smart applications include smart ticketing system, parking system, smart driving systems, smart traffic control and management systems, traffic light controlling, accidental and health care systems, disaster management, and precautions systems.

Thus, the chapter describes some solutions for the state of art methods as an accumulation of problems that are raised, which makes the smart city application most ineffective. The various problems arise are taken into account and the considerable solutions demanding the efficiency is provided. The forthcoming section describes the solution for most of the problems discussed earlier in this chapter. In addition to that some of the improvement in the smart transportation is also discussed here.

2. Related works

The proposed in which AI applications that illustrates some real values is the vehicle monitoring happening in network of transport (Rusul et al., 2019). An Automatic Vehicle Location (AVL) system was implemented for increasing the competence of public transport operations that could maintain the operational controls, which can improve the overall quality that can be enabled for public transport services (Barabino, Di Francesco, & Mozzoni, 2015). The system is capable of collecting data from tracking the units of transport in real-time by means of GPS signals, detecting issues informing vehicles of any changes, and managing alternate routes. It also provides passengers with information on their mobile devices, through APPS. Article (Barabino et al., 2015; Tilocca et al., 2017) show a good case study of the AVL-based GPS system that is implemented in the Italian city of Cagliari. At every instance of time the real-time data of the vehicle location has been updated to the AVL control center which is used for purposes like enhancing the reliability of the transport services that are needed publicly. The movement of the vehicles at the traffic signal is given priorities and the status of the vehicle is updated via short messaging service to the passengers for providing the reliable sharing of information. The details about various challenges in transportation and how can it be able to resolve those problems and make it reliable as well as an efficient transportation and smart urban locality design by reducing the impact on environment and community are well represented by Sadek (2007). Pečar and Papa (2017) presents some transportation issues of the regular (brilliant) city and focuses to their likely arrangements. A portion of these worries are as of now somewhat fathomed, yet open to more innovative work so as to discover far superior arrangements. One of the most basic parts of city availability is versatility (traveler or cargo) that ought to be empowered by Intelligent Transport Systems (ITS). Initially lots of drivers are working for schools and colleges on continuous regular work. The congestion is predicted to be more because of the increase in the strength of vehicles approaching the road at the same time. Also at the same time the garbage collection vehicles and the sanitation companies are moving in the road always and the persons who deliver the goods, i.e., the home delivery persons are moving always with the road. The various levels of shared economy is continuously interrupting or disrupting the mobility market now in current years (Pečar & Papa, 2017). This is also for the synchronization of the traffic light and also the traffic control center virtually.

The most important method for modeling the dynamics of traffic flows to ensure proper traffic light synchronization is the traffic network simulation. And it helps identify the positions and causes of congestion when applied with enough knowledge of the actual network. The various solutions are suggested for the determination of the various traffic conditions. The traffic flow is predicted for ensuring the road safety measures in the most useful way. The necessary assessment for the congestion of traffic is assessed for determining the congestion over the most effective decisions used for planning (Litman, 2015). The congestion in the traffic taking place is the most known time which is done irrespective to the end of the week. The conditions include, for example, the bad weather, poor maintenance of the road, traffic signal malfunctioning, local and minor accidents that occur in vehicles due to certain delays (How to: minimize freeway & road travel times in Los Angeles, 2015). The increase in the travel time and the variation in the consumption of the fuel viz., the increase in the fuel usage which makes the efficiency to be lower and reliability to be higher. Moreover, the flow of traffic related information that gathers from all the roads and the lanes that are separated might allow for the vast area for synchronizing the traffic lights. Based on the information the optimum light segment for the traffic are calculated dynamically for allowing the smooth and optimized form of traffic globally. Traffic managing centers typically use several sensors, often inductive loops and cameras. These sensors can be fairly costly to mount and maintain. The virtual traffic center for controlling has recently been established in Morlion (2016). The virtual sensors are used, which acquire various data from social medias, such as Facebook, Google, and so forth, and aggregate those floating vehicle data. This information can be utilized for different purposes: from age of street-arranged data sets without any preparation (Kores, Pavlič, Pečar, & Novak, 2017), to deciding the vehicle density as well as traffic patterns of any urban or rural roads and traffic conditions that occur in real-time. It used an autonomous system that can alert registered users to take a path that is already congested and the intimation is obtained for avoiding the area. There are a lot of methods used for enhancing the traffic using an advanced scheme. The suitable detour is the most recommended scheme for the identification of the traffic flow. Accessible by smart maps, alternative routes are often useless because of incorrect alternate best routes (Abraham, Delling, Goldberg, & Werneck, 2010) that serve as the multiobjective of suitable substitutions. The set of rules emerging (Duque, Lozano, & Medaglia, 2015) is good at discovering some alternate routes, but in some cases the need for the prediction of traffic is taken over for a long time and that is somewhat tough to proceed. Moreover, if all cars are moved to the similar bypass, there could be congestion, and hence there is a necessity for the right diversion and vehicles that is distributed along the possible and most alternative route for reducing the unexpected delay due to waiting time.

A system known as smart transportation in addition with the driving that might include the vehicles that are equipped with some capabilities for the sensing of information and the processing of communication, distributed computing, and the processing by means of enhancing the safety, performance along with the quality of service for drivers along with passengers is been proposed. By avoiding unnecessary cabling, an intravehicular wireless sensor network (IVWSN) the weight of the vehicles is reduced (Lin, Talty, & Tonguz, 2015). An IVWSN can work with bluetooth low energy (BLE) or ZigBee and must meet the accompanying objectives: sensible transmission rate, low idleness, fixed sensors, and strength to impediments in the sent condition. BLE has been demonstrated to be savvy by an achievability concentrated into potential remote correspondence advancements for an IVWSN. Similarly, it has a higher information rate than different advances, flaunts low postponement, and has a decent interoperability with convenient gadgets (Smart city system design). have suggested several solutions such as ITSs and urban planning for addressing those challenges and have come up with a new well-nourished smart city infrastructure by incorporating the machine intelligence in modern implementations. The idea behind their work is to parcel savvy city frameworks into applications, detecting, correspondence, security, and information planes and put noticeable quality significance on the information plane as the fortification of processing and information stockpiling. In Sundaramoorthy, lango, and Tamilarasan (2018) proposed a system for counting the passengers who were traveling on the bus by the use of Infrared (IR) sensors. In this system the smart bus passenger information and safety management is considered and it became another area of importance. For this a system using the Global System for Mobile Communication (GSM) module and IR sensors is studied. In these two IR sensors are used to count the boarding and deboarding of the passengers in the bus and one sensor is used to check the door statuses of the bus whether it is open or closed. The Liquid Crystal Display (LCD) device is used to display the number of people traveling in a bus and to display the door status. The information is continuously monitored and transferred to the higher authority manager of the bus, which helps give the authority to validate ticketing the data with number of passengers boarded on and deboarded from a bus and to check the compliance of safety standard. Bernini, Bombini, Buzzoni, Cerri, and Grisleri (2014) proposed a system that is used for calculating the passengers who were moving in and out the bus. This is an autonomous system, and it counts the details automatically. The artificial intelligence-based camera is used for capturing the image of the patients who were moving in and out the bus. Karaduman, Mucahit, and Eren (2017) analyzed the importance of smart vehicles since they are considered to be the most important part of the smart city. The accident care system that is implemented in the smart cities are enlightening the human lives. The driver's condition is analyzed periodically for the safety atmosphere creation. The sensors are used for predicting the condition of the driver. Kuge and Nobuyuki (2000) proposes a behavior recognition model for the driver by using AI and ML methods. This was enabled to find the driving methods and the driver behavior that can be predicted in an advanced way. The recognition is accompanied with the modeling. Oliver and Pentland (2000) designs a smart car that is used for the acquisition of the data and the machine learning models are used for the prediction of driver behavior. The designed car consists of certain sensors placed at various points and that help to identify the typical movements of the driver. The automated car that is engaged with a human-machine interface along with the driver helping system that provides most realistic support in some critical situations. Kazi, Bagasrawala, Shaikh, and Sayyed (2018) provides a method for ticketing scheme in the buses. The system consists of an intelligent application that will distribute the seats to the passengers automatically and the ticket reservation is done digitally. Here the mode of payment for the ticket is done through cashless mechanisms. Pečar and Papa (2017) presented some related problems in the smart city in association with mobility and transportation. Most of the problems are tackled, but there are some issues related to the research and the development for more effective approaches. The important problem in the implementation of this mechanism in the smart city and the technology adopted for the improvement of the efficiency of the smart transportation is somewhat critical from the user's point of view.

3. Proposed system

The proposed system consists of a smart bus that includes some of the advanced facilities using blockchain technology. The concept of ML and the AI is added as an additional feature to enhance the improvement in the smart transportation system. The advanced facility that gets incorporated in the smart bus is discussed here and limitations of state of the art techniques are dazed.

3.1 Blockchain-based smart vehicle for public transportation system using artificial intelligence

The system proposed for the smart transportation system is the use of smart vehicles. Regardless of the owned vehicles most of the peoples use the public transport. The usage of bus for public transport is most inevitable. The role of AI plays a vital role in accessing the beneficiaries of ITS that uses a smart bus with lot of smart facilities. The proposed system consists of some added facilities in-built in the bus, that will enhance the smart transportation mechanisms. Most important need is to tally the persons or passengers who were boarded and deboarded from the bus. Also, this system monitors the behavior of the passengers inside the bus. Fig. 3.1 shows the block diagram of the first module of the proposed system. The explanation of the components utilised is given below.

3.1.1 Arduino Uno

The Arduino Uno is one of the microcontrollers belongs to Atmel megafamily. This consists of 14 pin input and output pins where only six pins can be used for getting pulse width modulation (PWM) outputs. The resonator, USB connection, and other buttons are available, which helps to run the motor. This kind of device can be made usable with both hardware and software. Hence this can be implied effectively on the embedded platform, which is made flexible. The Arduino Uno can be programmed in such a way that it can receive the inputs from all the available sensors placed. Fig. 3.2 displays the Arduino Uno microcontroller used in this proposed system.

The programming on Arduino is totally based on its wiring modes. The Arduino can act as a standalone circuit and in turn can manage communication with the running software within the computer hardware like processing unit

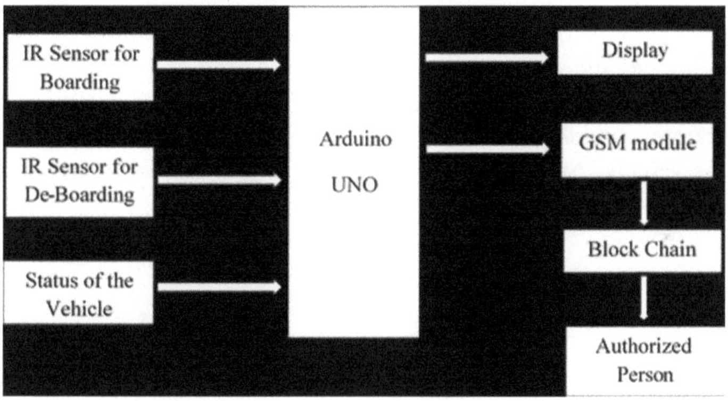

FIGURE 3.1 Block diagram for passenger information.

FIGURE 3.2 Schematic of arduino UNO.

and the flash memory. While comparing to all other processing units the rate of this is considerably low. The Arduino software uses the platform to run on the Linux operating system.

3.1.2 IR sensor

The IR sensor is used for the gathering of information during the boarding and deboarding of the passengers from the bus or any other vehicle, and the same can be used for predicting the status of the door whether it is open or closed and it also identifies the possibility of occurrence of faults in the doors. This will check the interruption of any persons or objects by tracing the IR radiations while falling over the device. The IR sensor consists of a couple of sections, i.e., the recipient circuit and the producer circuit. If there occurs any intermediate points between the input and the output circuit, the count will gets increased and if there is no intermediate points occurs means the count gets decreased. Fig. 3.3 shows the IR sensor unit used in this proposed system.

This type of sensor can also be called a photo coupler. The LED is considered to be the circuit for the emitter and the circuit used for the receiving purpose is the photo diode. The LED emits certain lights and the light emitted by the LED is much sensitive to the light. The emitter which receives the light is proportional to the output voltage. This sensor can be used for measuring the characteristics of the environments just by detecting and emitting the IR radiation. The hardware counter might use a bidirectional counter and it is the decade counter, which is used for counting the persons who is entering and leaving the vehicle. Another sensor used is to identify the faults in the door and also to check the status of the vehicle door.

FIGURE 3.3 Sensor.

3.1.3 GSM module

The communication in between the mobile device and the embedded device is GSM. In the GSM device, the modulation and demodulation of the signal takes place from any type of network that in turn permits internet connectivity. This device contains modules with the SIM card, analog-to-digital converter, and the power supply unit. Time division multiplexing is used in this system for communication purposes. The main purpose of this device is to reduce the size of the data by digitizing it and send the same to the channel with some gushes of customers. The implementation domain in the system is responsible for the variation in the cell. The system used might contains a mobile phone with transceiver section and is controlled by a SIM card that operates throughout the network. This will be acting as an interface in between the mobile station and the network. The diagram of the GSM module used in this proposed system is displayed in Fig. 3.4.

The transceiver section might comprise of some radio receivers for effectively handling the mobile communication by means of needed protocols. The controller for base station and the handset controller is added with this device in addition. The portable stations will have some essential system in association with it. The main purpose of the system used is to focus on the administration, i.e., for giving access to the other system. The additional task performed by the system is to keep a list of sending profiles. The information regarding the call list is stored in the cloud. Each device is enhanced with maximum security by providing the IMEI number.

FIGURE 3.4 Schematic of GSM module.

3.1.4 LCD device

LCD is one of the most widely used device in many of the projects and circuits. This device can display both digits, characters, and numbers. This consists of a set of serial and parallel pins that are much easier for interfacing with the Arduino. Lots of various sizes are in LCD. The number and character inside the LCD is displayed with rows and columns. For this, it consists of 16 pins. The column number is represented by X and the row number is represented by Y. The formula for the design of the LCD display is denoted as (X * Y).Totally our proposed design consists of 32 columns and eight rows in an LCD display device. The ST7066 controller is used for building the alphanumeric displays for the 32 characters and 8-line display. The power supply used in this device is of 5−7 V. It avails the support of an LED display with various different color textures. The LCD character finding is somewhat very well-known compositions. Normally the character recognition of LCD is similar for the pin-out composition. It might not have the control option for back light as the total pin count dedicated to this is 14. sometimes it may have the backlight facility and sometimes not. For controlling the backlight of LCD some extra pins are used instead of already existing pins. For displaying the data, the basic devices are used, which has cathode and anode. The LCD is fixed near the door end and the boarding and deboarding of the passenger count is being displayed. Two entries in total are in the smart bus: one for entry and the other for exit. It uses some IR sensors for counting the number of people entering the bus and leaving the bus. The sensors used at the door steps will also provide some security enhancements to the environment. The sensors output gets displayed in the LCD screen that is placed nearer to the authorized person particularly the driver which in turn visible to the checker. The closing and opening of the door is intimated periodically with a voice over for alerting the passengers to know about the stoppings. Also the intimation of the stoppings are messaged as short text to the passengers via GSM module. The message sharing by using the GSM module is enhanced by the use of blockchain. The information transferred might need some privacy and security, hence the privacy is enabled by the use of blockchain-based technology. The whole process is done automatically using the AI mechanism.

3.2 Smart ticketing system using blockchain technology

The system using electronic ticketing is ruling the smart bus ticketing system. This reduces the manpower requirements and eliminates lots of issues faced in the normal ticketing system. It is nothing but a paper ticket in digital form. In normal airlines and some of the other transportation system uses the same electronic ticketing system in the form of travel card. But here in this smart city public bus transportation, normal ticketing is advanced to the electronic ticketing scheme. The parameters used for normal travel are included in the ticket including location, count of the tickets, and the starting point, stopping point, etc. Smartphones are used for the ticket generation and also the tickets can be obtained using normal short messaging service. As there exists some advancement in the smart bus, the need for changing the ticketing sector is mandatory in terms of centralized activity. The blockchain technology plays a vital role in the generation of the electronic ticket. It creates a decentralized environment which enables a secure transmission of money for getting the tickets and thus enables a safety environment and ease of accessibility in satisfying the smart traveling criteria. By using the blockchain-enabled ticket distribution the time for waiting at the stops is reduced, the wastage of paper is reduced, and the seat availability is streamlined.

This is achieved by introducing a scheme of online registry of the user at the beginning, and an application is generated and it allows the user to select the tickets and allocate to them for all mode of traveling. Once the ticket is generated, the details such as a starting point and end point is recorded in the system with the help of blockchain-based cloud computing technology. The usage of the blockchain will enhance the security by safeguarding the secret information of the passenger. The application consists of a list of buses for the same route and the availability of the seats for the points. After the registration the user is allowed to fill the destination alone for time saving purposes. The arrival of the vehicle at the stop is notified earlier with a message. Once getting on the bus the indication of the end point is notified with the help of short messages periodically to alert the passenger. The GPS-powered device is available with the driver and the location of the travel and the intimation of the stops is initiated periodically to the passenger. The allocation of the seats to the bus is done with the help of a generated algorithm that implicates a queue-based approach. The first passenger in the queue is given preference and when the seat got vacant, then the lined passenger will get the seats allocated. The transactions are done in a safety basis with the use of blockchain technology, and a needed amount is deducted from the passengers' accounts directly based on the ticket cost. The acknowledgment for the deduction of the amount is sent to the passenger via short messages. The cancellation of tickets facility is included in the smart transportation system.

If the passenger does not use a smartphone, then the passenger can book the ticket using the device mounted at the bus terminals or bus stops. The ticket generation is done through normal short messages. The role of the conductor is just to check the ticket and the allotted seat number. At the initial stage the users get logged on to the system and the system will verify the users authentication device which is installed at the bus terminals for allowing the users to book the ticket. The source of the passenger is determined by the system using the authentication device. The whole information about the passenger is stored at the cloud server using IoT. The real-time tracking of the information is made possible by altering the destination of the user from the time he entered the bus. The bus for the various route numbers are the same and so all the entries of the bus is based on the bus number and the number plate. The diagram corresponding to the proposed system is sketched in Fig. 3.5. This kind of electronic ticketing system enables the smart transaction with the reduction in the cash usage thus by supporting the cashless environment as the transaction is done automatically by secure mode using blockchain.

3.3 Traffic mode prediction using machine learning

Traffic mode prediction is one of the most important aspects that gets included in the proposed system. The smart bus proposed here in this system have the facility of detecting the traffic in prior by using ML techniques. Normally the traffic flow data includes four major features including problems occurring due to noise, correlation in between the temporal and spatial parameters, dynamic sequential property, and incompleteness. The aforementioned issues might make the management of the traffic system more complex. Here a model based on median rank dynamic mode decomposition (MRDMD) is proposed, which can solve problems related to traffic issues, thus by predicting the flow of traffic by using the state transition matrix that personifies the similarities between the neighboring elements of traffic flow with median rank regularization. Many studies have been conducted for the prediction of the traffic flow at various time constraints by mentioning the existing methods.

At present all the cities in and around the world might have a predicted count of vehicles (approximately 40 million motor vehicles) running throughout the year. As the vehicles are increased year by year the condition of the traffic is becoming so complicated and the congestion on traffic is occurring frequently. The brilliant transportation system management effectively utilizes the flow of traffic prediction and it itself has a spread over applications in the guiding of paths, planning of paths, detection of events,

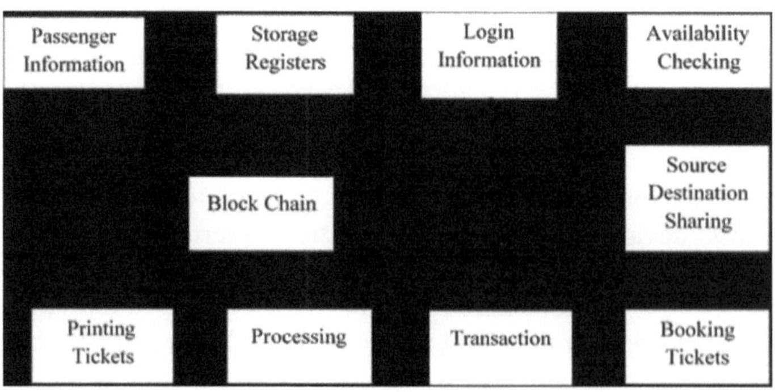

FIGURE 3.5 Smart ticketing system.

and so forth. The efficiency and the accuracy of the flow of traffic is predicted much closer to the city traffic management. The prediction of traffic at the earlier mode can give reliable indication for smooth transportation. Normally the prediction of traffic at low latency is a challenging task. The critical parameters like congestion, velocity and count of the vehicles are the parameters considered for the research. These are collected from the sensors and cameras placed at the road side and at the street lights placed near to the road. For analyzing some parameters, the sensors are fixed on the road. The sensors and the cameras are the most renowned detectors used most commonly for the determination of the traffic in some critical areas. These are installed with low cost and in most convenient manner. Instead of using these two devices we include the use of loop coil detectors that are fixed on the road at about 3–4 cm depth. This form of installation will safeguard the device from the natural calamities like rain and wind. Especially this kind of device might provide information about the congestion in the traffic and the distance between the vehicles.

Basically, the prediction of traffic can be done by using three methods: ML methods, Fuzzy logic methods and the statistical methods. The statistical method is used to implement statistics analyzing tool for the prediction of traffic. But in case of ML methods the prediction is done by training massive amount of obtained data to find the unknown data approximately. The statistical and the fuzzy based methods provide some of the rules for determining the traffic, but the short duration prediction is not possible in this case. But the ML methods are suffering from lot of issues by getting enough amount of data's needed for training. The proposed MRDMD for the low-term prediction of traffic flow uses a state transition matrix that characterizes a median rank relation between the nearby elements of congested area.

The idea used for predicting the most relevant information related to traffic is done by using ML concepts. Initially some of the sample information is taken in advance and that information is trained to the system by using some relevant algorithms like support vector machine and Hidden Markov Model. The samples are trained and make the system most comfortable to accept all the relevant traffic related issues. The CNN method is used for the effective training. This is achieved by establishing the differential equation thereby using the data sequence obtained originally. This provides a good prediction accuracy when compared to the previous methods. Since the pedestrians are considered as one of the factors for the traffic management, the genetic algorithm is used that includes the genetic mechanism. A model based on neural network is build which is based on the genetic algorithm for predicting the traffic flow. In general, the methods using ML for predicting the higher accuracy is done by training the large amount of data's which can even reduce the computational complexity in the prediction of traffic. In turn, it provides an effective traffic system modeling by controlling the traffic lights by producing the efficient outputs for the simulation. Also, one of the additional features added up in the system is the usage of ML algorithm for estimating the consumption of fuel along with the traffic flow prediction.

3.4 Intelligent driving mechanism

Seeking for driving safety in smart vehicles are mandatory in smart transportation that uses AI. The enhancing of comfortability and safety is much needed in the building of a smart environment. The proposed system consists of a safety mechanism that protects the driver and passengers from meeting accidents. The analysis of the driver's mentality and physical condition is checked periodically. The driver section consists of a smart device that includes certain sensors and some minor devices that brings out some parameters that helps the driver in driving smoothly. The driving assistant systems that implemented in the smart bus have some facilities can check the driver behavior frequently. For obtain this there need some arrangements in prior. That is the placement of sensors where the environmental and the biological data are retrieved from the driver's conditions. The behavior of the driver is analyzed by using the advanced learning mechanisms. The information's stored and processed must be stored in the cloud for further processing. The proposed system for the driver condition analysis is described in Fig. 3.6.

The recognition of the driver profile and the information related to the driver conditions and the environmental aspects are obtained at the time of driving. That obtained information is stored in the appropriate format. The data obtained are analyzed by determining the possible events that occurred from steering wheel. The detected events are perceived by the data that is

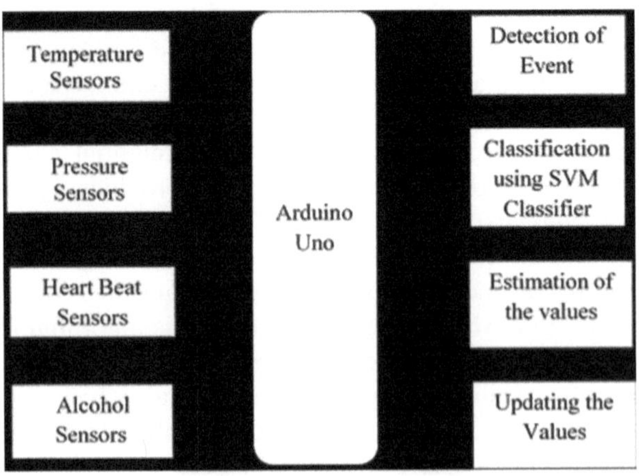

FIGURE 3.6 Alert system.

obtained at the time of driving. The events of the class are determined by the use of various estimation methods. The classification of the events will provide the analysis of the continuous driving data. Like this the overall behavior is predicted. The updating of the system is done periodically from the data obtained during real-time. The comparison of the instant data and the previous data is obtained. In addition to that the health monitoring of the driver is also considered which was taken from the sensors. The proposed system consists of various factors that combines the human and the vehicle data. The resultant driver conditions such as drowsiness, unsafe road conditions, and illness are identified and updated to the authority via short messages.

4. Results and discussion

The proposed system implements a smart bus that includes some applications like smart passenger's prediction system, smart ticketing, and brilliant traffic management systems. The installation of effective sensors and the implementation of the algorithms helped to achieve the predicted results in the way for finding the smart city environment. The application of blockchain and ML in the system enhanced the smart city applications by eliminating the security and privacy related problems that occur most commonly. The transfer of message and the transaction of money is made easy and effective by the implementation of a decentralized network using the blockchain. Initially the smart bus consists of some of the facilities like intimating the status of the passengers who were boarded and deboarded from the bus and also the door condition (whether it is closed or opened) is displayed inside the bus by an LCD display device. This information is updated to the cloud server and the intimation is given to the authority through the GSM module. Table 3.1 represents the passenger number and the status of the door (either closed or opened).

Thus the electronic ticketing system in the smart bus shows many advantages in supporting the cashless economy and reduces the maximum issues related to the security and privacy in the transactions. The proposed bus structure is shown in Fig. 3.7.

The seat 2, 8, and 17 are vacant; the black-marked seats are the allotted seats; and the gray-colored seats are blocked seats. Here the passengers having the nearest location will be allotted first. After the first passenger allotted seat is vacant then the second passenger will be allotted with that seat automatically. Figs. 3.8 and 3.9 displays the user interface and related status of the ticketing system.

The application of ML in the analysis of the traffic prediction consists of certain aspects assuming in the effectiveness of MRDMD on ITS. The analysis is done in four aspects: positive error, negative error, neutral error, and prediction against congestion. We know that the various kind of errors will create various effects in the systems implemented for transportation for MRDMD

TABLE 3.1 Status.

Passenger number	Door status
16	Open
18	Close
2	Open
17	Open
6	Open
3	Close
27	Open
30	Close
1	Close
32	Close
25	Close
20	Close
9	Close

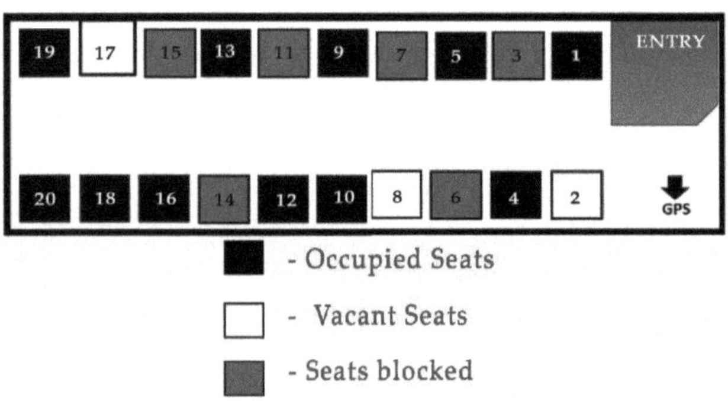

FIGURE 3.7 Occupancy status.

methods. The positive RMSE, Neutral RMSE and the negative RMSE is given in Table 3.2. The graph corresponding to the prediction error is given below in Fig. 3.10.

From the aforementioned table and graph, the details related to MRDMD value performs better. The value determines at various time intervals that the morning peak condition, the moderated peak condition, and the evening peak condition are determined effectively. The best prediction shows that the

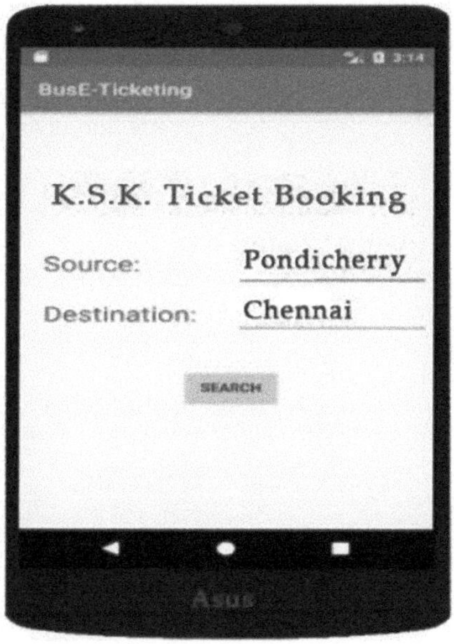

FIGURE 3.8 User interface window.

FIGURE 3.9 User interface window.

TABLE 3.2 Congestion.

Time at which the congestion occurs	Prediction error
6.20 a.m.	18
8.30 a.m.	30
9.30 a.m.	40
10.20 a.m.	25
12.00 p.m.	24
1.30 p.m.	38
3.30 p.m.	32
4.00 p.m.	20
6.20 p.m.	26
8.30 p.m.	32
10.10 p.m.	18
11.30 p.m.	19

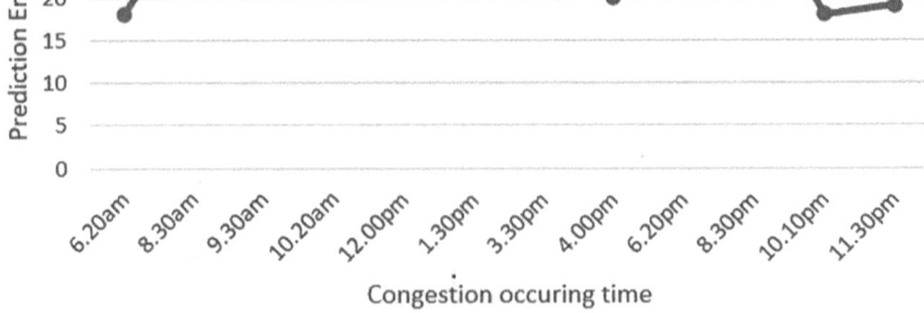

FIGURE 3.10 Congestion status.

variations in the morning peak and the evening peak are the root causes for problems in the traffic condition. The above diagram in Fig. 3.10 shows the traffic condition at specific time intervals. While the traffic prediction at different road regions or sections, the condition is given in Figs. 3.11 and 3.12.

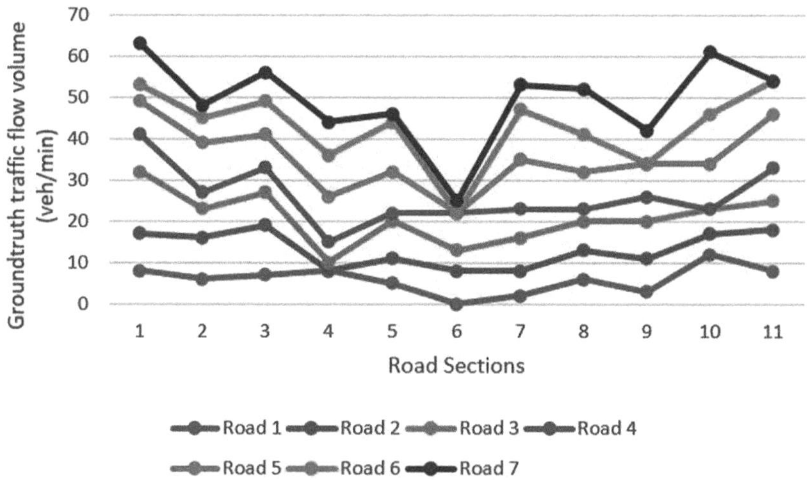

FIGURE 3.11 Traffic prediction at various roads.

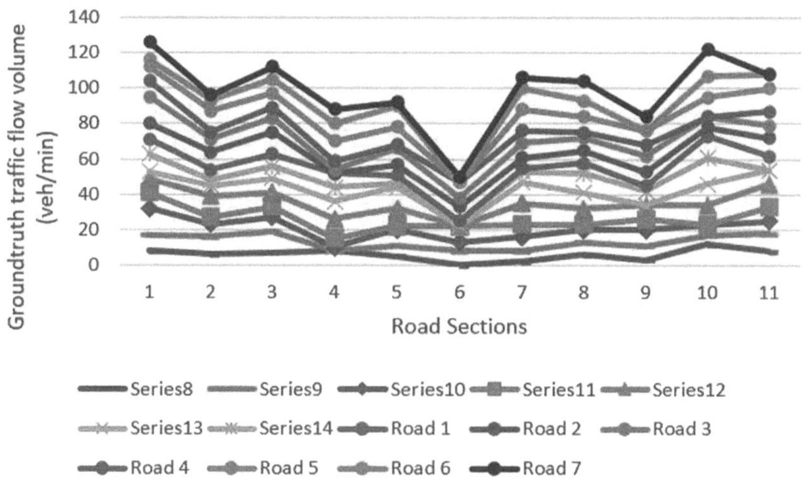

FIGURE 3.12 Comparison of traffic prediction.

The different road sections are taken into consideration for the analysis of the traffic status in both best and the worst conditions. Totally 11 different sections are taken into consideration. From the analysis the immediate changes are determined which can give clear prediction of real-time traffic flow. The MRDMD is considered better than LRDMD which is already existed.

Then considering the driver condition determination there have a great demand in this area as many accidents are occurring because of the drowsiness of the driver. The system is designed with the obtained needs. The biological

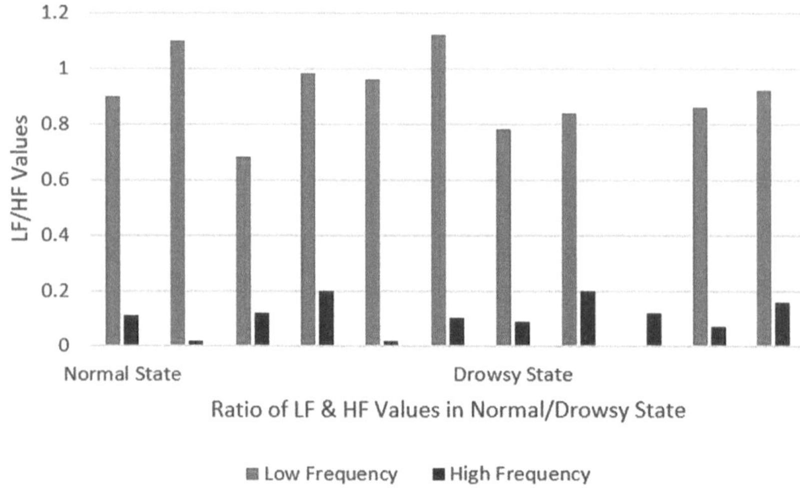

FIGURE 3.13 Driver monitoring.

details are determined from the system using some intelligent technologies. The gathered information is integrated into the system by using some smart mechanisms. Hence the driver helping systems are built by exhibiting the identical behavior to the real driver conditions. This form of systems reduces the risk of accidents. The conditions for the normal state and the drowsy state variation are depicted in Fig. 3.13.

These values could be different for determining against day time and night time. The low frequency and the high frequency values determine the actual conditions regarding drowsiness. The analysis reveals that the different values obtained at the normal state and the drowsy state conditions are used for the analysis of the driver's conditions predicted in real-time.

5. Conclusion

This chapter proposes an artificial system that uses ML concepts in association with blockchain technology. In this system some facilities are included for enhancing the facility of the smart bus. It consists of an automatic passenger monitoring and counting system that is enabled by using the IR sensors. The status of the boarding and deboarding of the passengers is displayed frequently using the LCD screen. The driver section in the bus consists of some sensors for determining the health conditions of the driver and analyses the drowsiness state. The condition before the drowsiness and after the drowsiness is determined using the HF/LF values. This helps to prevent the accidents. Then the ticketing system for the smart bus is advanced by the use of an electronic ticketing scheme. The additional facility added in this system is the traffic prediction; it determines the condition of the traffic in real-time. From the overall analysis the smart bus provides advanced facilities that are enhanced by the use of smartphones. The improvement in the technology helps to enhance the ease of living for people in smart cities.

References

Artificial intelligence in transportation. In *E-C113*, (pp. 72–79). (2007) (pp. 72–79).

Abraham, I., Delling, D., Goldberg, A. V., & Werneck, R. F. (2010). Alternative routes in road networks. In *Lecture notes in computer science (including subseries lecture notes in artificial intelligence and lecture notes in bioinformatics)* (Vol. 6049, pp. 23–34). https://doi.org/10.1007/978-3-642-13193-6_3

Australian Infrastructure Audit. (2015). *Infrastructure Australia. Our infrastructure challenge.* https://www.infrastructureaustralia.gov.au/sites/default/files/2019-06/australian-infrastructure-audit-volume-1.pdf.

Barabino, B., Di Francesco, M., & Mozzoni, S. (2015). Rethinking bus punctuality by integrating Automatic Vehicle Location data and passenger patterns. *Transportation Research Part A: Policy and Practice, 75*, 84–95. https://doi.org/10.1016/j.tra.2015.03.012

Bernini, N., Bombini, L., Buzzoni, M., Cerri, P., & Grisleri, P. (2014). An embedded system for counting passengers in public transportation vehicles. In *Mesa 2014 – 10th IEEE/ASME international conference on mechatronic and embedded systems and applications, conference proceedings*. Institute of Electrical and Electronics Engineers Inc. https://doi.org/10.1109/MESA.2014.6935562

Caterini, A. L., & Chang, D. E. (2018). Recurrent neural networks. In *SpringerBriefs in computer science* (pp. 59–79). Springer. https://doi.org/10.1007/978-3-319-75304-1_5. Issue 9783319753034.

Duque, D., Lozano, L., & Medaglia, A. L. (2015). An exact method for the biobjective shortest path problem for large-scale road networks. *European Journal of Operational Research, 242*(3), 788–797. https://doi.org/10.1016/j.ejor.2014.11.003

Du, M. N. S., & Swamy, K. L. (2014). Recurrent neural networks. In *Neural networks and statistical learning.*

How to: Minimize Freeway & road travel times in Los Angeles.(2015). http://mdscodelivers.com/tag/inrix-urban-mobility-scorecard-annual-report/.

Jin, K. H., McCann, M. T., Froustey, E., & Unser, M. (2017). Deep convolutional neural network for inverse problems in imaging. *IEEE Transactions on Image Processing, 26*(9), 4509–4522. https://doi.org/10.1109/TIP.2017.2713099

Karaduman, M., Mucahit, H., & Eren, H. (2017). In *2017 5th international istanbul smart grid and cities congress and fair (ICSG).*

Kazi, S., Bagasrawala, M., Shaikh, F., & Sayyed, A. (2018). Smart E-ticketing system for public transport bus. In *2018 international conference on smart city and emerging technology, ICSCET 2018*. Institute of Electrical and Electronics Engineers Inc. https://doi.org/10.1109/ICSCET.2018.8537302

Kim, P. (2017). *Convolutional neural network; MATLAB Deep learning.* A Press.

Kores, A., Pavlič, M., Pečar, T., & Novak, T. (2017). *Method, device and system for modeling a road network graph.*

Krizhevsky, A., Sutskever, I., & Hinton, G. E. (2012). Image net classification with deep convolutional neural networks". *Advances in Neural Information Processing Systems, 60*, 1–9.

Kuge, N., & Nobuyuki, T. (2000). A driver behavior recognition method based on a driver model framework. *SAE Transactions*, 469–476.

Lin, J. R., Talty, T., & Tonguz, O. (2015). On the potential of bluetooth low energy technology for vehicular applications. *IEEE Communications Magazine, 53*(1), 267–275. https://doi.org/10.1109/MCOM.2015.7010544

Litman, T. (2015). *Comprehensive evaluation of traffic congestion costs and congestion reduction strategies*.

Morlion, P. (2016). *Traffic management as a service*. https://www.linkedin.com/pulse/traffic-management-service-pieter-morlion.

Oliver, N., & Pentland, A. P. (2000). Driver behavior recognition and prediction in a SmartCar. In *Proc SPIE int soc opt eng, 4023*.

Patterson, D. (1990). Introduction to artificial intelligence and expert systems. In *Introduction to artificial intelligence and expert systems*.

Pečar, M., & Papa, G. (2017). Transportation problems and their potential solutions in smart cities. In *Proceedings of international conference on smart systems and technologies 2017, SST 2017* (Vol. 2017, pp. 195–199). Institute of Electrical and Electronics Engineers Inc. https://doi.org/10.1109/SST.2017.8188694

Rusul, A., Hussein, D., Sohani, L., & Asadi, B. S. (2019). Applications of artificial intelligence in transport: An overview. *Sustainability, 189*. https://doi.org/10.3390/su11010189

Sadek, A., et al. (2007). Artificial intelligence in transportation. *Transportation Research Circular, E-C113*, 72–79.

Smart city system design: A comprehensive study of the application and data planes. *ACM Computing Survey, 52*.

Sundaramoorthy, R., Ilango, Y., & Tamilarasan, A. K. (2018). Smart bus passengers information and safety management system. *International Journal of Pure and Applied Mathematics, 119*, 795–800.

Tilocca, P., Farris, S., Angius, S., Argiolas, R., Obino, A., Secchi, S., et al. (2017). Managing data and rethinking applications in an innovative mid-sized bus fleet. In *Transportation research procedia* (Vol. 25, pp. 1899–1919). Elsevier B.V. https://doi.org/10.1016/j.trpro.2017.05.184

Zhang, X. Y., Yin, F., Zhang, Y. M., Liu, C. L., & Bengio, Y. (2018). Drawing and recognizing Chinese characters with recurrent neural network. *IEEE Transactions on Pattern Analysis and Machine Intelligence, 40*(4), 849–862. https://doi.org/10.1109/TPAMI.2017.2695539

Chapter 4

Blockchain architecture for intelligent water management system in smart cities

S. Sundaresan[1], K. Suresh Kumar[2], T. Ananth Kumar[2], V. Ashok[3], E. Golden Julie[4]

[1]*Department of ECE, SRM TRP Engineering College, Trichy, Tamil Nadu, India;* [2]*Department of CSE, IFET College of Engineering, Villupuram, Tamil Nadu, India;* [3]*Department of CSE, Rajalakshmi Institute of Technology, Chennai, Tamil Nadu, India;* [4]*Anna University-Regional Campus, Tirunelveli, Tamil Nadu, India*

Chapter outline

1. Introduction

In recent years the need for fresh water has increased; it is now considered to be more a precious commodity. Recent analysis suggests that about 60% of the world population is suffering due to the scarcity of fresh water. The need is increasing day by day as the population is increasing further. Normally more water management methods are proposed by various researchers, as they have invented some advanced systems like supervisory control and data acquisition systems (SCADA). In this system there are certain limitations such as the

Blockchain for Smart Cities. https://doi.org/10.1016/B978-0-12-824446-3.00006-5

installation of the system is considered to be more complex, and there are issues with maintenance. Because of the maintenance increases, the cost is becoming higher. The introduced systems like SCADA and other systems are not providing sufficient support in getting the demand of the fresh water because of certain limitations. Hence the form of smart city implementation is somewhat difficult because of these challenges. In order to face the growing demand for fresh water, authorities have decided to encourage solutions that can provide effective and efficient method of handling the waste water and make it usable for the household uses. The solution also gets accompanied with the reduction of water waste. The implemented system for smart waste water management is meeting the increased demand of fresh water use in smart cities by the method of detection of available water sources and preventing the contamination of water from chemicals, and also predicting the overflow that occurs in sewage. These processes of water management are done by using sensors, which is based on the Internet of things (IoT) (Vakula & Kolli, 2017).

The combination of the biological and chemical process is used in the treatment of wastewater. The technology used here is working without the intervention of electricity. The maintenance needed is and manpower needed is lower. The most important thing to consider is if this system is self-sustainable and ecofriendly. The adaptive technology is an effective technology for water pollution control. It is being used as a pollutant sink because it sinks the nutrients and metals from the water sources.

Many mechanisms are involved in the treatment of wastewater in wetlands. The important measures to be used will consist of sedimentation and filtration mechanisms. The bacterial actions and the uptake of nutrients are also involved in the analysis. The mechanism consists of a set of independent cells that can support the transformation of various sizes of gravels and stones in which the anaerobic digestion takes place in the cell, which is made of different layers that support the gravel as like the advanced filter cell used in wetlands. The technology would treat the water wastes by natural treatment methods by not adding any chemicals. The accomplishment of biological factors involved in the usage of water-using plants and semiwater-using plants are taken into consideration for the treatment. The engineering parameters are utilized in an effective way for the treatment of the waste water using biological factors (Lim, Kim, & Jung, 2018). The wetland which is filterable will be planted with amphibian and additionally the semioceanic plants where the wastewater will get streamed is processed vertically for better water power and sufficient maintenance period. Primary and secondary treatment of the water wastes from the municipality and the sludge management units will make use of it. The landfill leachates will be responsible for the treatment of the agricultural effluents and the industrial sewages.

Abnormal climatic conditions for local environments have been established; the changes occur globally. In context to changes, smart cities now tend to establish the response in case of emergency to ensure the local residents a lifeline. This has involved greater efforts for undertaking the activities of water management by considering the philosophical effects in which water has improved human life. The additional expansion considered in case of the urban areas are associated with the urban population and the industrialization terribly increasing its worst conditions. The maximum rise in areas of the urbanization yields increased results in case of straight out flows that in turn results in the reduction of the water resources availability which includes the variable count in damage occurring due to flood which inhibits the water resources supply (Lundin & Morrison, 2002; Wintgens, Salehi, Hochstrat, & Melin, 2008). Imbalances occur in the water cycle such as ground water depletion created by increases in areas that are impermeable because of urbanization, supply water scarcity, water pollution occurrence like floods, and so forth (A Synthesis Report on the U-City Testbed Construction, 2008). These kind of issues are resolved with the help of urban systems for water cycle treatment, which can improve the effectiveness of the water cycles. The techniques used for the managing the recycled waste water is used from sources like rain-water harvesting plant, ponds, and reservoirs (Korean Society, 2010).

Thus, the blockchain technology for the management of water resources are analyzed, and the key requirements that enable the most secure technology are discussed in this chapter. The remaining part of this chapter shows various directions in the field of water management using IoT and blockchain technology.

2. Blockchain in water management

The available water resources consume of about 85% of the agricultural usage. This was considered to be the most predominant uses in the total need. The sustainable management of water is indeed an imperative necessity, as it is the most important source of any kind of biological need (Gutiérrez, 2013). The water policy standards followed in the European countries would emphasize that the most well organized and precise controlling of water is necessary for achieving the sustainable need and the resource management of the planet. Hence the requirement of water to the service providers directly should be measured from the total consumption. The emergence of the IoT can possibly embed the energy-efficient mechanisms that are connected to the devices within the system used for the distribution of the water. The consumption of the water is monitored much accurately and automatically. The technologies like the low-power wide area network (WAN) might provide the pervasive and reasonable connectivity including the service providers and the stake holders (Adelantado et al., 2017).

The combination of blockchain advances into IoT frameworks is extensively being researched in several areas (Salek et al., 2019). Specifically, for overseeing assets, for example, water and vitality, such coordination would help with several business strategies in, for example, book-keeping, charging, and conveyance (Lazaroiu & Roscia, 2017). The trust-less idea of a blockchain-based framework empowers the consistent incorporation of a few outer entertainers that, because of the straightforward nonrenouncement of the information by which the public blockchain might store the information that could straightforwardly confirm and even reward end-clients dependent on the genuine utilization of the observed asset. The minor analysis that has been implemented in blockchain technology might determine the available IoT systems for agriculture. The deployment will get favored with the low cost and the efficiency of power (Khutsoane, Isong, & Abu-Mahfouz, 2017). These are normally interpreted by the sectors of the device integrators, and the hardware device choice is made interact directly with the infrastructure, which is based on blockchain. Considering all the mentioned cases, the sensor nodes are not listed as direct actors into the system, which is based on the blockchain, may not manage the identities of the blockchain. Subsequently the involvement of third parties and the intermediate persons who were marking the transactions regarding the sensor nodes which introduces the security measures within the corresponding system (Lin et al., 2017) thereby reducing the dependability of the assimilated data. This is the reason by which the accessibility of the IoT devices are dealing with the requirements in participating the direct actions directly in between the system which is based on blockchain. For improving the distribution of the water resources the intelligent meters are used along with the devices which are used for sensing. This is mainly for the reduction of energy consumption. Normally a blockchain gives an exchanging stage to trading tokens that lessen the expense of energy. A friendly game-based approach (Cristina & Giacomo, 2017) is proposed for the analysis of behaviors that occur socially in case of water and the energy needs by which the infrastructure using the blockchain might provide a more secured storage area for the storing of the obtained data by the involvement of IoT devices that are owned by companies' utility that are dedicated to the user's need and how they consume water resources most efficiently and effectively. The obtained data is then utilized by the gamification cycle, unmistakably profiting by the blockchain straightforwardness to forestall exploitative practices. Generally, like our proposition, as of late, the creators of (Bordel et al., 2019) present a framework to productively oversee and arrange the utilization of water by a water system network where a few people assemble for better admittance to the asset. For this situation, a blockchain empowers trust among network individuals dependent on the data gathered by some IoT gadgets.

2.1 Industrial water management

Industrial applications might consume some reasonable amount of water from the overall usage. The most efficient usage of water resources will make the storage areas and application units to maintain an effective ecosystem. Normally the industrial outlet water gets polluted badly because of the addition of the chemicals and other toxic substances. Hence the disposal of the wastes is mandatory for the effective utilization of water. IoT will aid in an effective mechanism that enhances the purity of the water resources using various sensors and actuators, especially for analyzing and processing data. For each modern activity utilizing the previously mentioned data correspondence advancements, data is legitimately shared with the cloud (Xie et al., 2020). The vulnerable situation arises during the information sharing that leads to the treats in the security and the proper strategy for the security is required. Along with the security of the data, the other key challenge in the information sharing is the backlogs, which exist in addition to the scope for the data analysis.

For handling the encounters that occur in the information sharing, blockchain innovation fills in as an expected arrangement; furthermore, the research is being directed in this space. It examines zeroing in on various parts of blockchain innovation, and industiral internet of things (IIoT) have been distributed (Bai, Hu, Liu, & Wang, 2019). For basic IIoT issues, noticeable examinations center around the joining of blockchain innovation inside IIoT applications and the improvement of vigorous systems (Reyna, Martín, Chen, Soler, & Díaz, 2018; Moin et al., 2019). Blockchain innovation includes applying three advancements such as the internet, cryptography, and agreement instruments. Cryptography is a private key that has been utilized in blockchain for satisfying the prerequisites of validation (for example, the copyright). The blockchain-based modern wastewater engineering contains four layers: information obtaining, edge-figuring frameworks, a distributed computing stage, and an application layer (second layer). In information procurement in the first layer, different hardware (for example, stream meter, water meter along with level measure, stockpiling tank for water, in addition to a wastewater treatment plant [WWTP]) related to the business remain furnished along with IoT-based brilliant item, which is fit for detecting, checking, preparing, gathering, and imparting data related to the water conditions in the water stockpiling level, measure of utilization of water, wastewater age measure, along with the modern wastewater, which is preserved by WWTPs.

The gathered information were moved to edge-figuring frameworks (for example the edge hubs/savvy entryways) that occur in the second layers by means of the accessible correspondence advancements (for example, the wireless fidelity and some network generation types) for constant handling, continuous dynamic, information accumulation, and furthermore, transitory capacity of information blocks that consists of a predefined measure of time. The amassed information blocks are confirmed by means of agreement utilizing the other edge hubs and added to the blockchain in the wake of executing the brilliant agreements. The edge hubs forward the confirmed information squares that is added to the water's blockchain dwelling on cloud-oriented servers (third layer), which are fit for handling, putting away, overseeing, breaking down, and taking care of blockchain-related tasks with a lot of information created by different IIoT-based businesses identified with water utilization, mechanical wastewater age, and wastewater treatment.

Applications to screen and oversee assignments identified with water work on the application layer (i.e., the fourth layer). All the remaining partners (e.g., brilliant businesses along with government organizations, natural checking offices and nonadministrative associations, water providers, and appropriations organizations) that are engaged with mechanical wastewater observing and the board will have the option to get to/inquiry the put away information in the blockchain through savvy contracts; as they might have the option to view, process, and substantiate the information for which they have consent.

3. Proposed system

The system for the management of the water is mandatory for the resource handling phenomenon in smart cities. The water to be managed is either drinking water or wastewater. The various systems are proposed for the effective handling of the water, but some drawbacks are identified in the management of the same. The usage of the internet is increased day by day and the involvement of internet in the smart city environment handling is yielding better results. The introduction of blockchain in smart cities made the process easier. In the proposed system the process of waste management in the smart cities is enhanced by the implementation of the blockchain technologies.

3.1 Water management in smart cities

The proposed system describes a model that estimates the consumption of water by each and every resident in the smart cities. In addition to that the generation of the electrical power is also done by using this model. For the calculation of the water utility by each house in the smart city we are using blockchain-based mechanism. This device performs better in doing some activities that effectively manage the utilization of the water. The proposed architecture for water management using blockchain technology is given below in Fig. 4.1.

The system consists of a control panel that is used for controlling and monitoring the flow of water from the reservoir. The whole system is administered by the use of blockchain technology associated with IoT and is indicated by the short messages or with an alarm. The autonomous monitoring

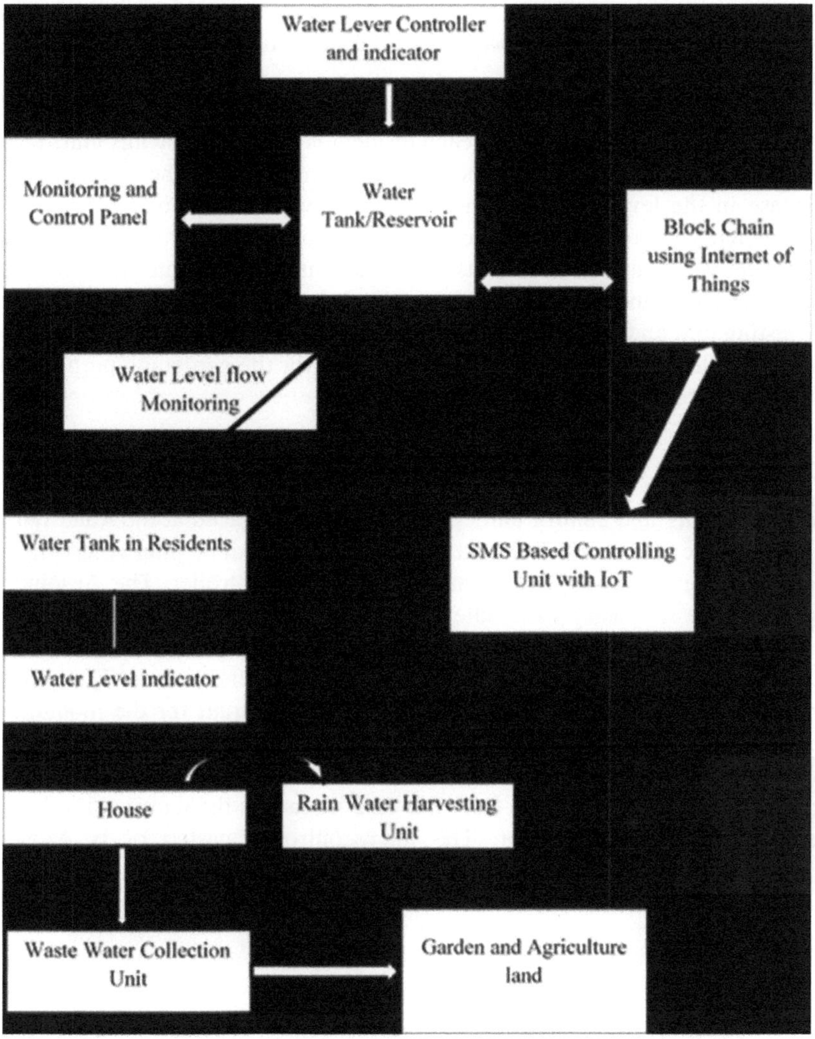

FIGURE 4.1 Block diagram of water management.

and the control system provide a most effective solution for the management of water with maximum accuracy. The major component used in this control unit consists of certain control systems. That includes the controlling of the water flow, the pump control, water level in the reservoir or water tanks, and the SMS-based alarm control unit. The mechanism used for the operation of the control system is coated with an electrical shock-proof material to avoid electrical shock. The control system that is proposed is enhanced by the three state switches, which are designed and implemented. This mode of setup will facilitate the manual operation in addition to the automatic stage and the IoT state. The control panel used here is the most reliable one, which gives an acceptable solution for the management of water for domestic and agricultural usage.

The reservoir is included with the water-level controlling unit and the limitation of the units of water to each resident is controlled by this unit. There is a limited supply of water based on the requirements of the residents. The increase in the level of the water is indicated with an alarm. Each house is constructed with a smart water tank consisting of a level controlling unit that periodically indicates the residents about the status of the water tank with the SMS and alerts them about the usage. Each house consists of a rain water harvesting unit and the collected water is transferred to the reservoir. The used water is collected in the reservoir and is reused for the gardening purpose and the agriculture land usage.

3.2 Intelligent control and monitoring unit

The monitoring and control unit consists of sensors placed at the water tanks, pipes, and the reservoirs. The data processing can be done after receiving the data from the sensors by using an Arduino microcontroller. The Arduino is used because it is easy to handle the interfacing of the hardware with the software. The multiple formats of the inputs like lights, biometrics, texts and images are handled effectively with the help of Arduino boards. The outputs produced will also be more accurate. The block diagram for the monitoring control unit is given below in Fig. 4.2.

The controlling unit consists of a monitoring circuit that guarantees a reliable connection within 2 s from the time it reaches the required level after deactivated by the processor. The microcontroller casted of is Arduino

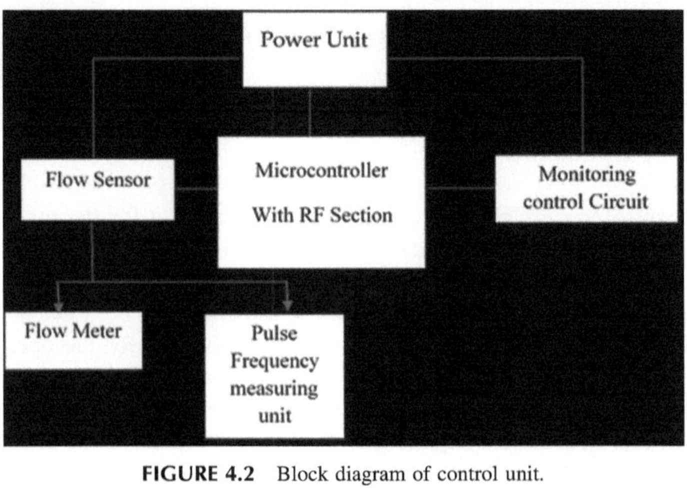

FIGURE 4.2 Block diagram of control unit.

processor and it used to send a message to the authorities for every second through the RF section. The water-level variations are noticed periodically by the sensors placed inside the reservoir. When the level of water reaches the threshold level then the processor will indicate it with an alarm and the information is passed to the authorized person. When the water tank overflows, automatically the sensor will send a signal to the controlling unit and the motor will set back to the turns off position. The flow sensor will sense the level of water and it consists of two parts viz., the flow meter and pulse frequency measuring unit. The flow meter and pulse frequency measuring unit will keep track on the typical variations noted frequently from the reservoir or the water tank.

The power supply for the smart water tank is provided by the smart solar system shown in Fig. 4.3. The system is build using two mechanisms one is the installation of the phase change materials (PCM) and another is the capacitor control unit. The open loop circuit consists of a solar panel setup that comprises of a capacitor control section and the battery for storing the generated charge. The PCM unit is fixed in series with the water tank, and the storage system that uses the PCM system is used for optimizing the matching circuit. The whole solar system model is constructed by the use of transient system in association with some program software. Here the load connected will be acting as a heat sink. The combination of the phase charge materials in the system will permit to store lot of energy by keeping the similar volume and the problem that arises in the PCM unit in the form of some shortcomings due to the low discharge rate is solved. This happens at the time when there occurs a conjunction of PCM system with the water tank because of the large requirement of the heat. If we place the PCM system before the storage unit,

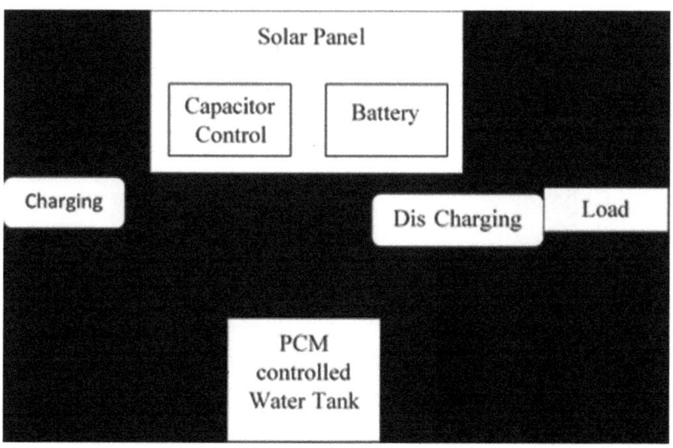

FIGURE 4.3 Block diagram of solar powered water tank.

then it will reduce the temperature of the inlet. The solar panel consists of a capacitor control unit and battery for storing the charge. The most important capability noticed in this solar panel is the facility of fault tolerance.

3.3 Short Message Service (SMS)-based control unit using Internet of Things (IoT)

The information regarding the water overflow and the water-level variation in the tank is indicated to the authorities via short messages. Here an IoT-based Global System for Mobile Communication (GSM) model is used for transmission purposes. The structure used for this transmitting device is simple. It consists of signal gathering unit, the code conversion unit, the encoder and the GSM module. The setup is displayed in Fig. 4.4. The code generation block is used and the block has 1024-bit linear-feedback shift register (LFSR) and it is alerted when the timer expires and at the time when short messages are transmitted.

The proposed device used for receiving consists of an integrated circuit-based setup for controlling. The connection for the GSM network with IoT is enabled by the third-party modem. The interfacing of the modem is done by a Universal Asynchronous Receiver/Transmitter (UART) device, as the interface using modems are available for ease of use. The device used for the receiving of the information via SMS is traced in Fig. 4.5. The code transmitted from the transmitter is compared with the received value and the difference is noted. Both the transmission and the reception is done through blockchain-based IoT. The blockchain usage might keep track of the information sharing and the security of the information is guaranteed by the use of fog computing. The transparency of the information is made possible through the use of blockchain. Due to the decentralized nature of the blockchain, there occurs some form of secure transmission in the flow of information.

The timing variation is validated by the real-time clock. The sending and receiving of the code are based on the time constraints marked. In this module an asynchronous type interrupt is used, as the device for receiving will wait for

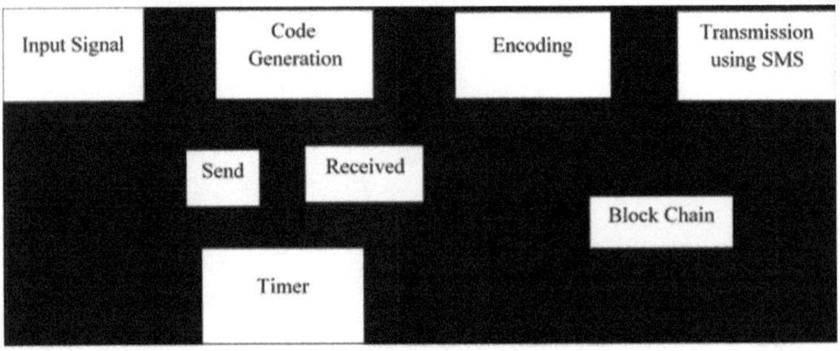

FIGURE 4.4 Generation of information.

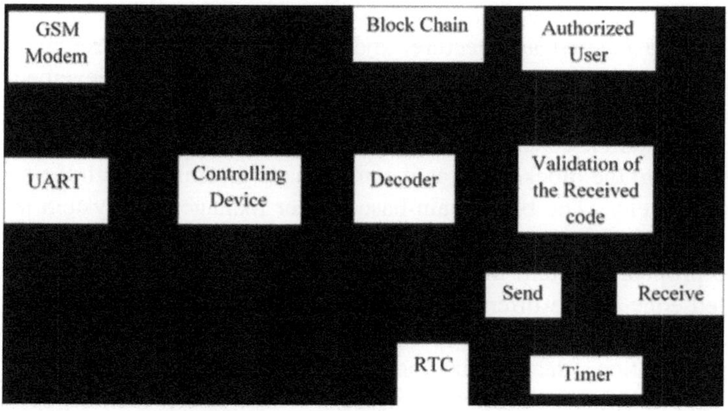

FIGURE 4.5 Block diagram of communication.

the incoming short message formerly acting. The encoding and decoding process are considered to be symmetric and that should be made to perform on the nonsimilar devices. The device used for the transmission and the reception is to be initialized by the similar protocol. The two devices, which is kept under pairing, must share the key. For suitability the transmitting device will generate the key automatically by using the random number generator. The transmitting device gets displayed with the key and it is handled manually by the receiving device. At one point the transmitting and the receiving device have the same key that allows the incoming messages to get decoded. After the pairing gets completed the device for receiving must wait for the transmitter messages. The automatically generated code is responsible for the effective sharing of information. For making the process easier the automated code is enabled by the timers within certain time period for the transmitter and the receiver. The message is transmitted successfully within the termination period and hence there is a chance for the security issues because of the transmission delay. In most critical applications the security threat can be reduced by choosing the transmission time to be a most feasible value.

3.4 Blockchain using Internet of Things (IoT) in water management

The water requirement is the major problem that is facing as the most recent problem in the country. The most advanced application for the management of the water is made possible through the use of IoT and blockchain. For this, a decentralized solution using aqua chain is being used that includes the traceability using blockchain. IoT is being used for the consumption of the digital information within the chain. For accessing the aqua chain effectively, the use case is to be determined classically within the provided vertical domain, i.e., to the buyer from the supplier. The traceability can be achieved by deploying the use case. The proposed system for the management of water might have a layered architecture, and that is able to keep counted on the blockchain in association with the IoT technologies for the achievement of the transparency of the stored information in a nontrusted environment. The usage of the blockchain is considered with the aqua chain by permitting it to be much independent to the blockchain, as it can be made assembled with the available software systems. The blockchain-based water management system is shown in Fig. 4.6.

The proposed system might compensate by increasing the capabilities of advanced edge device offerings that might be unswervingly cast off as wide-ranging nodes of our implementation of the blockchain, thereby encompassing the confrontation, decentralization, safety, and confidence of the whole network. The important components of the aqua chain include application programming, interface, and the controller. The application programming interface (API) exposes the abilities of the aqua chain with other applications by the highest level of abstraction by allowing the familiar integration of the available software systems. The controller is responsible for transforming the highest level of function calls addicted to the corresponding low level. The individuality constraints and modern industry water requirements pose the foremost challenges to accomplish an auditable, transparent, and reliable blockchain management.

Most of the challenges are remaining at the heterogeneity of the users and the WiFi. The users at different levels of interoperability might be available within the involved systems at the lack of data governance. The blockchain-based distributed technology consists of the business logic that is implemented through the blockchain with smart contracts using the proper gateway. Based on the designated blockchain, the module will vary with the complexity that is varied according to the software abilities with the client interfaces.

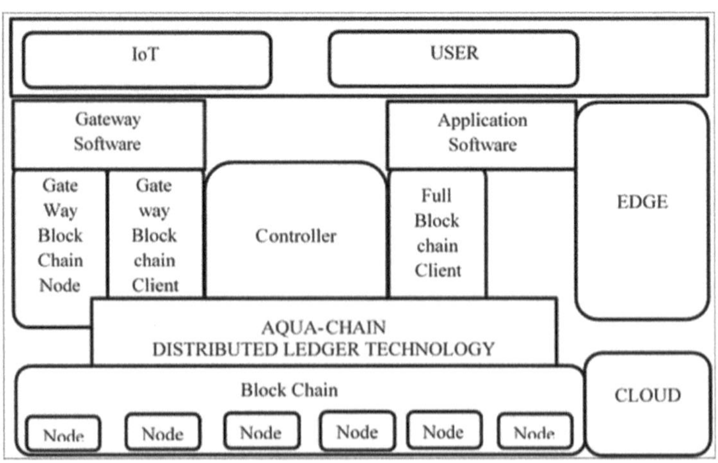

FIGURE 4.6 Architecture.

3.5 Blockchain-based smart water quality monitoring system using internet of things (IoT)

The water quality management system which is proposed comprises of three sensors, i.e., pH sensor, ultrasonic sensor, and DHT sensor. The Arduino Uno is used as the main processing unit. The major role is played by the microcontroller. Here Arduino Uno is used because of its facilities like reliability, Low cost and the power consumption is low. The solar panel setup is used for driving the power. From the three sensors, two sensors are used for the collection of data in the form of continuous time signals. The microcontroller unit has an analog-to-digital converter which renovates the obtained analog signals to its corresponding digital form. To obtain the analog output from the sensors the sensors output will be linked with the microcontroller unit using analog pins. The other sensors are linked directly to the digital pins of the microcontroller unit. All the data that is processed by the microcontroller unit with updates are done at the PC server by using the WiFi module. The diagram for the water quality management system is given in Fig. 4.7. The software used for the design of the whole system is Embedded C, and for Arduino, it is written using Arduino IDE. For collecting the data regarding the pH value, level of the water, temperature, humidity, and the turbidity of the atmosphere, the sensors are used by the system. Blockchain technology is used and hence the authorized users can access these data by using the individual ID provided. The real-time gathering of information is done and the collected information is stored and is transmitted at the needed time. Here ESP8266 is used as WiFi module because of its low cost.

The code boots from outer flash legitimately throughout the preparation of the program by expanding the framework execution and the capacity prerequisites because of their advanced reserve limit. ESP8266 utilizes transmitter and receiver for sequential handset pins for sending and getting

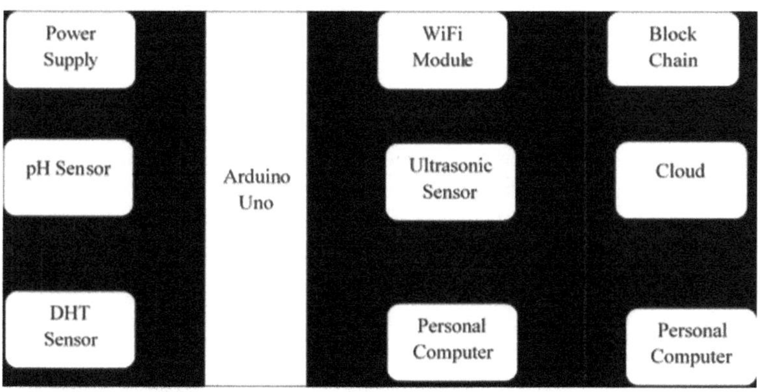

FIGURE 4.7 Block diagram of quality management.

information, for changing remote module settings, for changing sequential inquiry orders. Two pins (transmitter/receiver) are required to convey, yet just joined, between a WiFi module and a microcontroller yet associated oppositely. It uses the transmitter and receiver pins for sending and receiving the information. It is anything but difficult to set up an IoT application through a WiFi module by means of SPI and UART.

3.5.1 Role of blockchain module

The blockchain module consists of an intelligent contract that exists in between the application that is decentralized in case of water management systems. There are two different kinds of contracts that occur in smart manner for providing the interaction mode of the platform as depicted. The application of the blockchain used in Fig. 4.7 typically corresponds to the communication functioning of the wireless module in accordance with the storage area. Along with the normal maintenance of the sensing device the interface is also provided in between the applications of the smart sectors that shows the logic of the applications mentioned. The Arduino must be programmed in such a way to accommodate the specific functions mentioned to the interfacing sectors.

4. Results and discussion

Water management in smart cities might be enhanced by the use of blockchain with IoT. In this setup there are a number of sensors used for the level indication and controlling the flow level. The flow sensors are attached to the adapter made of rubber. Normally for the purpose of why the water management system is used in smart cities is to regulate the water supply by limiting the usage of water by each resident and to reduce the wastage of the water. The users are provided with the estimate that shows the permitted daily usage. The individual storing capacity is allotted to the water tanks and the information about the allocation and the usage log is stored in the electrically erasable programmable read-only memory. By using this device, the values obtained from the analysis are generating impact while considering the daily consumption of the water. The amount of water that is distributed from the water tank to the device that is installed and the amount that is stored at the location pointed out in the memory device. The values are noted from the memory location and are erased at the end of the week. Table 4.1 shows the data that are gathered at the time of running for the whole week. The total water needed is obtained in liters, and the water wasted is calculated, which is used for indicating the residents about their part of wastage of resources. The water dispensed is also calculated, and the information is stored, which can be kept as the reference for identifying the individual usage. Fig. 4.8 shows the collected data for the running mode.

TABLE 4.1 Record for one week.

Day	Water required (volume in liters)	Water wasted (volume in liters)	Water dispensed (volume in liters)
Day 1	50	60	75
Day 2	60	62	54
Day 3	54	50	56
Day 4	52	54	54
Day 5	67	67	56
Day 6	68	45	54
Day 7	60	70	66

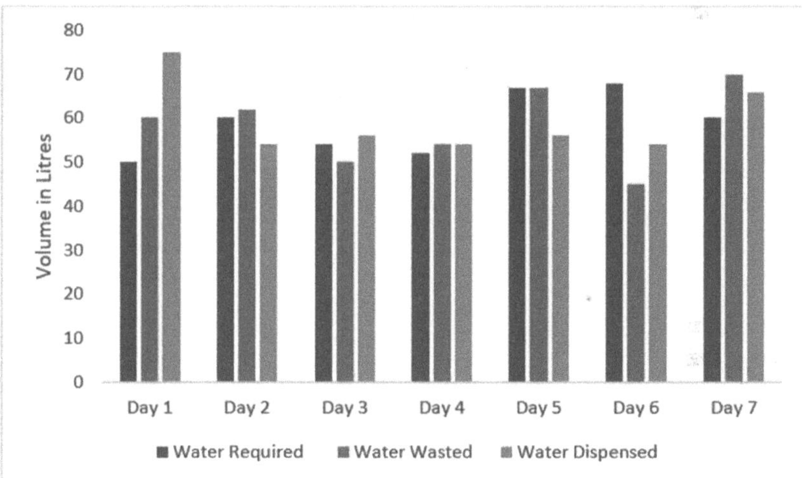

FIGURE 4.8 Bar chart for water quantity.

The reservoir that is fixed at some other location, which is far away from the current location, is noted periodically, that is, throughout the week, and the status of the tank is to be intimated to the authority time by time. Table 4.2 records the filling mode status of the reservoir and the water tank. The water that is intended to be filled is calculated in liters. Here also, the wastage of the water is marked as it may cause due to the overflow of the tank or by some other issue. The graph corresponding to the filling mode information after the data is being collected is shown in Fig. 4.9.

TABLE 4.2 Data for a week.

Day	Water to be filled (volume in liters)	Water wasted (volume in liters)
Day 1	110	45
Day 2	90	52
Day 3	98	40
Day 4	96	56
Day 5	87	55
Day 6	92	56
Day 7	120	60

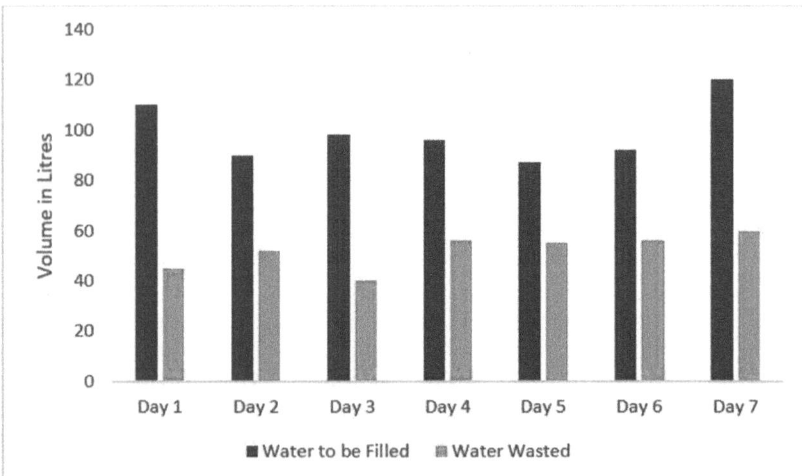

FIGURE 4.9 Bar chart of filling mode.

The above two modes, viz., the running mode and the filling mode, are calculated for a week, and the combination of these two are obtained for a month, and it is used for the calculation of the usage cost and the wastage penalty. Table 4.3 shows the corresponding values of the running mode and the filling mode that was taken earlier. The data for the values collected for a month in both the modes, the running mode and the filling mode are shown in Fig. 4.10.

TABLE 4.3 Data for a month.

Day	Water dispensed (volume in liters)	Water used (volume in liters)	Water wasted (volume in liters)	Day	Water dispensed (volume in liters)	Water used (volume in liters)	Water wasted (volume in liters)
Day 1	216	116	62	Day 16	222	136	84
Day 2	222	124	64	Day 17	228	134	66
Day 3	210	150	52	Day 18	216	132	68
Day 4	285	130	58	Day 19	291	125	56
Day 5	262	128	59	Day 20	268	133	62
Day 6	218	126	63	Day 21	224	159	63
Day 7	224	119	70	Day 22	230	139	67
Day 8	212	127	82	Day 23	218	137	74
Day 9	287	153	64	Day 24	293	135	86
Day 10	264	133	66	Day 25	270	128	68
Day 11	220	131	54	Day 26	226	136	70
Day 12	226	129	60	Day 27	232	162	58
Day 13	214	122	61	Day 28	220	142	64
Day 14	289	130	65	Day 29	295	140	65
Day 15	266	156	72	Day 30	272	138	69

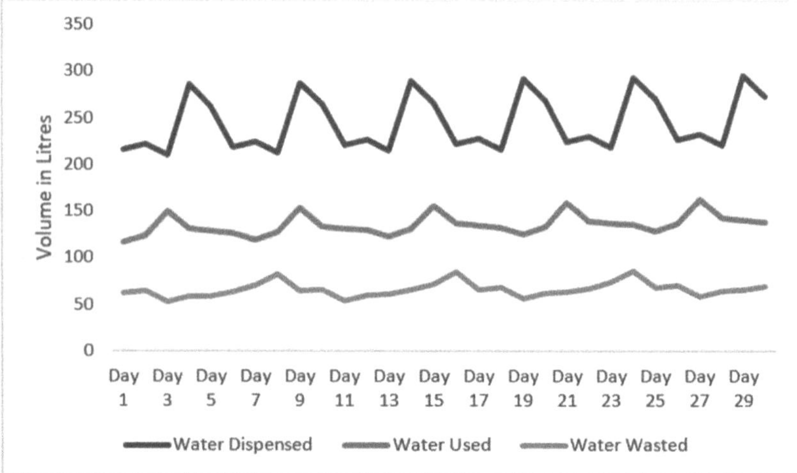

FIGURE 4.10 Graph for a month.

The additional feature added up in this system is the water quality management. The neighboring temperature of the environment and the update of it with the cloud server are sketched in Fig. 4.11 and Fig. 4.12.

According to the timing variations the value of the temperature and the humidity is being updated. The temperature is measured in Celsius and the measurement of the humidity is done with percentage. The other sensors fixed at the device like the pH sensor and the turbidity sensors might give much accurate values while calculating. The value of the temperature and the

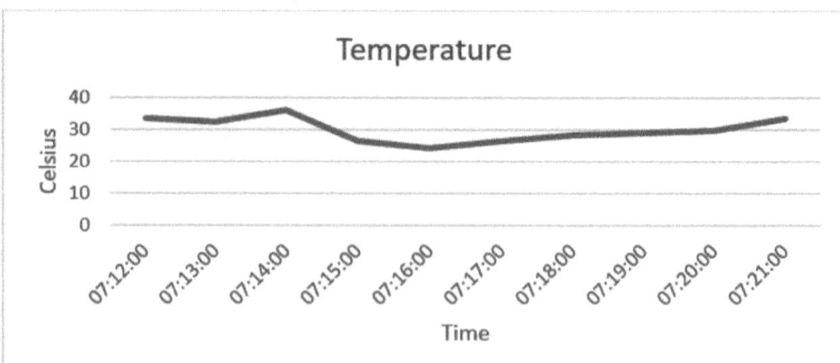

FIGURE 4.11 Temperature measure.

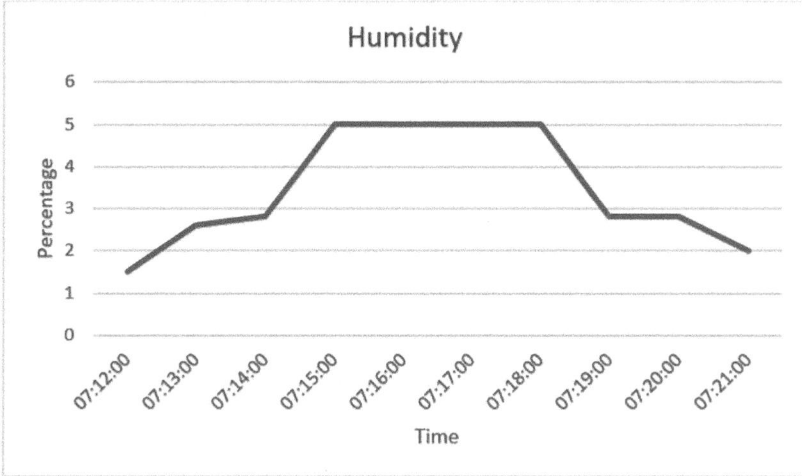

FIGURE 4.12 Humidity measure.

moisture value is monitored by the sensors and the information is recorded at the Arduino Uno. Then the pH value of the hydrogen ions is calculated by doing the acidity balancing test. Otherwise the same can be determined by the calculation of the alkaline content in the water. The pH value must be in the range of 6.5–9.5 values and it is considered to be safe for drinking. The increase in the pH value decreases the hydrogen ion concentration. The sensors used consist of an electrode and the hydrogen ion content in the water is sensitive to the electrode measurement which might have the potential that is directed toward the ion concentration. The value of the pH sensor is shown in Figs. 4.13 and 4.14.

The ultrasonic sensor delivers a measurement range in between 3 cm and 10 m. The sensors are installed on a module consisting of a transmitter and the receiver in addition to the control circuit. The frequency generated by the sound wave is calculated, and the value is measured in between the signal transmission end and the receiving end that in turn determines the object distance. The ultrasonic sensors valued for distance and time is shown in Fig. 4.15. When we analyze the data transmission done in between the decentralized node with the microcontroller and the sensor node, the research was made to attach the potentiometer-based device to the sensor node inputs, i.e., the analog-to-digital converter and hence different values of the voltage levels could be implied, and the same is measured and sent to the cloud server by permitting the data transfer in between the decentralized node and the sensor node. The sensor node voltage applied here is about 5 V. The important consideration to the device is the extraction of data from the Arduino Uno which includes the processing mechanisms in the decentralized node that gets

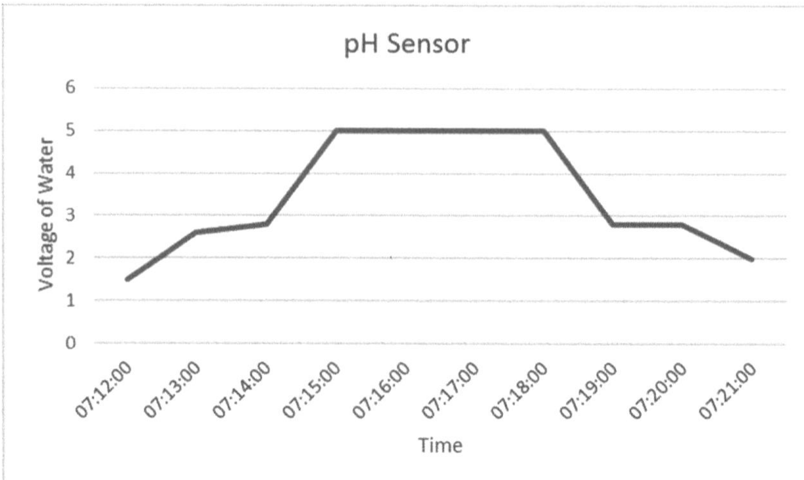

FIGURE 4.13 Voltage measure.

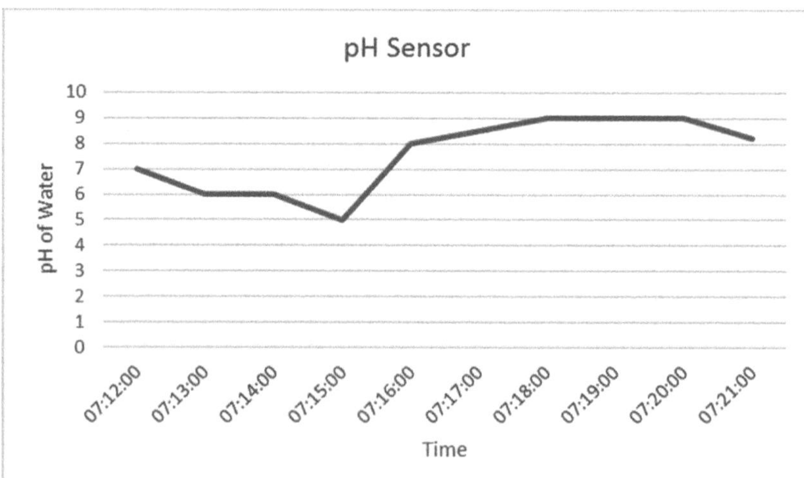

FIGURE 4.14 pH measure.

optimizes and filters the received data. The flow-sensor values at various intervals are shown in Table 4.4.

The graph showing the various sensor values in accordance with the different timing intervals are shown in Fig. 4.16. Since the device consists of various external attached devices for performing certain unified operations, the power consumption is analyzed properly to safeguard the usage. The complete

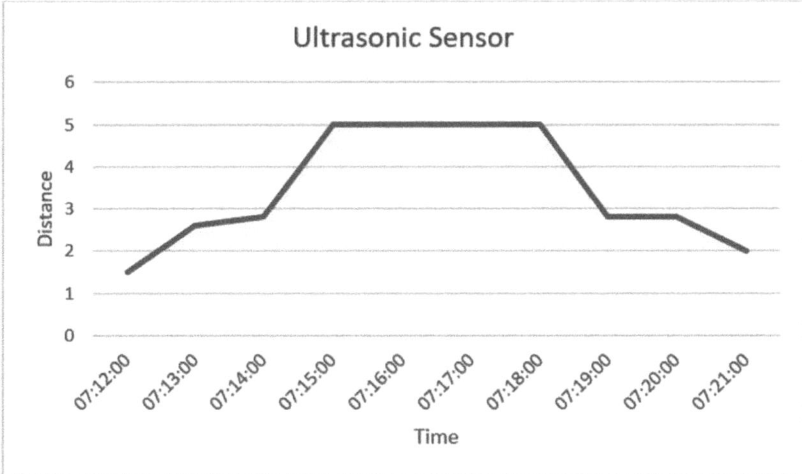

FIGURE 4.15 Ultrasonic sensor measures.

TABLE 4.4 Flow sensors measurements.

Time (seconds)	Flow sensor 1	Flow sensor 2	Flow sensor 3	Flow sensor 4
0	5	5	5	5
5	12	24	33	18
10	28	40	49	34
15	120	132	141	126
20	124	136	145	130
25	125	137	146	131
30	126	138	147	132
35	122	134	143	128
40	128	140	149	134
45	126	138	147	132
50	128	140	149	134
55	135	147	156	141
60	140	152	161	146

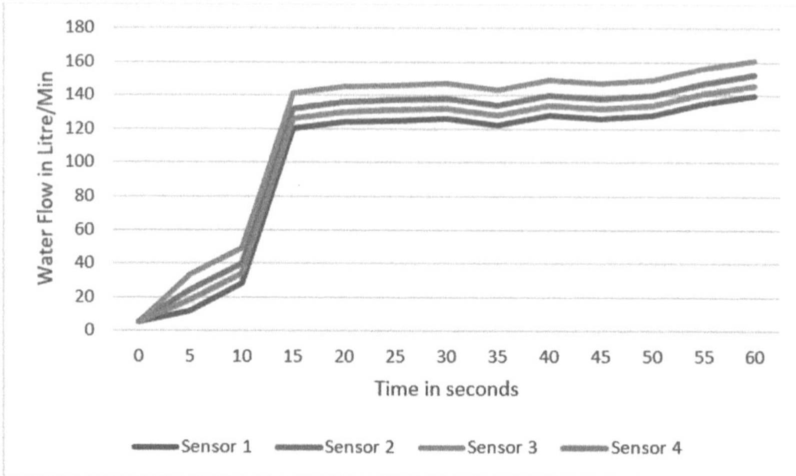

FIGURE 4.16 Flow sensors values.

device module consists of certain control units with the help of microcontrollers that uses Arduino Uno and the consumption of the energy of the proposed system is measured at regular intervals and at various stages including the idle stage. Table 4.5 gives the average consumption at various processes in the modules like sensing, transmitting, receiving, encoding, and decoding.

The obtained results show that device for management of water is affordable and the designed device can directly interact with the blockchain

TABLE 4.5 Utilized power.

Device status	Average energy consumption in joules
Idle case	1825
Sensing	1774
Encoding	235
Hashing	354
Compressing	1652
Transmitting	225
Decoding	368
Receiving	924

technology. The operations corresponding to the blockchain implementation might limit the various functionalities in the system. The aforementioned operations such as sensing, transmitting, receiving, encoding, and decoding are the blockchain operations. The energy consumptions can be varied according to the microcontrollers used.

5. Conclusion

The proposed system can effectively manage the beneficiary utilization of the water by reducing the wastage up to 73% for a month and serves as an efficient system with reduced costs. The facilities inside the system can make it easy to handle, and the two modes specified by the operation of the device can be motion-specific. The unauthorized usage can be easily tracked and transmit the information to the authorized person through short messages. This system is implemented in a way to considerably help in extenuating the worldwide crises about the water and it can provide a sustainable atmosphere in the management of water worldwide.

References

A synthesis report on the u-city testbed construction. (2008). National Information Society Agency.

Adelantado, F., Vilajosana, X., Tuset-Peiro, P., Martinez, B., Melia-Segui, J., & Watteyne, T. (2017). Understanding the limits of LoRaWAN. *IEEE Communications Magazine, 55*(9), 34–40. https://doi.org/10.1109/MCOM.2017.1600613

Bai, L., Hu, M., Liu, M., & Wang, J. (2019). BPIIoT: A light-weighted blockchain-based platform for industrial IoT. *IEEE Access, 7*, 58381–58393. https://doi.org/10.1109/ACCESS.2019.2914223

Bordel, B., Martin, D., Alcarria, R., & Robles, T. (2019). A blockchain-based water control system for the automatic management of irrigation communities. In *IEEE international conference on consumer electronics (ICCE)*. https://doi.org/10.1109/ICCE.2019.8661940

Cristina, R., & Giacomo, V. (2017). A privacy-friendly gaming framework in smart electricity and water grids. *IEEE Access*, 14221–14233. https://doi.org/10.1109/ACCESS.2017.2727552

Gutiérrez, J, et al. (2013). Automated irrigation system using a wireless sensor network and GPRS module. In IEEE (Ed.), *1. IEEE transactions on instrumentation and measurement* (pp. 166–176). IEEE. Issue 1.

Khutsoane, O., Isong, B., & Abu-Mahfouz, A. M. (2017). IoT devices and applications based on LoRa/LoRaWAN. In *Proceedings IECON 2017 – 43rd annual conference of the IEEE industrial electronics society* (Vol. 2017, pp. 6107–6112). Institute of Electrical and Electronics Engineers Inc. https://doi.org/10.1109/IECON.2017.8217061

Korean society of civil engineers, 58(12), (2010), 18–24.

Lazaroiu, C., & Roscia, M. (2017). Smart district through IoT and blockchain. In *2017 6th international conference on renewable energy research and applications, ICRERA 2017* (Vol. 2017, pp. 454–461). Institute of Electrical and Electronics Engineers Inc. https://doi.org/10.1109/DISTRA.2017.8191102

Lim, H., Kim, W., & Jung, J. (2018). Integrated water cycle management system for smart cities. In *Proceedings — 2018 2nd international conference on green energy and applications, ICGEA* (Vol. 2018, pp. 55—58). Institute of Electrical and Electronics Engineers Inc. https://doi.org/10.1109/ICGEA.2018.8356311

Lin, J., Yu, W., Zhang, N., Yang, X., Zhang, H., & Zhao, W. (2017). A survey on internet of things: Architecture, enabling technologies, security and privacy, and applications. *IEEE Internet of Things Journal, 4*(5), 1125—1142. https://doi.org/10.1109/JIOT.2017.2683200

Lundin, M., & Morrison, G. M. (2002). A life cycle assessment based procedure for development of environmental sustainability indicators for urban water systems. *Urban Water, 4*(2), 145—152. https://doi.org/10.1016/S1462-0758(02)00015-8

Moin, S., Karim, A., Safdar, Z., Safdar, K., Ahmed, E., & Imran, M. (2019). Securing IoTs in distributed blockchain: Analysis, requirements and open issues. *Future Generation Computer Systems, 100*, 325—343. https://doi.org/10.1016/j.future.2019.05.023

Reyna, A., Martín, C., Chen, J., Soler, E., & Díaz, M. (2018). On blockchain and its integration with IoT. Challenges and opportunities. *Future Generation Computer Systems, 88*, 173—190. https://doi.org/10.1016/j.future.2018.05.046

Salek, A. M., Massimo, V., Miguel, P., Koustabh, D., Fabio, A., & Husain, R. M. (2019). Applications of blockchains in the internet of things: A comprehensive survey. *IEEE Communications Surveys & Tutorials*, 1676—1717. https://doi.org/10.1109/comst.2018.2886932

Vakula, D., & Kolli, Y. K. (2017). Waste water management for smart cities. In *2017 international conference on intelligent sustainable systems (ICISS)*.

Wintgens, T., Salehi, F., Hochstrat, R., & Melin, T. (2008). Emerging contaminants and treatment options in water recycling for indirect potable use. *Water Science and Technology, 57*(1), 99—107. https://doi.org/10.2166/wst.2008.799

Xie, S., Zheng, Z., Chen, W., Wu, J., Dai, H. N., & Imran, M. (2020). Blockchain for cloud exchange: A survey. *Computers and Electrical Engineering, 81*. https://doi.org/10.1016/j.compeleceng.2019.106526

Chapter 5

Blockchain-based energy-efficient smart green city in IoT environments

K. Suresh Kumar[1], A.S. Radha Mani[2], S. Sundaresan[3], T. Ananth Kumar[1], Y. Harold Robinson[4]

[1]*Department of CSE, IFET College of Engineering, Villupuram, Tamil Nadu, India;* [2]*V.V. College of Engineering, Tirunelveli, Tamil Nadu, India;* [3]*Department of ECE, SRM TRP Engineering College, Trichy, Tamil Nadu, India;* [4]*School of Information Technology and Engineering, Vellore Institute of Technology, Vellore, Tamil Nadu, India*

Chapter outline

1. Introduction

For improving the Internet of Things (IoT) every day, intelligent devices including sensors, cameras, and a lot of smart home appliances are connected together. Radiofrequency-based innovations in the sensor network have been promoted in most of the intelligent home IoT applications (Helin, Arokiaswami, Wen-De, Chen, & Xianzhong, 2020). Smart home consists of comfortable inhabitants from past to future. In this view, the intellectual home

Blockchain for Smart Cities. https://doi.org/10.1016/B978-0-12-824446-3.00007-7

can be tenacious. Its automation has smart devices which consist of wired and wireless technology for integrating the home. Also, network vulnerabilities and attacks will happen in a smart home. The method of communication faces many obstacles, such as intellectuals' security and privacy room addressed with combat variations and methodologies of obviation. Depending on the machine's shape and level of attack, simplified threat-related future attacks are categorized.

Blockchain or else Distributed Ledger Technologies is specifically premeditated in the direction of allowing decentralized transactions by eliminating security mechanism. Cryptocurrencies also probably alleviate decentralized network issues. Blockchains are decentralized data systems or ledgers which might store transactions using digital platform safely deprived of a centralized authority. Most significantly, blockchain technologies allow smart contract automation in peer-to-peer networks (Red, 2017). Additionally, this might be interpreted as repositories allowing several users to change the directory concurrently, leading to multiple iterations of the chain. Each member of the network holds a duplicate of recording chain instead of handling the ledger with one trusted hub. Therefore, this finds consensus on the actual ledger condition. The exact consensus-building approach is an evolving field of research that can vary in many implementation areas. New transactions require previous cryptographic transactions and thus stable, efficient blockchain networks. Any network consumer can check if transactions are real, transparent, and trustworthy.

Blockchain potential in the energy industry has only begun to occur in conjunction with use cases in other sectors as demonstrated by means of increasing the number of startups which in turn includes the pilots, trials, in addition with the research projects (Ahmad & Khaled, 2018). Normally the blockchain is considered to be a digital data network which is a distributed one with open database with constantly evolving transaction log and clock order. In other words, the data structure is a booklet with digital transactions, documents, and executable (Yong, Yannan, Junfeng, & Jianwei, 2018). Transactions might get accumulated which is hooked on bigger structures named blocks that remain time stamping which is cryptographically connected to earlier blocks that form a sequence of record carrying database which define the event or blockchain sequence order. Blockchain Terminology which can be used in the recordical proceedings for defining the applications and architectures, algorithms, or networks-control blockchain domains. By using these networks, the data transfer is similar to the copying of data from one location to another. Thus, it yields in copying digital coins from one consumer to another. In this context, the system has to ensure whether bitcoins are spent only once or not, as spending twice is not possible. A traditional approach consists of using a central authority, like reserve bank, that can be acting as a reliable agent among the transacted parties and has the purpose of preserving and retaining the register's legal status. When many parties will write concurrently in the ledger and the role of the central authority is to process the request, it must also implement price controls and make adjustments in the headline (2019). Often central management can be excessive or inefficient because it increases average costs and encourages the trustable users of the network for trusting a third party for running the system. Thanks to a single failure point, centralized networks have significant drawbacks, which renders the most vulnerable to technological attacks in addition with the malicious attacks. Blockchain technology's primary goal is to eliminate the need for such intermediaries by replacing them with a centralized, digital user network by collaborating the authorized users for verifying the transactions thus by safeguarding the ledger's trustworthiness.

From recent knowledge, the implementation using blockchain is considered to be the critical technology replacing the Bitcoin technology and a candidate solution in IoT systems to tackle security, cyber assaults, and privacy concerns. The blockchain is a ledger with various data maintained. When the block is complete, it is modified with the blockchain. It uses a method called the mining method where different nodes joined together for attempting to solve some difficult cryptographic puzzle that consumes resources. The adaptive block might get added to the blockchain after mining. This is considered to be a distributed system in which all the nodes involved in the chain will have a chained copy (Dorri, Kanhere, & Jurdak, 2017). Blockchain's great thing is that its decentralized nodes must be part of the chain. Two key elements of blockchain technology exist user-generated network activity, known as transactions, and all the completed transactions will get recorded in the form of blocks in which the blocks might hold the correct sequence which remains unaltered. The advanced blockchain technology is capable of a highly decentralized environment of securing IoT users' privacy. Besides, blockchain technology can assist in defining and implementing various rates of access policies to reduce unauthorized IoT-generated data operations.

Blockchain technology is a health record innovation that company and analysts have foreseen as a developing technology which could act as a crucial role for monitoring and controlling, particularly for protecting the IoT-based devices. The authors will describe a roadmap that integrates IoT and blockchain technology for sharing IoT resources and services, allowing cryptographically automated time-sensitive workflows (Alex et al., 2019). This research examines methods and workarounds for mutual use of blockchain and IoT. For example, the data communicated for the IoT-based devices will always be proved cryptographically by ensuring the sender's signature of a specific pair of keys, ensuring the security along with the integrity of the transmitted data. Additionally, the public ledger documents all IoT device transactions, which could be traced back safely. Already the blockchain appears to be a solution for addressing privacy-related issues in current centralized systems. Normally there are lot of researches even hurdles by preventing their incorporation into new IoT networks. For computers with extreme resource constraints, most of the consensus algorithms that are used by existing blockchain-based systems which will not get optimized. Proof of work is considered to be the first consensus algorithm for the network using blockchain which have many current applications in use. This extends decisions to all nodes, including miners. This method, called mining, involves enormous computing capabilities. This chapter starts with incorporation of blockchain and smart city growth consisting of IoT, Big Data, and Networks. First, the critical role of the Smart City IoT, Big Data, and Energy Internet is addressed with the blockchain's basic concepts, core innovations, and domestic status, then summarizes many technical IoT, Big Data, and Energy Internet bottlenecks, discussing similarities and discrepancies between the blockchain and them. Through examining the characteristics and advantages of the blockchain, it reveals that the technological capabilities of the blockchain are best suited for Internet infrastructure and can also fix the issue of networking and large data centers, such as high centralized database maintenance costs, privacy leakage, terminal equipment upgrading. This chapter will act as a conceptual guide to help adapting blockchains to the future of smart cities.

2. Related works

Distributed energy systems play a significant role as they are mainly equipped for renewable energy production, storage, and distribution. Taking advantage of these benefits is essential, particularly as regards the current climate change stoppage debate. Various innovations, such as smart algorithm waste heat recovery, can improve energy efficiency and save substantial resources (Mylrea & Gourisetti, 2017). Considering the other side, the usage of the blockchain technology in the field of consumable energy markets will encourage renewable energy usage and provide a transparent mechanism in order to monitor the real-time production of energy in addition with the consumption of information. Blockchain technology will also allow the most independent agent exchange thus by creating a healthy energy community. The proposed system demonstrates how the blockchain could be used to build and leverage energy networks and energy transaction management techniques and allow a safer trading environment for consumers and producers. The authors proposed a blockchain-based energy network for facilitating the energy exchanges in a prosumer community (Ioan, Masoud, Yacine, & Rana, 2020). Another approaches like Blockchain Reputation–Based Consensus scheme where a node should have a credit score above the network-based trust point before inserting a new chain block. A jury of judges is randomly chosen to watch each consensus node's behavior and change the node ranking. They simulated a blockchain network to show scalability and malicious resistance to various network situations and malicious actions (de Oliveira et al., 2020).

Blockchain offers support in the energy distribution system (DES), which highlights operational challenges, control of electricity generation, trading and sharing of energy, financial flows, energy sources, carbon trading, etc., which also describes more about the effects of blockchain in DES (Manoj, 2018). Customer recognition is similarly known to understand company practices with both private and licensed ledgers. Validator nodes are already identified and trusted, and no artificial rewards are needed to ensure the network's efficiency. Private and certified ledgers can be streamlined, scalable, and accurate at the cost including the immutability along with the censorship resistance (2020). A contract includes information about the number of coins that are exchanged by addressing with the mentioning the parties for transaction. Typically, all kind of transactions are encrypted through the receiver's cryptographic key which is digitally authorized by the sender, and submitted to the network using the bitcoin. Particular nodes will get aggregated with all the outgoing transactions in one row for the past 10 min. These nodes also finalize the authentication protocol for validation and inclusion in the blockchain for around 10 min (Dwyer, 2015). Various solutions have been proposed, such as sharding, sidechaining, and the use of payment networks. Sharding is said to be a most advanced technique which is used to validate each transaction with another subset of validator pool nodes. Security features can be preserved when multiple validating nodes are used (On Sharding Blockchain, 2018). This will potentially permit parallel processing in transaction verification and speed up the transaction, but the methodology is still being developed. Another work gets developed a roadside unit for reducing the consumption of energy in addition with the cost of deployment. The specification for vehicles is required for end-user service quality (Premsankar, Ghaddar, Di Francesco, & Verago, 2018).

Fernández-Caramés, Fraga-Lamas, and Suárez-Albela (1798), have developed a multilayered enterprise for edge computing platform, which in turn focused on Augmented Reality (AR). Effectiveness of the proposed model was demonstrated by the real-time implementation. Schneider, Rambach, and Stricker (2017), has technologically advanced with a live remote access system with supported architecture. The video compression optimized the bandwidth. An integrated blockchain solution for LoRaWAN network servers is proposed in order to introduce a transparent, trustworthy, decentralized, and deceptive program. An undisputed mechanism is given to check the transaction data that existed on the network at an exact time. The authors note that blockchain technology is taking advantage of this research, which is the first to combine blockchain with LoRaWAN IoT technology (Lin, Shen, & Miao, 2017). Polyzos and Fotiou (2017), researched IoT dealer blockchain prospects. Essential safety requirements are addressed and how blockchain helps overcome them. A blockchain-based IoT application delivery system is developed to investigate how existing security frameworks can be strengthened using blockchain technology. Samaniego and Deters, 2017, proposed a lightweight blockchain-based IoT architecture. The proposed design would reduce the overall burden of conventional blockchain systems with equal security along with the privacy. The authors might validate the corresponding proposed architecture to prove efficiency in a smart home environment. Simulation tests indicate that the new solution substantially reduces packet and overhead capacity relative to traditional blockchain technology. The Smart City project aims to encourage the cities that provide crucial services and provide a decent quality of life, a secure and healthy climate, and the use of "smart" solutions. The goal is to create a sustainable and equitable environment, providing a replicable model for other emerging cities.

3. Proposed system

The systems and techniques already available in the implementation of the smart cities can be enhanced by using the proposed blockchain-based technology that might compensate the latency, computational cost, management of resources, and the secure transmission. The total architecture comprises of blockchain technology along with the edge computing platform. The privacy and security are much guaranteed in various applications pointed out in the smart cities. The primary application of the smart city is being enhanced by the process of transparency and effective connectivity. The transmission of information in the smart city environment is handled effectively and securely. The proposed methods can improve the scalability and reliability in the various application of the smart city. For some of the applications, the deep learning techniques are used for the connection purpose and the communication purpose.

3.1 Proposed energy management system using blockchain technology

The proposed system focuses more on smart city energy management system as energy consumption is one of the essential needs in the implementation of the smart cities. The environmental changes might impact energy conservation most effectively in many areas. Most suggested and readily available challenges in the management of energy are by generating an efficient structure based on sustainable energy, as it consumes the benefits of the IoT technologies. The changes noticed in the climate and global warming might enforce a standard move in the manipulation of the resources, and that gives the techniques related to the efficient management of energy resources. The most integral part of the idea in the efficient management of energy is the delivering, manufacturing, and utilization. The transition of the energy in the smart city can be done with the use of smart grids. The smart grid technology using blockchain might guarantee flexibility, and it will permit all the consumers to involve in the energy system. The managing and the monitoring of electricity transportation are exploited by the use of smart grids since it used advanced digital technology to provide the information's quickly and most effectively. The use of distribution grids can improve the reliability of the networks. The storage is complementing the energy from the renewable sources during the time at which the microgrids might help the same by reducing the power failure by providing the local distribution of the energy. Considering the power system already available which is of a single directional one that allocates the electricity that is developed from the power plant intended to the consumer, the use of the microgrid will get fulfilled with the local power supply unit, and there exists a storage mechanism that is cornered as a nondependent mode of power sources. The availability of the energy network in the existing power system can get connected with the available energy network, which has the needed amount of energy like heat from the electricity using the distributed power sources separately. This, in turn, provides the owners with the capability to produce the energy by their own as the microgrids. It may minimize the addiction on the providers of the energy by helping them to reduce the costs and to eliminate the end usage charges. The microgrid will produce the profits by satisfying the enormous need of the power, that could be sold to the distributor of energy.

The IoT technologies used in the microgrids are acquainted with the realizing of the smart systems that are able to schedule the system faults automatically, and it improves the efficiency of the consumption of the energy. The smart microgrid energy management scheme is introduced for using renewable sources (i.e., solar energy and wind energy). This will reduce the usage of energy arriving from the power grid. The smart IoT technologies will use a scheduling algorithm that is able to control the power loads based on the needs of the consumer (Nayanatara, Divya, & Mahalakshmi, 2018). The smart power system uses a low power consumption scheme which is of reduced cost, and the system is accomplished with the facilities to alert the consumer at the time of faults occurring due to overcurrent and short circuit. Here we are using an automated system, and hence the need for human intervention is completely eliminated.

The smart microgrids that are used in this system used technology to monitor the flow of energy, and it will change according to the change in the demand of the energy which provides and flexible supply in real time. The enhancement made in this proposed system uses blockchain technology in addition to machine learning (Jui-Sheng, Shu-Chien, Ngoc-Tri, Chih-Wei, & Chia-Chi, 2019) and artificial intelligence (Bose, 2017). It uses a decentralized network with smart meters for efficient handling of electrical energy that can normalize the electrical bills at a low cost. This can gather, process, and transmit the various electrical energy-based information that is focusing mainly on the smart energy control of the consumer. Fig. 5.1 given below shows the proposed system that uses a blockchain controlled smart energy meter for efficient handling of energy.

Unlike the smart meter proposed in Rosario, Claudio, Gaetano, and Chandra (2017), which uses the current transducer and the voltage transducer, this new smart meter consists of blockchain control, and this may help to enhance the utilization of energy in a well-efficient manner. It uses the Edge computing technology for making connections with the smart energy meter to

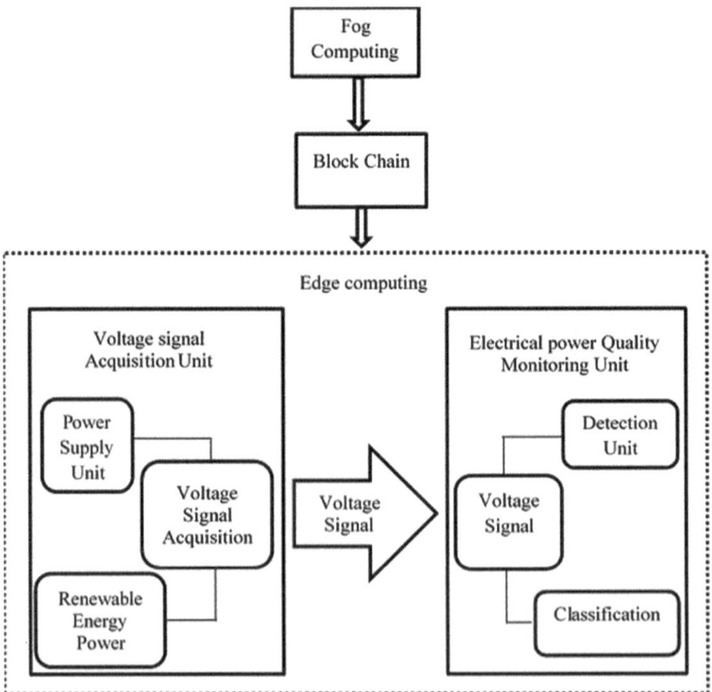

FIGURE 5.1 Smart meter.

other smart energy meters, and the Fog computing is used for establishing connections with the smart meter to the base stations where the information is about the power usage is shared. The blockchain feature used in this system can permit individual residents to access the system effectively. This can keep track of the usage status alive, and each usage can be predicted in real time with accurate data.

3.2 Proposed smart waste management system using blockchain technology

The process of managing the domestic waste (soil, water, and air) is an essential aspect in the management of smart cities. The improper management of waste leads to a harmful effect on humanity and the environment. Many methods are proposed for the effective handling of the wastes. In Al-Masri et al. (2018), there introduced an IoT architecture which is of serverless, for identifying the wastes before the process of separation. The identification of the hazardous wastes from the units and separating them can be made cost-effective using this technology. Totally air waste, water waste, and the soil wastes are taken into consideration. The segregation of the wastes is done at the base level, thus by reducing the interaction of humans. The use of the blockchain in the waste management system can reduce the effect of centralized network intervention.

4. Domestic wastes collection and management

Buildings in the smart cities are installed with an autonomous collection box at separate points. This is pretended to operate with the help of a mobile application. The setup consists of a robotic approach that can separate the degradable and the nondegradable wastes, which consists of an IoT-enabled structure for disposing of the unwanted wastes and for sterilizing the organic wastes. The proposed system consists of two major control units like user identification unit and smart system for garbage collection. The whole unit consists of a waste management system that includes the blockchain unit. The autonomous waste collection box used in this system has sensors which can sense the waste category within the boxes and the robotic arm in the collection box separates the biodegradable and the nonbiodegradable wastes. The use of blockchain provides a decentralized network that can make the residents use the collection box effectively. Another facility added up in this system is it can alert the residents regarding the status of the box via a beep signal if the box gets filled up. The sensors used in this system is fixed at the houses and other waste collection areas. The sensors can sense the availability of the wastes and intimate the residents and the waste-collecting authority about the status. This is made possible by the microcontroller placed in this smart bin. This system also includes a user identification unit, and this uses the ultrasonic sensors that will sense the availability of the persons from a particular distance aside by using the sound transmitting and receiving capability. The transmission and reception capability are already provided to the ultrasonic sensors. The various pulses generated by the sensors are not listenable to the human ear, as it is beyond the listening range. Totally three ultrasonic sensors are placed at the box. One is placed at the interior side of the box, and it will indicate the level of wastes collected. The other one which is placed at the external side is used for detecting the users or personnel's when they are near to the collection box. The third sensor is placed at the household points to intimate the residents about the availability of the wastes. The block diagram is given below in Figs. 5.2 and 5.3 demonstrates the proposed architecture for solid waste management.

4.1 Waste food management system

The food wastes are much responsible for the generation of greenhouse gases to some extent. The review states that there is a possibility of generation of

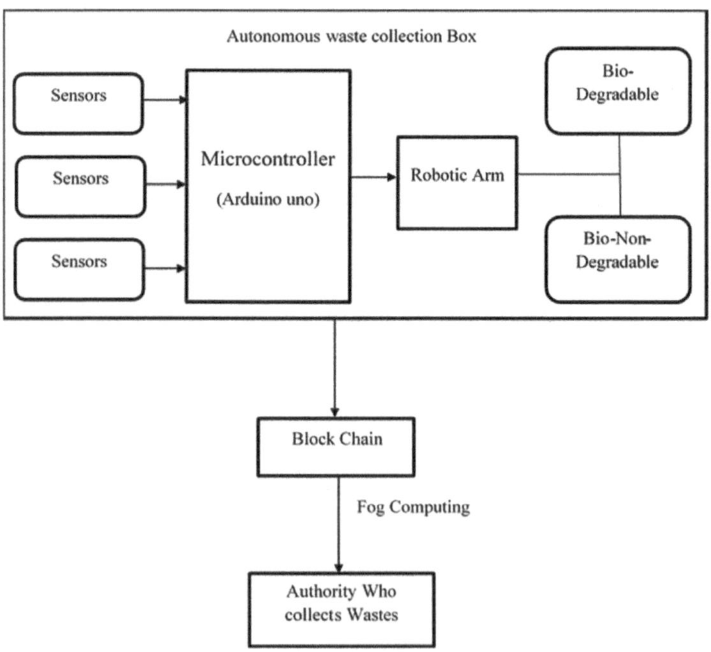

FIGURE 5.2 Smart garbage collection.

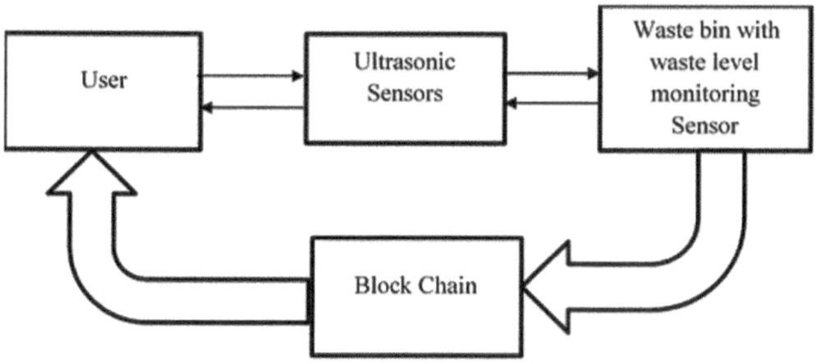

FIGURE 5.3 User identification unit.

6% of the global greenhouse gases emitted from the food wastes (Food waste is responsible for 6% of global greenhouse gas emissions, 2020). From the refrigerators outlet, there occurs the emission of hydrofluoric carbons. The collected food wastes can generate carbon dioxide gas, nitrous oxide gas, and methane (The climate impact of the food in the back of your fridge, 2018).

The food wastes are one of the essential domestic wastes that often creates problems because of inadequate disposal mechanisms. Handling of the food wastes is mandatory in the smart city, as the waste food will act as a living place for the flies and some worms which seem to be a disease generating agents.

The biodegradable food substances will generate the NH3 (ammonia) emission. This can be effectively handled to make it as a fertilizer and leaving it as such can create many issues in the environment. Thus, a model is proposed for the management of the food wastes in the smart city using the IoT-enabled blockchain technology. Fig. 5.4 demonstrates the proposed food waste management system. In this system, Arduino UNO is used for the processing of the various gases obtained from the wastes and it consists of NH3 sensor, CO_2, sensor, CH_4 sensor, and the HFC sensor. The sensors are attached in the smart dust bin. The smart dust bin consists of pollution level indication unit connected with the microcontroller. The composting materials will compost the various food waste according to the category. The meat has separate composting materials, and the vegetables have separate composting materials. The thermometer fixed to this will identify the temperature level in the mixture, and the detail is transmitted to the responsible authorities via blockchain technology. The use of blockchain technology will bring the transparency of the information is transmitted. Generating messages and sending them in quick mode using this technology can able to reduce the effect of the greenhouse gases emerging from the smart cities.

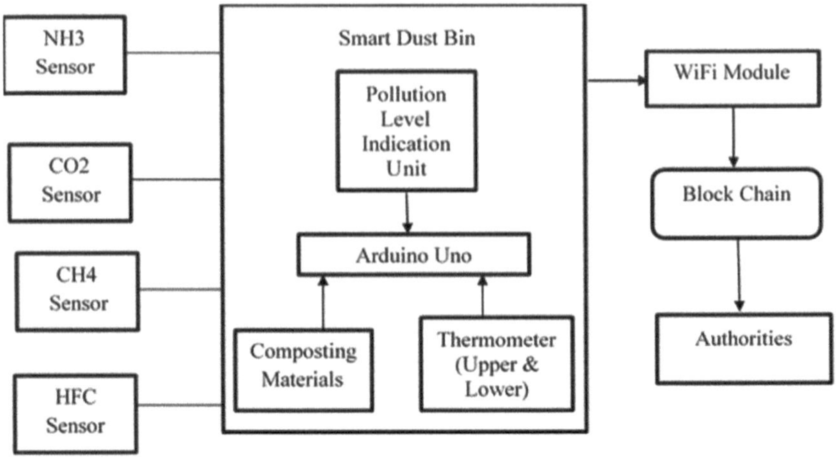

FIGURE 5.4 Waste food detection system.

4.2 Air pollution management system

The changes in the ecosystem will create the hardness in the air by polluting it by the generation of the effluents like chemicals and the polluted particles which can emerge out from the houses, vehicles, and the industries outlets. Instead, pollution can affect the climate very severely and now become a global concern that affects the lifestyles of many living beings. The usual situation available is making contact with the air since it seems unavoidable. The polluted air will create a significant impact directly to the humans, plants, and animals, as they are respiring it. Hence, the need for pollution control scheme is mandatory. Basically, there are two various kinds of air pollution available. The first one is the pollution created by the primary pollutants such as by the atmosphere, like the fire created in the forest and due to volcanic erosion. Moreover, the other is the pollution created by the secondary sources like the chemicals, like CO_2 emission, carbon monoxide, CFC, etc.

Another critical effect of air pollution is the emergence of nitrogen oxide that could generate from the vehicle effluents and from the factories. These are a part of the greenhouse gases, and that may yield ash and dust too. The burning fuels and the burning plastics may generate these kinds of gases. This polluted air can create many troubles to humanity like contributing the diseases like cancer and also some heart diseases when it is being inhaled. The secondary air pollution that gets created while exposed to solar energy, i.e., the photochemical smog. These can affect the oxygen ion bond and when inhaled can create a lot of troubles. These emerging gases also affect the environment by affecting the ozone layer. Hence, there requires a solution for the effective handling of the air wastes. The proposed system imposes an idea of using the IoT technology in combination with the blockchain for monitoring and detecting the polluted air from the houses and industries outlet. This can also suggest implying on the vehicles in the congested areas. In this system, technology is used that will detect the polluted air from the specified area and notifies whether the level exceeds the suggested limit, i.e., the accepted level. After the arrival of the accepted level, the system will send a prompt message to the server to indicate the authority about the severe cases regarding the pollution in that area spotted. On seeing the messages regarding the severity of the pollution, the person responsible can able to take the corrective action like excavating the peoples from the spotted areas. An advanced feature is added with this system, like a minimum level purifier is implemented with the help of a water-based system that can reduce the effect of air pollution to some extent. This also involves some kind of pollution test in the vehicles automatically. The system used a blockchain-based technology that can provide a decentralized network, which in turn helps to access with more security. Fig. 5.5 shows the architectural diagram of pollution management system.

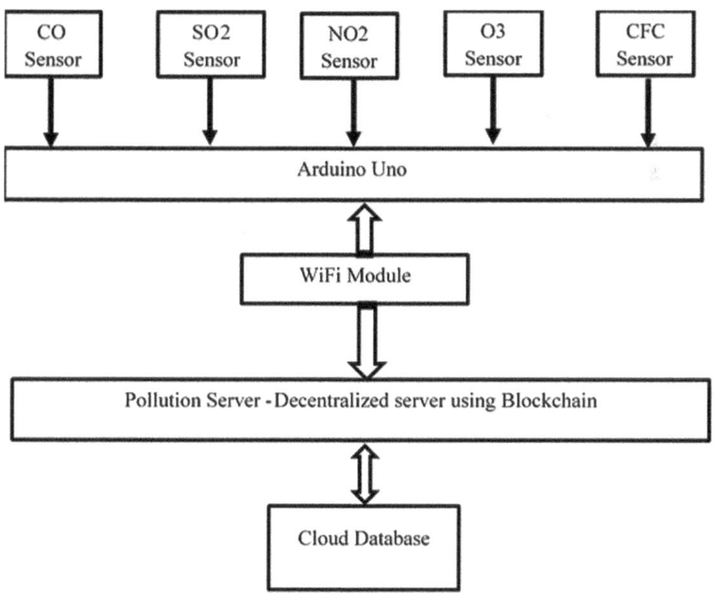

FIGURE 5.5 Pollution monitoring system.

The proposed system has a detection system for polluted air identification that consists of the IoT-based technology. This is mainly focusing on the renewal of energy sources. The detection system proposed might use the power setup from solar energy, as it consists of a system that can convert solar energy into electrical energy. This includes a messaging system that can alert the user or the authorities about the increase in air pollution in that particular monitored area. So that the authorities can rush up there for the immediate rescue by identifying the cause of the pollution, the corrective actions can be taken further. Mostly the factories and vehicles are responsible for the maximum amount of pollution causes. This system can test the vehicles and factory outlets during the running time, and a real-time prediction of the information is obtained. This will have the threshold value, and the predicted value is then tested with the available threshold value, and the deviation is noticed, and the variation ratio is notified to the authorities.

4.3 Soil waste management system

Contamination of the soil is the most crucial challenge to the environment. In the implementation of the smart cities, the usage of plastics, chemical washing liquids, and detergent wastes might damage the quality of the soil. Moreover, the handling of wastes in the soil is essential. Fig. 5.6. shows the block diagram of soil waste management system.

This will increase the productivity in the smart agriculture field. The properties of the soil like the pH value, electrical conductivity, and the organic content in the soil are calculated by using some standard methods, and the periodical fetching of information is being recorded. If the value of the abovesaid parameters is not in the available range, then that is to be adequately treated. If it fails, the soil quality will get damaged, and it will make the soil not to fit for any kind of irrigation purposes. The chemical analysis is conducted periodically for determining the acid mixture in the soil. The organic waste materials that can be used instead of the chemical materials can improve the soil quality.

The plastic and chemical materials modify the pH value of the soil. The deficiency characterizes the acidic nature of the soil in the nutrition level. This happens because the pH value affects the concentration and the movement of the ions in the soil texture. When the plant takes the nutrient value from the soil, its productivity might get reduced. Thus, the proposed system uses a quality indicator that will check the quality of the soil in three aspects like physical, chemical, and biological matters. The level indicator indicates the contaminated level, and the organic matter is separated into organic and nonorganic wastes, and a recycling strategy is implemented to enhance the

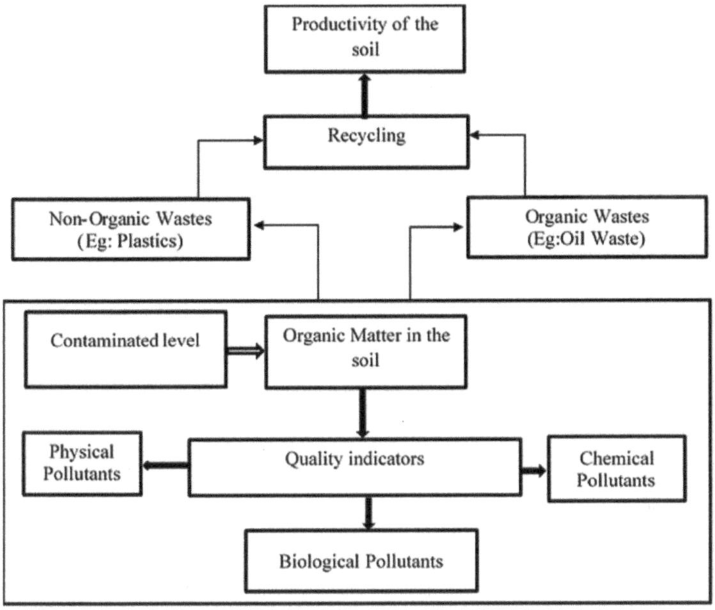

FIGURE 5.6 Soil waste management system.

quality of the soil, thus by helping to improve productivity. In addition, the contaminated area is indicated to the residents with the help of a text message. This also provides transparent communication with the help of blockchain technology.

5. Overview of proposed architecture design

The proposed architecture for the blockchain-based smart city is shown in Fig. 5.7. In this, the layered architecture is used for checking the connection between the IoT-based devices and the devices used for the multiinterface network. The data are overwhelming at the adaptation layer by the Fog network. This is initiated by the use of the blockchain along with the SDN. At the adaptation layer, the raw data which are filtered are directed toward the Fog node. This can improve the connection ability and reduces computational complexity. The fog node having the actuators is mainly responsible for the analysis of the data and the conversion of one format to another format in the pointed time.

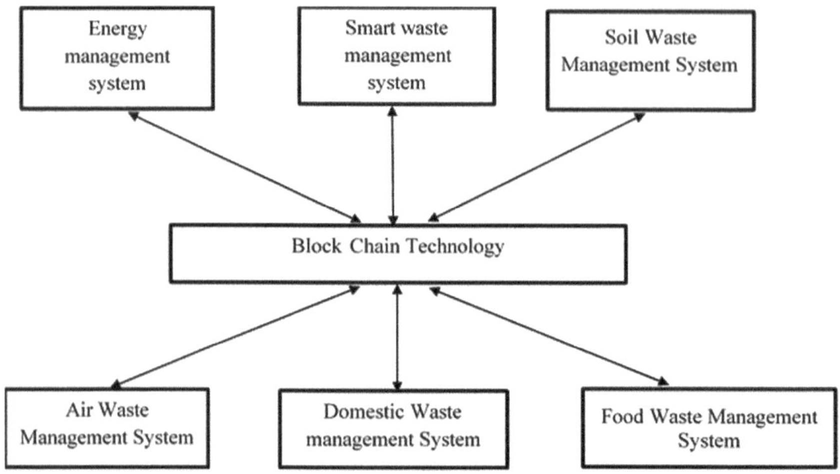

FIGURE 5.7 Secure smart city.

The combination of the blockchain and the SDN might provide privacy in addition to security at low resources and cost. The SDN controller having the actuator node is getting connected with the distributed mechanism by using the blockchain method. The usage of the SDN will enhance security and latency.

6. Results and discussion

In this proposed system, the use of blockchain provides a decentralized network with the improved scalability and latency for the smart management systems discussed. The scalability is improved for the smart management system for the effective utilization of the energy. The computational latency is comparatively high when compared to the previous methods. The smart management systems used in this proposed approach is showing better performance when compared to the previous energy management techniques. The performance that is calculated effectively with minimum latency and the maximum scalability of the novel smart systems implemented. The smart energy data are utilized that consist of some samples related to the different categories of energies considered. The collection of data used in the consumption of energy is controlled by the smart meter. The analysis obtained over all these might consist of some advanced techniques specifically for the smart cities.

TABLE 5.1 Latency versus data source.

	Data source				
Latency	Energy management system	Soil waste management system	Air waste management system	Domestic waste management system	Food waste management system
101.599	2.08595	2.41389	1.89512	2.24455	2.18981
100.334	3.28824	3.87198	4.66396	3.45911	4.00807
102.735	4.69414	5.01901	5.50863	4.50331	5.30705
95.3642	5.38993	6.41725	7.35503	5.18682	6.14302
90.4238	6.63929	7.50595	9.70537	6.38809	7.17904
83.0583	7.18466	8.80493	11.2627	6.6531	8.18026
62.2153	7.96947	9.65984	12.7745	7.51874	8.3936
66.477	8.67499	11.518	14.0832	9.10883	8.65452
76.2488	9.3851	13.9763	15.8391	10.2569	9.25924
94.5798	10.4533	15.0251	16.915	11.4643	9.85322
70.6483	12.0378	16.7876	18.1326	12.1217	11.7973
69.1834	19.1062	18.7333	19.1175	12.6763	13.3184
80.1592	20.419	20.5428	20.5658	15.8478	16.7646

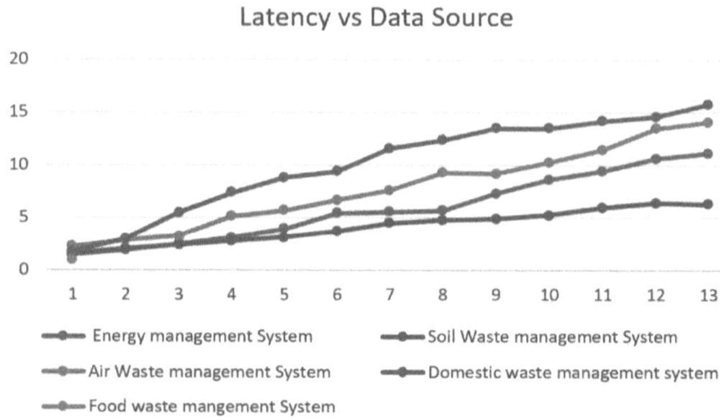

FIGURE 5.8 Latency values.

The proposed scheme also calculates the scalability by using the deep learning method by building an LSTM like smart network holding the input data. The overall scalability is based on the smart systems input data. Since the blockchain technology is adopted and the SDN is enabled using the fog node through the deep learning cloud. The performance proposed in the smart management system shows some variation in the latency while considering it with all the smart management systems used. Table 5.1 shows the variation of the latency in accordance with the data source predicted for the systems used.

The graph corresponding to the above Table 5.1 is given below in Fig. 5.8.

The secure foundation delivered by the blockchain technology introduced a transparent, secure transmission with some technology enhancements regarding the transfer of data straight into the reliable communication method. Another essential aspect of calculation is the consideration of the scalability values. The most critical issue to be considered is the scalability value prediction for all the smart management systems. Table 5.2 given below shows the various scalability values of the smart management system proposed.

The graph corresponding to the above Table 5.2 is given below in Fig. 5.9.

Table 5.3 describes the various attributes performance against the smart management system proposed.

The response time is the main criteria to be considered as the system used SDN along with the blockchain for the security and reliability-related purposes. The variation of the data shows the various attributes that has already taken against the proposed smart management systems in consideration with the previous systems.

TABLE 5.2 Scalability versus data source.

Scalability	Data source				
	Energy management system	Soil waste management system	Air waste management system	Domestic waste management system	Food waste management system
101.599	1.48225	1.59378	2.24302	1.8148	1.40755
100.334	1.93861	2.06139	2.8329	2.95108	3.42228
102.735	2.45585	2.38013	3.26317	5.50044	4.02137
95.3642	3.11275	2.76742	5.12901	7.33354	5.6795
90.4238	3.90319	3.15215	5.7015	8.78959	6.6751
83.0583	5.44467	3.73487	6.67561	9.41222	8.0851
62.2153	5.55262	4.51457	7.62209	11.5912	8.62792
66.477	5.70713	4.78112	9.25311	12.3611	10.3382
76.2488	7.31615	4.90134	9.1825	13.4545	11.2888
94.5798	8.63508	5.28761	10.2681	13.4365	13.2984
70.6483	9.45212	6.01461	11.4459	14.1589	14.1942
69.1834	10.6109	6.40906	13.4933	14.5928	15.8038
80.1592	11.1184	6.29906	14.116	15.7813	16.9544
111.441	12.5765	8.06873	15.2589	18.5992	19.0668
113.26	13.7732	9.33189	15.6702	20.4262	20.5745

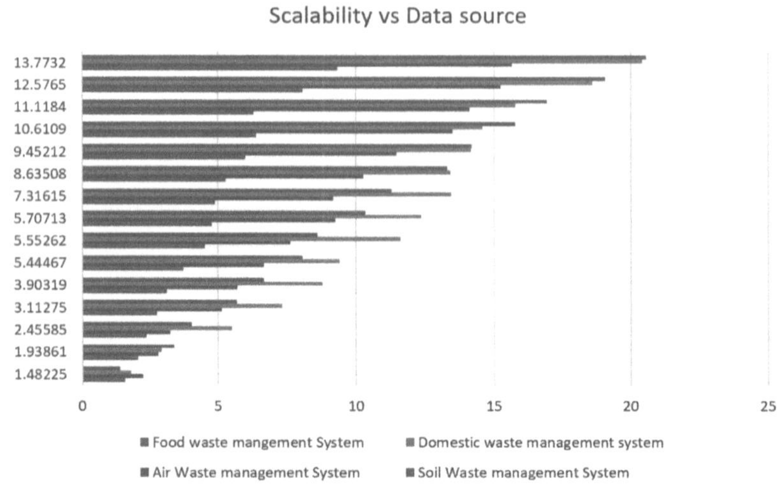

FIGURE 5.9 Scalability versus data source.

TABLE 5.3 Different attributes performance against the proposed smart management system.

Attributes	Performance %				
	Energy management system	Soil waste management system	Air waste management system	Domestic waste management system	Food waste management system
Latency	86	85	78	90	94
Bandwidth	88	87	80	88	90
Delay	76	83	84	79	90
Handover	62	63	58	66	70
Resource consumption	78	78	65	80	80
Power consumption	85	73	67	86	92
Mobility	81	72	62	81	91
Scalability	80	78	70	87	92

7. Conclusion

In this chapter, the various smart management methods are integrated into the computation of some sensing devices with smart applications in the smart city. The proposed blockchain-based distributed environment was enhanced with the use of SDN along with the Fog node. The probable architecture mitigates the decentralized network by providing transparency over the data provided with the smart environment. The applications like electrical energy management system and waste management system in the smart city are enhanced with the use of blockchain in order to provide security and privacy. The latency and scalability of the proposed smart management systems are taken into consideration for analysis. The deep learning method is used at the cloud, and the improvement in the performance is handled easily by the decentralized network.

References

Ahmad, K. M., & Khaled, S. (2018). IoT security: Review, blockchain solutions, and open challenges. *Future Generation Computer Systems*, 395–411. https://doi.org/10.1016/j.future.2017.11.022

Alex, R., André, da C. C., Rodrigo, da R. R., Ferreira, da S. V., Roberto, G. J., & Schmidt, D. C. (2019). Analyzing the performance of a blockchain-based personal health record implementation. *Journal of Biomedical Informatics, 95*, 103140. https://doi.org/10.1016/j.jbi.2019.103140

Al-Masri, E., et al. (2018). A serverless IoT architecture for smart waste management systems. In *IEEE international conference on industrial internet (ICII).*

Bose, B. K. (2017). Artificial intelligence techniques in smart grid and renewable energy systems—some example applications. *Proceedings of the IEEE*, 2262–2273. https://doi.org/10.1109/JPROC.2017.2756596

Dorri, A., Kanhere, S. S., & Jurdak, R. (2017). Towards an optimized blockchain for IoT. In *Proceedings - 2017 IEEE/ACM 2nd international conference on internet-of-things design and implementation, IoTDI 2017 (part of CPS week)* (pp. 173–178). Association for Computing Machinery, Inc. https://doi.org/10.1145/3054977.3055003

Dwyer, G. P. (2015). The economics of bitcoin and similar private digital currencies. *Journal of Financial Stability, 17*, 81–91. https://doi.org/10.1016/j.jfs.2014.11.006

Fernández-Caramés, T. M., Fraga-Lamas, P., Suárez-Albela, M., & Vilar-Montesinos, M. (1798). A fog computing and cloudlet based augmented reality system for the industry 4.0 shipyard. *Sensors, 18.*

Food waste is responsible for 6% of global greenhouse gas emissions.(2020). https://ourworldindata.org/food-waste-emissions.

Helin, Y., Arokiaswami, A., Wen-De, Z., Chen, C., & Xianzhong, X. (2020). Learning-based energy-efficient resource management by heterogeneous RF/VLC for ultra-reliable low-latency industrial IoT networks. *IEEE Transactions on Industrial Informatics*, 5565–5576. https://doi.org/10.1109/tii.2019.2933867

Ioan, P., Masoud, B., Yacine, R., & Rana, F. O. (2020). Blockchain for energy sharing and trading in distributed prosumer communities. *Computers in Industry, 123*, 103282. https://doi.org/10.1016/j.compind.2020.103282

Jui-Sheng, C., Shu-Chien, H., Ngoc-Tri, N., Chih-Wei, L., & Chia-Chi, T. (2019). Hybrid machine learning system to forecast electricity consumption of smart grid-based air conditioners. *IEEE Systems Journal*, 3120–3128. https://doi.org/10.1109/jsyst.2018.2890524

Lin, J., Shen, Z., & Miao, C. (2017). Using blockchain technology to build trust in sharing LoRaWAN IoT. In *ACM international conference proceeding series* (Vol. 130655, pp. 38–43). Association for Computing Machinery. https://doi.org/10.1145/3126973.3126980

Manoj, K. N. (2018). Blockchain: Enabling wide range of services in distributed energy system. *Beni-Suef University Journal of Basic and Applied Sciences*, 701–704. https://doi.org/10.1016/j.bjbas.2018.08.003

Mylrea, M., & Gourisetti, S. N. G. (2017). Blockchain for smart grid resilience: Exchanging distributed energy at speed, scale and security. In *Proceedings - 2017 resilience week, RWS 2017* (pp. 18–23). Institute of Electrical and Electronics Engineers Inc. https://doi.org/10.1109/RWEEK.2017.8088642

Nayanatara, C., Divya, S., & Mahalakshmi, E. K. (2018). Micro-grid management strategy with the integration of renewable energy using IoT. In *7th IEEE international conference on computation of power, energy, information and communication, ICCPEIC 2018* (pp. 160–165). Institute of Electrical and Electronics Engineers Inc. https://doi.org/10.1109/ICCPEIC.2018.8525205

de Oliveira, M. T., Reis, L. H. A., Medeiros, D. S. V., Carrano, R. C., Olabarriaga, S. D., & Mattos, D. M. F. (2020). Blockchain reputation-based consensus: A scalable and resilient mechanism for distributed mistrusting applications. *Computer Networks, 179*. https://doi.org/10.1016/j.comnet.2020.107367

On sharding blockchain.(2018). https://eth.wiki/sharding/Sharding-FAQs.

Polyzos, G. C., & Fotiou, N. (2017). Blockchain-assisted information distribution for the internet of things. In *, Vol. 2017. Proceedings - 2017 IEEE international conference on information reuse and integration, IRI 2017* (pp. 75–78). Institute of Electrical and Electronics Engineers Inc. https://doi.org/10.1109/IRI.2017.83

Premsankar, G., Ghaddar, B., Di Francesco, M., & Verago, R. (2018). Efficient placement of edge computing devices for vehicular applications in smart cities. In *IEEE/IFIP network operations and management symposium: Cognitive management in a cyber world, NOMS 2018* (pp. 1–9). Institute of Electrical and Electronics Engineers Inc. https://doi.org/10.1109/NOMS.2018.8406256

Red, V. A. (2017). Practical comparison of distributed ledger technologies for IoT. In *Proceedings of SPIE - the international society for optical engineering* (Vol. 10206)SPIE. https://doi.org/10.1117/12.2262793

Rosario, M., Claudio, D. C., Gaetano, F., & Chandra, M. S. (2017). A smart power meter to monitor energy flow in smart grids: The role of advanced sensing and IoT in the electric grid of the future. *IEEE Sensors Journal*, 7828–7837. https://doi.org/10.1109/JSEN.2017.2760014

Samaniego, M., & Deters, R. (2017). Internet of smart things - IoST: Using blockchain and CLIPS to make things autonomous. In *Proceedings - 2017 IEEE 1st international conference on cognitive computing, ICCC 2017* (pp. 9–16). Institute of Electrical and Electronics Engineers Inc. https://doi.org/10.1109/IEEE.ICCC.2017.9

Schneider, M., Rambach, J., & Stricker, D. (2017). Augmented reality based on edge computing using the example of remote live support. In *Proceedings of the IEEE international conference on industrial technology* (pp. 1277–1282). Institute of Electrical and Electronics Engineers Inc. https://doi.org/10.1109/ICIT.2017.7915547

The climate impact of the food in the back of your fridge.(2018). https://www.washingtonpost.com/news/theworldpost/wp/2018/07/31/food-waste/.

Yong, Y., Yannan, L., Junfeng, T., & Jianwei, L. (2018). Blockchain-based solutions to security and privacy issues in the internet of things. *IEEE Wireless Communications*, 12–18. https://doi.org/10.1109/mwc.2017.1800116

Chapter 6

Applications of blockchain in smart cities: detecting fake documents from land records using blockchain technology

R. Dinesh Kumar, V.N.S. Manaswini
Siddhartha Institute of Technology and Sciences, Hyderabad, Andhra Pradesh, India

Chapter outline

Blockchain for Smart Cities. https://doi.org/10.1016/B978-0-12-824446-3.00017-X

1. Introduction

In the past, most land documents, in India were marked through a village map boundary or through paper documents in which it included the occupant's details. Because of this method, there have been many issues such as illegal land possession, land litigations and property scams etc. One of the main challenges which were faced is illegal land possession.

Illegal land possession is a kind of ownership which is legally authorized, and imposed due to illegal occupation and it is simply obtained with the occurrence of illegal occupancy by generating fake documents, statements, or through bad influence, and by enforcing on legal owners of the land property. To overcome this issue, the Indian government launched the Digital India Records Modernization Program (DILRMP) (Digitization of land records, 2020) in August 2008. The main objective of this program was to reduce the scope on land property disputes and improve the transparency of maintaining land documents. But today, maintaining digitized records also leads to some problems of integrity. Cyber-attacks on digitized records are gradually increased day to day over the past few years. Considering the breach level index (BLI), approximately 3 million land records were in land property disputes in the first six months of 2018. By using these e-governance systems, cyber attackers hack the information of owners, which results in an increased number of land property disputes and social economic risk factors. To resolve these issues, in this paper we are using blockchain technology.

1.1 Blockchain key features

Recent advancements in online storage and access deals with the security and privacy issue. To provide better security and unauthorized access is prevented by introducing the blockchain concept. The blockchain key features are given in Fig. 6.1.

The blockchain key features are listed below.

1. Increased Capacity
2. Better Security
3. Immutability
4. Faster Settlement
5. Decentralized System
6. Minting

1.1.1 Increased capacity

The foremost key feature of blockchain is the increased capacity of networks. It can handle heterogeneous types of networks where many devices are connected to provide high computing power in centralized and decentralized things.

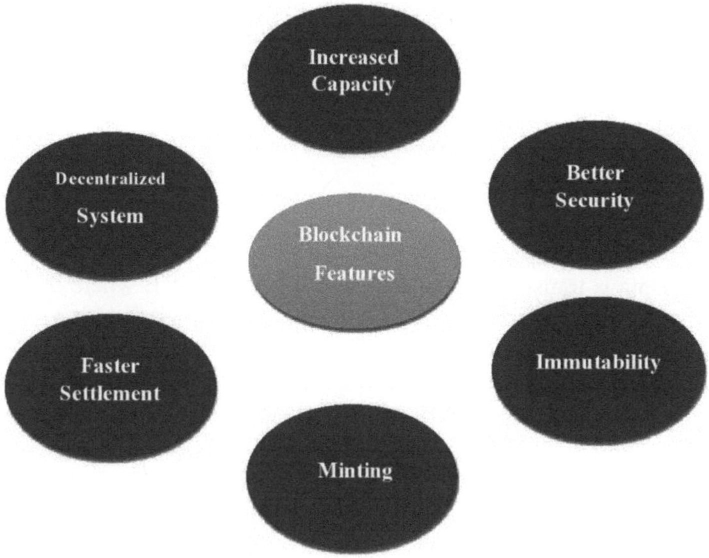

FIGURE 6.1 Blockchain features.

1.1.2 Better security

All the banking transactions need more security. The blockchain concept provides better security where shutting down the system is not possible. The blockchain includes a number of computers where each transaction is recorded in the ledger record, which built high security.

1.1.3 Immutability

The centralized database can be hacked and requires a third-party to manage security. The blockchain creates ledgers that cannot be immutable. The old ledgers remain unchanged and it preserves the original form.

1.1.4 Faster settlement

In the earlier banking system, settlement is made at a slow speed. To improve more settlement and transaction, blockchain provides a better way for the customer to update to the new banking system.

1.1.5 Decentralized system

Decentralization of the system improves easy storage and access with the support of the internet. The public key holds the control and process transaction. Minting blockchain provides a solution to the minting of process of manipulation. The trust level is increased on blockchain.

1.2 Blockchain architecture

In blockchain architecture, every block consists of following elements.

Index.

Timestamp.

Previous Hash.

Hash.

Data.

The following architecture is given in Fig. 6.2.

Data: Data is transferring between node A and node B where the data that stores in each block which is depended on blockchain type and it performs the data transactions like certificate transfer, electronic fund transfer etc.

Previous Hash: The hash is also be a fingerprint where the hash of a previous block is a tamper-proof sequence, while executing the transactions the hash is generated and broadcast with the help of cryptographic hashing algorithm (SHA 256) to network and this algorithm allows simple hash and dehash options which is why Hash Trees are common to choose.

Hash: The final hash element within the block is stored in the body of previous block, and it is the main hash element for security of blockchain architecture.

Timestamp: Time takes for generating block.

Index: It optimizes the performance of blockchain process.

2. Related survey

Blockchain based framework for educational certificates verification (Saleh et al. 2020) were identified some security themes for verifying documents using blockchain technology, and this paper also identified some gaps and loopholes with present certificate verification solutions. Finally, in this paper

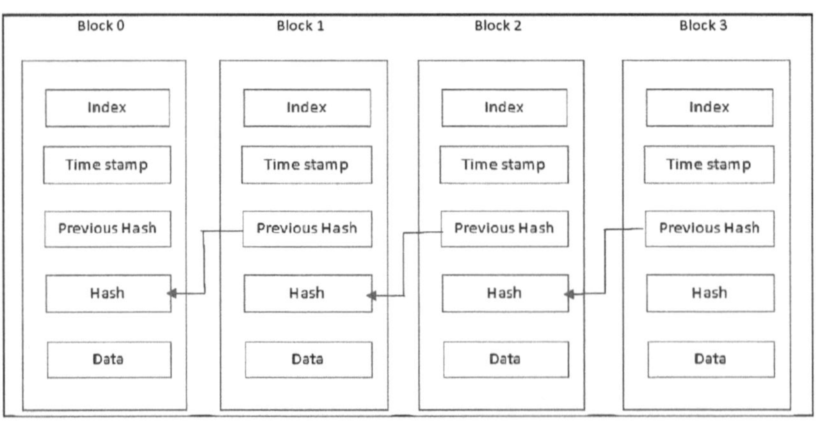

FIGURE 6.2 Blockchain architecture.

they focused on various themes like authentication, authorization, confidentiality, privacy and owner ship by using Hyperledger Fabric Framework.

Blockchain land registry platform—reducing frauds and delays (2020) were proposed on land registry platform using blockchain technology, and it is used to reduce frauds and delays with land property. So, in this paper he introduced one platform called LeewayHertz platform so this is successfully created for land registry to get selling and buying land properties with the help of Hyperledger sawtooth where Hyperledger sawtooth is open source platform where it supports both permissioned and permission less blockchain networks.

India's poor risk loss of privacy, land in drive to digitize records (2020) was said that in India after digitizing the land records also some states are not going for survey still some land records are not stored in digital land records more than a century and it may causes some data privacy issues also due to the accessibility with digital land records also there is to hacked by cyber attackers It leads to loss of data. And she also intimated that India is not having personal data protection law and it is not clear that who will maintain or control the data.

This paper (2020) proposed a system which are implemented on land records. They not only improving the security and integrity issues but also entangled the coordination of departments through this blockchain technology with this article they developed solution for getting the tamper-proof land records.

"Detecting insider attacks on databases using blockchains" (2017) introduced blockchain technology to the database because they detected insider attacks on the database and proposed a solution using tamper-proof resistance of blockchains. And the main intention of this paper was to make the database be robust.

Survey of India (SOI) used new technologies for mapping activities to get efficient records because so many lands are not registered in revenue records. It may lead to land disputes for this SOI-introduced GIS database and joined with the nation's Principal Mapping Agency and traces out the land property with the help of maps. It introduced two series of maps according to the National Topographic Database (NTDB); those are open series maps (OSM) and defense series maps (DSM) (Survey of India using new technologies for mapping and data generation, n.d.).

The maintenance of land records in paper form (2017) can deteriorate over decades. Various committees have been formed to build strengthen land management. To manage land records the conclusive titling system has been done. In this conclusive titling, the encumbrances of previous records to be made and legal policies to made in registering the land document in both central and state laws. This manages the ownership disputes and provides the evidence with updated land records.

The new term digitization (Lantmäteriet, 2016) is made for societal improvement by providing storage and easy access digitally. To support digitization, blockchain has been introduced in protecting documents.

The digital fingerprint for land documents can be used in the blockchain. The confidentiality of the data is not compromised in a digital fingerprint. The fingerprint placed in a different location helps to trace the document and avoid redundancy at maximum level.

One-third of the country's economy deals with real estate. Most construction projects have been stopped due to ownership issues. The aforementioned problems can be solved using blockchain concepts with strong security and easy access of exact ownership details.The Ethereum blockchain with smart contracts have been proposed for the proof of history and ownership of land records (2018).

With advancements in medical fields the storage and access of medical records in a timely manner with more security and privacy Xu Z. et al. (n.d.) is built with a blockchain concept. The key features, advantages, and safe transfer of medical records in health care management and advancement in pharmaceutical research in respect to oncology have been proposed.

The research (n.d.) in terms of security identities for blockchain provides secured data storage and transfer in distributed systems. Personally identifiable information with public key cryptography proposed to provide secure data storage. The analysis of blockchain on different fields like Internet of Things (IoT), artificial intelligence (AI), and cyber security have been made, which leads to new advancement in technology.

3. Existing land registration framework

In existing processes for buying or selling land property, first we have to make clear the idea of titles and check whether it is involved in land disputes or other illegal issues such as generating fake documents, and mortgage issues like creating double, general power of attorney, government lands, encroachments, and so forth, and have better approaches for legal advisers to register land without any litigation.

3.1 Procedure for land registration

Step 1: First, we must verify the documents those are valid or not. For that we follow some steps.

(1) Draft sale deed

Check whether having correct deed or not. For instance, if the land is presented as gift then required gift deed. suppose the land is purchased then maintain sale deed where sale deed is nothing, but which includes contract, payment details, standard terms and conditions agreed by seller and buyer, payment period, etc. and also, we have to check Encumbrance certificate only the certificate which confirms that land is having not having any liabilities like loans, litigations and mortgages.

(2) Preparation of stamp paper

To execute transactions on property along with linked documents like conveyance deed, sale agreement and sale deed government charges some amount of money as fees to be paid. It is nothing but stamp duty. You may receive stamped paper from an authoritative vendor.

(3) Executing deed

The deed must be executed at the subregistrar's office, and both parties must attend, sign the documents, and complete some formalities. If anyone of the parties, either seller or buyer, is not able to attend, then a power of attorney can be generated with an execution.

(4) Property registration

After completion of all formalities by seller and buyer are found to be acceptable, then they perform last step, i.e., registration will be held on property.

There are some following challenges occurred with this existing land registration process:

(a) The involvement of third-party occurrence
(b) The increasing number of property scam cases
(c) Delay with time

4. Proposed framework

With the use of existing system, there is a chance to create fake documents, but in the proposed system by using blockchain technology, there is no chance to develop fake documents, because after uploading the data in blocks there is no way to change the data as we can observe in Fig. 6.3.

Stepwise approach for getting land registration using blockchain technology (Khan, Ansari, Sachadeva, & Jain, 2020).

4.1 Step 1: user creates an account

Users are those who created an account by registering into the blockchain land registration system. First, they enter their personal details and choose whether they are sellers or buyers. The identity information of the user gets a hash, which is stored into the blockchain system.

4.2 Step 2: sellers upload property details on the system

Firstly, the sellers must upload complete details about land property, like images, land possession certificate, various documents, and pin the location where the land is situated on the map. Seller's land property is recorded and it is accessed by buyers who are signed up into the system.

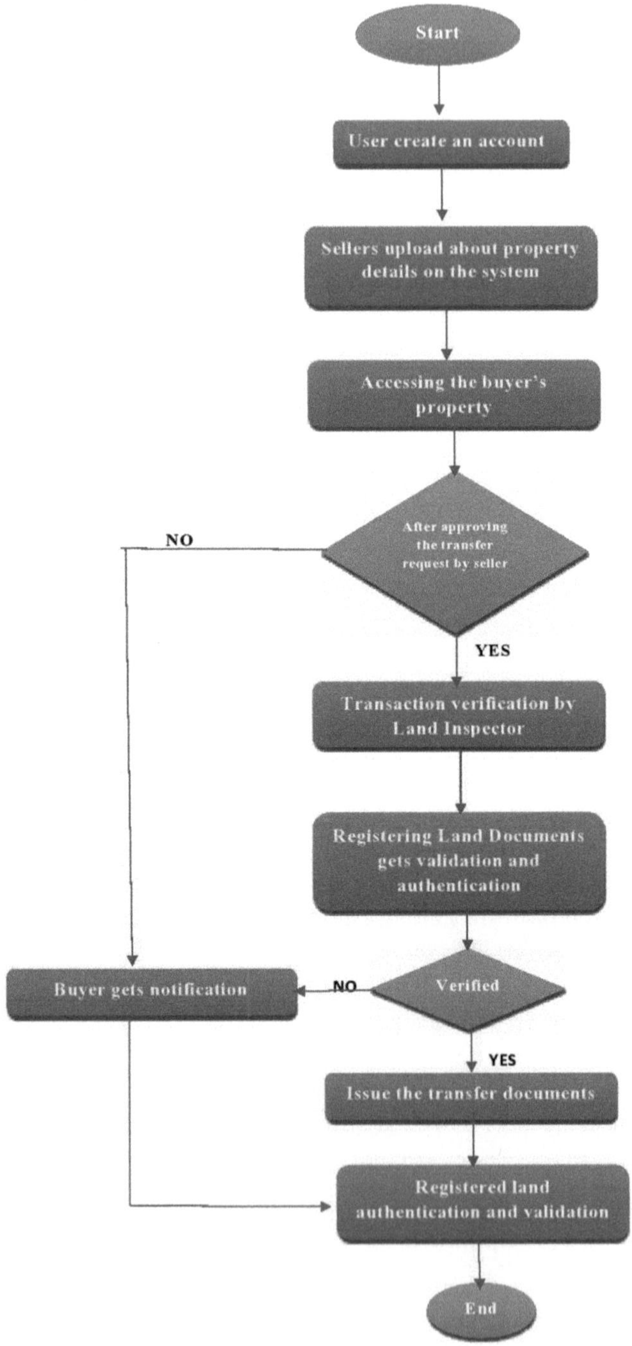

FIGURE 6.3 Flow chart.

4.3 Step 3: accessing the buyer's property

Buyer is able to access the recorded property, which is stored in the system and they can request the seller to give access to the property details. If sellers are satisfied with the buyer's acceptance, then it allows them to access the specifications of the property. If the seller accepted the request, then the buyer can be able to check land property documents and ownership records on land possession. If the buyer is satisfied, then requested to purchase and begin the process of transfer.

The blockchain then starts its functionality where the transactions of sellers and buyers that are related to land property is stored in blocks. By using the hash, the blocks of data are indexed for ensuring the data, which is to be authentic and traceable.

4.4 Step 4: after approving the transfer request by seller and then land inspector gets involved

If the seller approved the transfer request to the buyer, then the land inspector will initiate his work for transferring property. After verifying the land documents, they will conduct a meeting for transferring the ownership rights with seller to buyer. This meeting is also recorded in the blockchain for further land disputes.

4.5 Step 5: transaction verification by land inspector and performs land transitions

Land documents are verified by the land inspector after sellers and buyers completed some formalities in front of the land inspector online, and these documents are uploaded and the records stored in the blockchain land registration platform.

4.6 Step 6: registering land documents gets validation and authentication

After completion of uploading land documents into the blockchain with hash, it gets authenticated and validated and it will not get modified once it is uploaded in the blockchain.

5. Implementation

5.1 Technical workflow

This is the process of checking the authority of land details by the registration authority, and if it is clear they will approve registration, then store in the blockchain technology safely with a unique ID (Fig. 6.4).

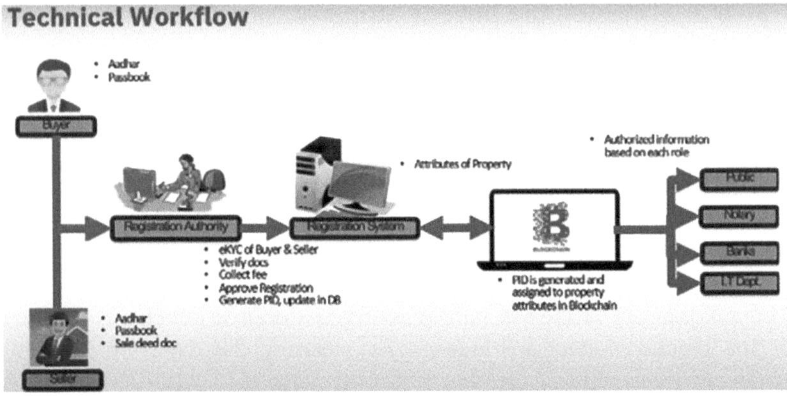

FIGURE 6.4 Technical workflow.

5.2 Creating buyer registry

In this step we are creating the register of the buyer (Fig. 6.5).

5.3 Creating seller participant

After creating buyer's details, the next process is creating seller's details (Fig. 6.6).

5.4 Property list asset registry

In this, we will maintain the property lists and their asset registry (Fig. 6.7).

5.5 Property for intent of sale

In this process we are taking the details of a property intended for sale (Fig. 6.8).

```
{
    "$class": "org.example.empty.Buyer",
    "Bname": "Romika",
    "Bemail": "romika@gmail.com",
    "Aadhar": 123456789012,
    "BankName": "Canara Bank",
    "BankAddress": "Janak Puri",
    "AccNo": 176132190876,
    "IFSCCode": "CNRB001234",
    "BBalance": 40000
}
```

FIGURE 6.5 Properties of buyer.

```
{
  "$class": "org.example.empty.Seller",
  "Sname": "Harshul",
  "Semail": "harshul@gmail.com",
  "Aadhar": 12343212321,
  "BankName": "ICICI Bank",
  "BankAddress": "Prem Nagar",
  "AccNo": 17613215432,
  "IFSCCode": "ICIB0004564",
  "SBalance": 50000,
  "SaleDeedDoc": "SD123456"
}
```

FIGURE 6.6 Details about seller.

```
{
  "$class": "org.example.empty.PropertyListing",
  "PLID": "1",
  "owner": "Mayank",
  "mktprice": 2000,
  "RegistrationDate": "14-9-2016",
  "PropertyType": "House",
  "Location": "Gurugram",
  "Status": "IntentForSale"
}
```

FIGURE 6.7 Property list according to assets.

```
{
  [illegible]
  [illegible]
  [illegible]
}
```

FIGURE 6.8 Property intent of sale.

5.6 Registered transactions

The sales transactions of seller and buyer is to be updated with the help of the Hyperledger platform using blockchain technology (Fig. 6.9).

```
{
  "$class": "org.example.empty.Registered",
  "PID": "3",
  "rpropertylisting":
"resource:org.example.empty.PropertyListing#2"
}
```

FIGURE 6.9 Registered transactions.

5.7 Generating certificates

cd first-network
export FABRIC_CFG_PATH = $PWD
cryptogen generate — config = ./crypto-config.yaml

5.8 Creating block and channeltx

configtxgen -profile TwoOrgsOrdererGenesis -outputBlock./composer-genesis. block.

5.9 Key can be found in -

"first-network/crypto-config/peerOrganizations/org1.example.com/ca/c7d8243x 5d76cb36bb1499edf1b9b256144753a458a8467ed1ff67607cef09179_sk".

6. Conclusion

In this chapter we are proposing a land registration system where it maintains the land records of seller and buyer while providing authentication and validation to the registered documents, which are stored in the blockchain system. In the blockchain system, the data is stored in blocks with hash values, where in the existing system the land registration is done physically in the sub registrar office (SRO) office, but with that system, we will face a lot of issues especially on land disputes and increasing frauds. Producing paperless documents on land registration will not be an easy process but are also very secure, and once we upload in blocks in the blockchain system, it is impossible to edit the information with this; thus we can reduce fraud cases and data lost. By using this platform our land properties can also be executed using the cryptocurrency technique and mapping with land records, which are created by a seller. Hence, the extent of use can be widespread.

References

Blockchain land registry platform – reducing frauds and delays.(2020). https://www.leewayhertz.com/blockchain-land-registry-platform/.

(n.d.). *Breach level index*. https://breachlevelindex.com/.

Detecting insider attacks on databases using blockchains. (2017). In *ACMSIGSAC conference on computer and communications security*. https://security.cse.iitk.ac.in/sites/default/files/db-blockchain.pdf.

Digitization of land records.(2020). https://realty.economictimes.indiatimes.com/realty-check/digitization-of-land-records/2966.

India's poor risk loss of privacy, land in drive to digitize records.(2020). https://ciso.economictimes.indiatimes.com/news/indias-poor-risk-loss-of-privacy-land-in-drive-to-digitise-records/74558835.

Khan, R., Ansari, S., Sachadeva, S., & Jain, S. (2020). Blockchain based land registry system using ethereum blockchain. *Journal of Xi'an University of Architecture and Technology, 9*, 3640–3648 (1006–7930) https://www.xajzkjdx.cn/gallery/355-april2020.pdf.

Lantmäteriet, Telia Company, ChromaWay, & Kairos Future. (July 2016). *The Land registry in the blockchain.* http://ica-it.org/pdf/Blockchain_Landregistry_Report.pdf.

(n.d). *Survey of India using new technologies for mapping and data generation.* https://dst.gov.in/survey-india-using-new-technologies-mapping-and-data-generation.

Saleh, O. S., Ghazali, O., & Rana, M. E. (2020). Blockchain based framework for educational certificates verification. *Journal of Critical Reviews, 7*(3), 79–84. https://doi.org/10.31838/jcr.07.03.13.

Chapter 7

Citizen e-governance using blockchain

Rishabh Setiya, Sneha Pandey, Ayush Kumar Singh, Deepak Kumar Sharma

Department of Information Technology, Netaji Subhas University of Technology, New Delhi, India

Chapter outline

1. Introduction

Governments store many types of information about their citizens such as their income, land holdings, employment, and government-issued documents like passport, voter card, driving license, and certificates of birth, death, and marriage. This information is used for providing as well as denying services.

Blockchain for Smart Cities. https://doi.org/10.1016/B978-0-12-824446-3.00010-7

The availability of personal data to the government is daunting to the citizens who are concerned about their right to privacy. It is felt that the access to citizen records should be controlled in a manner that prevents disclosure of information, which may be unnecessary for a particular purpose. Such systems can be designed using blockchain technology. Blockchain, in essence, is an immutable ledger which allows for transactions and data to be recorded, shared, and synchronized across a distributed network in a decentralized manner.

1.1 Introduction to e-governance

E-governance or electronic governance refers to the application of Information and Communication Technology (ICT) for providing government services, and exchange of information and communication between the government and the four major stakeholders of a nation: citizens, businesses, employees, and other government organizations. Fig. 7.1 shows these interactions.

As can be seen from Fig. 7.1 the concept of e-governance incorporates strategic institutional integration of the four major stakeholders. This enables the citizens to be more participative and active during decision-making.

E-governance improves the delivery and efficiency of the government services to all the stakeholders. The access of all the information to the citizens and increased transparency in the administration decreases the corruption and empowers the citizens. The major change that e-governance brings about is the reduction of bureaucratic paperwork ensuring a better management and

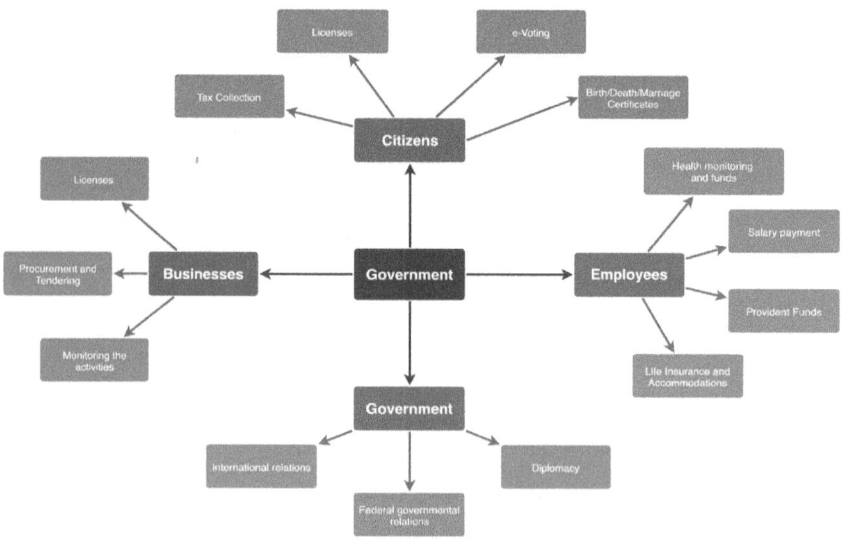

FIGURE 7.1 Interactions between the government and various stakeholders of the nation.

legitimacy of the government. This also reduces the stationery expenses, saves resources, and requires minimized workforce. The reduction of manual labor and use of electronic medium speeds up the provision of services and awareness of government policies among the citizens.

At the root of e-governance reside the ICTs, which contribute to good governance practices for development like

Automation: substituting present manually executed practices of processing information. Example: Automation of existing accounting or clerical functions.

Informatization: aiding present manually executed practices. Example: Aiding and promoting existing practices of decision-making and implementation.

Transformation: devising new ICT-executed practices. Example: Creating new approaches for Citizen Administration.

ICTs can thus provide cheaper and more efficient services. The benefits of ICTs show that e-governance has a key role to play in the current and future development of nations (Richard, 2001).

1.2 Traditional systems of e-governance

Many countries worldwide have implemented centralized e-governance systems for the development of efficient communication and interactions between the government and citizens. A centralized system of governance means that a central authority has the power to take decisions regarding the distribution of services to the citizens. The data required for these services are stored on centrally managed databases.

Some examples of governments which have made significant advances in the implementation of e-governance services are as follows:

Citizen friendly e-government of Kazakhstan

Using a comprehensive plan for establishing and developing e-governance, Kazakhstan has made use of ICTs to connect the citizens and build their trust with the government authorities. There have been four major stages in expanding and diversifying e-governance in the country since its inception (Fig. 7.2).

The government aims to provide services as per the citizens' requirements where the citizens do not even need to direct any requests for the same but only need to confirm those actions (E-Government, n.d.).

United States Social Security Administration

The program benefits the retired citizens, survivors, and the families whose wage-earner dies. The portal ensures providing the best facilities by maintaining the records and presenting a user-friendly interface. The implementation of such programs on a large scale without ICT would have been difficult (U.S. Social Security Administration, n.d.).

FIGURE 7.2 Stages in e-governance of Kazakhstan.

1.3 Challenges in traditional systems of e-governance

Centralized systems of governance essentially compel the citizens to trust their governments whether or not they are worthy enough of that trust. Every digital document must be verified by the central authority to be considered authentic. For work to go on ceaselessly throughout the region coming under a system, the system must always stay functional. However, this does not happen as systems are always susceptible to failures. System failures may be caused by internal or external hazards. Internal hazards include power outages, hardware failure, software error, network issues, and human error. External hazards include infiltration, ransomwares, denial-of-service attacks, and natural calamities.

Governments often outsource the e-governance systems to private companies. Entrusting the private sector with confidential information of citizens can be a threat to their privacy. Additionally, any disputes arising out of the contract could lead to a severe impact on the citizens. As an example, consider the Kochi government which outsourced its e-governance operations to Tata Consultancy Services (TCS). As per a report published in The Hindu on June 13, 2020, TCS was unwilling to provide further support to the government. It had completed all the exit formalities but had not received any communication in response. Its dues had not been cleared. It warned that it shall not be responsible for the disruption of services (TCS Backs out of Kochi Corporation's e-Governance System, 2020).

Such systems entrust a lot of power in the hands of a few people. The possibility of cyber-attacks increases if the concerned authorities do not have sufficient expertise in network security. Attack on a single central server can leak a substantial amount of confidential information. The software used may be built with backdoors which could make malicious activity undetectable. Further, the users are given little or no information regarding the security of their data (Qi, Feng, Liu, & Mrad, 2017).

The increasing dependence on ITCs due to development in e-governance has made such attacks progressively frequent. One notable example is the DDoS attack on Canadian government websites of June 17, 2015. The attack was so fatal that it completely crashed the e-mail services of government employees. Further, dozens of websites of government departments were rendered inaccessible (Chase, 2015). Another incident in which an e-governance service was affected was the hardware failure which the United States Postal Service (USPS) suffered from in April 2014. The hardware failure wiped off a Computer Incident Response Team database. Incidentally, the backup of the database was also stored in the same hardware which failed. This led the USPS to audit and implement corrective backup measures for other essential databases (Williams, 2014).

These challenges, which are faced by the traditional systems of e-governance, can be reduced by using blockchain technology.

2. Use cases

This section discusses various sectors of e-governance in which blockchain-based solutions have been explored. Some of these ideas have been successfully implemented as well.

2.1 Digital identity

Identity is the means of distinguishing a person from everyone else. Increasing use of digital mode in various domains calls for the creation of a digital identity of every person using these services. For example, many teaching platforms teach through prerecorded video lectures or live classes and the certificates are issued online too. For such services to provide trustworthy certifications, it is essential to authenticate the identity of the beneficiary of education. In other words, no two persons having the same name should be able to claim their right over the same certificate. Also, no student should be able to impersonate another student which could easily happen in the absence of a foolproof system of digital identity.

Usually, these educational services do not require one to provide legally valid identity proof and the name on the certificate reflects whatever the learner wishes. As a result, nobody is bound to accept these credentials as genuine which leads to them being regarded as merely showpieces. However, if they could be linked to a digital identity authenticated by the government, their value would increase considerably.

A digital identity involves two processes. The first being authentication at the time of its creation and the other being verification when a service is to be availed. In the abovementioned example, the authentication step will be performed by the government when a person decides to register for the scheme of digital identity. For this, a person will need to show proof of address, birth certificate, driving license, or any other government-issued document which can establish his/her identity. Verification will be required at the time of registration in an online course and also when the certificate issued is required for job or university admission applications (Sullivan & Burger, 2018).

The Illinois Blockchain Initiative is a digital identity framework which aims to establish the digital identity of a person from the time of birth itself. Government agencies would be responsible for cryptographically signing the identity attributes like legal name, date of birth, sex, or blood group. The permission to view or share them would be stored in a blockchain in the form of a decentralized identifier. Only the owner of the identity would be able to provide permission to access any details without which data would not be accessible.

This system will reduce the need of various state agencies and private companies to build and maintain their own identity databases. They will be able to request access to the information as per their requirement. This would decrease the chances of illegitimate usage of data as well as leakage of data in case of a breach (Douglas, 2017).

The uPort system of decentralized identity in the Municipality of Zug in Switzerland is a project which aims to provide a self-reliant blockchain-based identity to authenticate for e-government services and share personal data with third parties (David, Maciej, Lorenzino, & Francesco, 2019).

The biggest challenge in the application of blockchain for identity management is regulation and lack of standards. A framework must analyze risks and limitations in addition to the benefits offered.

2.1.1 Case study: United Nations World Food Program in Jordan

As of 2020, Jordan has over 600 1000 Syrian refugees (Farah & Soheib, 2020). Before the incorporation of blockchain, the United Nations World Food Program (WFP) used to supply cash to intermediaries like banks and financial service providers. The intermediaries were also provided a list of beneficiaries. Bank accounts of all beneficiaries were opened and the mode of payment was agreed upon. The beneficiaries could go to a local store to purchase whatever they needed and get their iris scanned at the billing counter for identification. This let the cashier know what benefits they were entitled to. The banks later reimbursed the local stores after verifying the transactions (Zambrano, Young, & Verhulst, 2018). The proof of concept of Building Blocks was conducted in January 2017 in Sindh Province of Pakistan with 97 people (Alden, n.d.; Building Blocks Blockchain for Zero Hunger, n.d.). It was built on the public Ethereum blockchain (Juskalian, 2018). In May of the same year, a pilot phase for more than 10,000 people was completed in Jordan (Blockchain Against Hunger: Harnessing Technology in Support of Syrian Refugees, 2017) and was eventually scaled to 100,000 people by 2018 (Juskalian, 2018). However, this was on a private, permissioned version of Ethereum as opposed to the public one used during the initial phase (Juskalian, 2018). It used the Proof of Authority consensus mechanism (Theresia & Novelo, 2019). This was done as the transactions on the public blockchain were slow and costly (Juskalian, 2018). The project has led to a reduction of 98% in expenditure as well as a significant decline in processing time (Farah & Soheib, 2020). The control over transaction data lies with the WFP (Juskalian, 2018). WFP maintains these virtual wallets to record how much it owes to each supermarket. These records are settled at the end of each month (Farah & Soheib, 2020). The WFP aims to connect Building Blocks to all refugees in Jordan and eventually every refugee in the world (Juskalian, 2018).

2.2 Education

Education is one of the strongest pillars of governance. It is one of the most important measures of development of a nation. But unfortunately, education systems in many nations lack aspects of transparency, effective communication between teachers and students, and proper management of data. Blockchain technology has a prominent solution to these problems. The effective management and storage of data as well as accountability through smart contracts can improve the system for a better future. Students often have conflicting opinions regarding the marks allotted to them or the attendance received by them as the system lacks transparency. An implementation of a system which makes use of blockchain can help the teachers and students update, access, and report anomalies in data conveniently.

This system would provide students and teachers unique Ids and all the data would be stored on a blockchain database. The smart contracts on the blockchain network can ensure a better understanding between the teachers and students. The contracts can provide students with marks or credits as soon as the work is evaluated. The credits would be recorded on blockchain nodes called wallets that are protected by cryptographic algorithms and hence immutable and secure. This also guarantees a reliable evaluation method based on the wallet currency (Guang, Bing, Manli, & Nian-Shing, 2018). Massachusetts Institute of Technology in association with Learning Machine launched Digital Diploma, a pilot program that provides digital diplomas to the students via an app Blockcerts Wallet that uses blockchain technology in 2017 (MIT Introduces New Digital Certification Program Powered By Blockchain Technology — Observatory Of Educational Innovation, 2017).

EduCTX is a global higher education credit platform based on European Credit Transfer and Accumulation System (ECTS) that uses blockchain. The platform is based on the ARK blockchain technology and processes ECTS tokens which hold the credits earned by the students. ARK is a decentralized company providing a flexible and open-source platform for implementing blockchain technology. The transactions information stored on blockchain nodes include the sender's official name, the anonymous student, credit score, and the course identification. It is based on distributed consensus protocol. Therefore, no computing power is required and prevents random miners from joining. An EduCTX blockchain explorer is available providing solutions to the university students regarding certificate management and verification (Muhamed, Marko, Kristjan, Marjan, & Aida, 2018).

The Open Source University is another decentralized platform for education based on blockchain technology. It provides verified certification to learners all over the world. Its "credentials wallet" or "competence passport" is an immutable digital credentials portfolio storing all of the learners' achievements at one place. Also the framework of its smart contract system makes it more effective and trustworthy. The framework consists of L2A (learner-to-academia) and B2L (business-to-learner) smart contracts. L2A smart contracts provide a database of the learners' accomplishments which reflects their ambitions and hence the academia suggests other courses and opportunities of their interests. Furthermore, B2L smart contracts become the database for businesses as the corporates can approach the learners with the required credentials more effectively. The "handshake" feature of B2L smart contracts is the key to recruitment when the terms of the contract are met. Hence, this system inculcates academia, businesses, and learners all together creating a perfect platform (What Is the Open Source University at OS.UNIVERSITY and Why Does It Matter?, 2017).

2.2.1 Case study: Malta

In September 2017, the Ministry for Education and Employment in Malta started its pilot project in association with Learning Machine Technologies, a blockchain company. The project was the world's first national pilot of Blockcerts and with this the Maltese government began its journey to become the blockchain island. In February 2019, the Maltese government extended the Blockcerts implementation to all institutions to create an education management across the nation. It became a platform to create, issue, verify, and view educational certificates using blockchain technology (Malta Begins Nationwide Rollout of Blockcerts Blockchain Credentials For Education And Employment, 2019; Malta Is First Country To Put Education Certificates On Blockchain, 2019). The Blockcerts app allows the students to create an account and then add their educational institutions as an issuer of certificate. The institution creates the certificate using its private key and it can also access the students' public keys. The assigned certificates are also added to the students' wallet and they can access it anytime (Fig. 7.3). This has effectively improved the system of sharing and managing academic records (Blockcerts, n.d.).

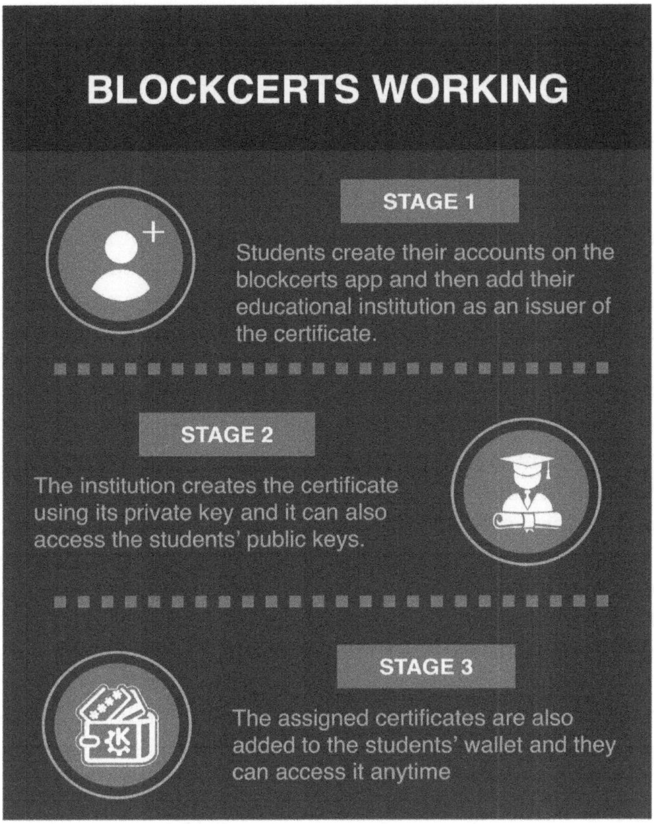

FIGURE 7.3 Blockcerts working.

The institutions such as the Ministry for Education and Employment, the Institute for Tourism Studies, the Malta College for Arts, Sciences, and Technology and the National Commission for Further and Higher Education have successfully established the Blockcerts certificate management system and are motivating other institutions to implement this soon (Malta Begins Nationwide Rollout of Blockcerts Blockchain Credentials For Education And Employment, 2019).

2.3 Healthcare

Provisions of good healthcare services and management have become a vital part of any nation's development. A structured healthcare industry consists of medical services and facilities, infrastructure and equipment, pharmaceuticals, medical research, patient management, and record keeping. Recent times have seen an increasing trend toward digitization of patient records and healthcare data, and smart wearable technology in healthcare. But these advancements come with the challenge of storing, secure sharing, and robust authentication of healthcare data without violating the patient's privacy.

Different healthcare providers implement their own centralized data storage. This leads to fragmentation of healthcare data among different sources. It is evident that a medical process involves interactions with different individuals and institutions like physicians, pharmacists, medical laboratories, and so on. This fragmentation makes authentication and processing of healthcare data a tedious task for all the medical professionals involved. Further, this also puts the patient's privacy at risk. This is an ideal scenario to make use of blockchain technology, as it enforces privacy, authenticity, data integrity, and easy availability by using decentralized storage (Agbo, Qusay, & Eklund, 2019).

A good example of this is the Gem Health Network launched by Gem. It is powered by Ethereum blockchain and aims to provide interoperability and data sharing between all the medical stakeholders. It will allow medical stakeholders to view their patient's history from a single source, saving resources and time during treatment. Moreover, the cryptographic algorithms of blockchain makes hacking a patient's data, or violating their privacy, practically impossible (Prisco, 2016).

Another example of an Electronic Medical Record (EMR) management system built on blockchain is Healthbank, a Swiss digital health company. Users can upload and maintain their own medical records and have complete ownership over them. These data can be made available to medical professionals with proper authorization. Healthbank takes its medical innovation one step further by giving the users the opportunity to contribute to medical research by sharing their data. In exchange for this, Healthbank compensates the users financially. The amount compensated depends on the quantity and quality of data shared by the users (Health Bank, n.d.; Mettler, 2016).

Healthchain is another EMR application that is built on IBM Blockchain's Hyperledger Fabric.

Yet another example of an EMR is OmniPHR, a proposed model to handle patient's health records which provides an elastic and scalable platform and promotes interoperability (Thomas, Raymond, Zhechao, & Debiao, 2019).

Since transactions in blockchain are immutable and traceable, one can instantaneously determine the entire life cycle of a product; its production and ingredients, and all transfers of ownership. This traceability can be used to control the production and supply of counterfeit medication, which is a serious problem in the pharmaceutical industry. If a medicine was contaminated or its dosage was altered, it could be determined when it happened and under whose ownership.

Counterfeit Medicine Project, launched by Hyperledger, aims to tackle the prevalence of counterfeit medication using blockchain. It plans to attach a timestamp, containing details about the manufacturer, to every medicine manufactured. This makes the medicine traceable to its origin and can control the production of counterfeit medicines (Mettler, 2016).

Blockchain has immense potential in healthcare and has the ability to make many of the present systems more secure, efficient, and reliable.

2.4 Utility supplies

Electricity and water are essential utilities in every house. They can be billed in multiple ways. One of the ways is that an employee of the electricity or water department visits every house and records the meter reading. This leaves a lot of room for corruption as the customers can bribe the reader and get their bill reduced by manipulation of reading as well as the meter itself. This is a common practice in countries like India. Some customers tip the reader even to get their actual bills instead of inflated bills. A slightly improved method is that the customers click photographs of the meter and send it to the water or energy board through any messaging application or a specialized application made by the board for this purpose. This method could cause huge losses to the utilities if there is any scope of uploading manipulated reading by the customer. The customers can digitally edit the clicked photograph or upload a photograph clicked in the past when the reading was lower. Minimizing such problems is necessary as neither can these services be returned if an inflated bill is produced nor can the authorities confiscate them if the user fails to pay for them.

Bill payment was one of the first applications of blockchain in the energy sector. BAS Nederland was the first energy company to accept payments in Bitcoins. BAS stands for Beneficial to All Stakeholders, which represents the aim of the company to gradually progress toward lesser and lesser dependence on nonrenewable resources and turn the customers into producers of clean energy (Marckx, 2014).

Drift, founded in 2014, is a Seattle-based startup which intends to bring about increased transparency in energy bills and reduce costs. It allows consumers to purchase electricity in a peer-to-peer (P2P) marketplace from small energy producers like wind farms, solar farms, and large buildings with excess power. Blockchain is used for these P2P transactions. Weekly bills are generated which include detailed information of the various sources from which energy was purchased. Till 2020, the company was operating only in New York (Drift | Find Your Power, n.d.; Drift Is A New Startup Applying Peer-To-Peer Trading To Retail Electricity Markets, 2017).

SunContract is another blockchain company with a model similar to Drift operating in Slovenia. In November 2019, it received a $2,300,000 USD grant from the European Union's European Regional Development Fund for a Research and Development project that will introduce artificial intelligence and automated trading in its platform (Energy Trading Platform | Suncontract, n.d.).

In case of water and gas too, smart meters incorporating blockchain technology can allow users to pay via cryptocurrencies and help them adopt sustainable consumption patterns.

2.4.1 Case study: Walenstadt, Switzerland

The lighthouse project Quartierstrom, started in 2017 (Die Entstehung Vom Projekt | Quartierstrom, 2019) is led by a team from ETH Zurich and University of St. Gallen and cofunded by the Switzerland Federal Office of Energy as a pilot and demonstration (Ableitner et al., 2019). The local electricity market started operating in January 2019 and since then 37 parties have been able to trade electricity in a peer-to-peer manner. 28 out of the 37 had their own solar power systems (prosumers) and the rest were pure consumers (Quartierstrom 1.0 - Das Projekt In Kürze, n.d.). Whenever excess power was generated, it was sold to Water and Electricity Works Walenstadt and when the locally generated power was insufficient, it was purchased from them. The smart meters installed with the prosumers functioned as full nodes in the blockchain (Ableitner et al., 2019). Raspberry Pis installed with self-developed modules acted as smart meters as no commercial smart meters were available which could be used for this purpose (Brenzikofer, Meeuw, Schopfer, Wörner, & Dürr, 2019). The permissioned blockchain allowed the users to set their sale and purchase price limits. A double auction algorithm was run on the nodes every 15 min and all resulting transactions were recorded on the blockchain (Ableitner et al., 2019). The feature of setting prices was replaced by automatic pricing based on demand and supply for a month and it was observed that the majority of the participants especially those who frequently used the portal preferred this system (Quartierstrom: Die Ergebnisse | Quartierstrom, n.d.). Although the pilot phase was completed in January 2020, the local energy market continued. The project continued in its second phase as Quartierstrom 2.0. The core team founded a startup called Exnaton to develop a scalable product based on the technology (Smart City Innovation Award Für Quartierstrom 2.0 | Quartierstrom, 2020).

2.5 Tax collection

The most important method of revenue generation for funding public expenditure is collection of taxes. Blockchain has the potential to completely reform and reshape the way taxes are collected (Blockchain 101 for Governments, A Note Prepared for The Committee of Experts on International Cooperation in Tax Matters, n.d.). Most current systems allow the citizens to declare their incomes at the time of filing income tax. This allows them to keep multiple sources of income undisclosed and hence untaxed. Even though transactions are done using cheque or bank transfer and have written records, lack of communication between various government departments allows such flaws to creep in.

Use of blockchain-based smart contracts in tax administration can greatly increase state revenue by ensuring real-time automated tax payments when any taxable transaction takes place (Blockchain: Taxation and Regulatory Challenges and Opportunities, 2017). In particular, payroll tax is an ideal use case for blockchain. Multiple regulatory agencies collect, store, and audit the tax-related data independently. This process can be carried out with much greater efficiency with the use of smart contracts. However, the potential remains completely unutilized due to a number of factors. The smart contracts required for this need to be skillfully designed. This requires a deep technical tax knowledge base. The ever-changing tax regulations add to the complexity. Also, there exists no cryptocurrency which governments accept for deposits/payments. Interested private companies are waiting for the establishment of a fiat cryptocurrency. Once this happens, they would be able to establish agreements with government revenue departments to act as payroll tax regulators (Ainsworth & Viitasaari, 2017).

Unlike Bitcoin and Ethereum, which are public blockchains, a tax application would require confidentiality. Private blockchains can serve this purpose. Though the data would be transparent to selected parties such as the tax administration and maybe banks, it would remain hidden from the public. This system could lead to a major reduction in transaction costs.

Although consideration of blockchain in payroll tax collection has taken a backseat, its application to overcome VAT fraud has been discussed in literature. The EU loses €50 billion in good-based VAT frauds every year. In this fraud, a seller in one country sells goods to a seller who further sells them in another country. The seller in the first country files for a return on the VAT paid and the seller in the second country is supposed to pay the VAT in the country he sells. If he does not pay the VAT, he is said to be the missing trader. And the fraud is known as Missing Trader Intra Community Fraud. The root cause of this fraud is that multiple tax jurisdictions are bound in a community and each insists on keeping its own centralized database. This fraud can effectively be solved if a decentralized database is used (Thompson & Andrew, 2016).

2.6 License management

Licenses are a ubiquitous feature of all governments. The aim of licensing is to ensure that all authorities and individuals are working under the norms set up by the government which ensures public welfare and judicial governance. The various licenses issued by the government include driving license, medical license, license for practising law, manufacturing goods, or for a business or trade. The processing of license for any authorization often takes about two or more months in countries like India and there also is a substantial chance of human error such as inputting incorrect name, date of birth, or any other information. The reason for this is that the systems used are years old and require major human intervention Also, the approval of licenses in many fields require the same documents such as ID card, birth certificate, or passport and therefore it becomes burdensome to present them again and again. Hence, there is a need for a centralized license management system that governs all types of licenses of an individual under a single database. This can be achieved by a distributive ledger that is blockchain.

Dubai Economic Department (DED) has developed Unified Corporate Registry that allows government authorities to issue all types of licenses to the citizens using blockchain network. Any license which is issued, renewed, modified, or suspended is updated at once and is accessible to all the authorities easily at one platform. This has helped DED in providing infrastructure and services that are secure, transparent, and distributed. Its implementation uses softwares such as PostgreSQL RDBMS and CouchDB. PostgreSQL RDBMS is used for initial uploading of data. The government authorities share the data in csv (comma separated values file) format which is imported into PostgreSQL and then added to the blockchain network using REST APIs. CouchDB is the database storing all the issued license data. The blockchain nodes of this system are managed by Hyperledger Fabric software. Hyperledger is an open source community developing softwares for the blockchain implementation in numerous fields. The system uses REACT, an open source JavaScript library, for developing the interactive user interface (Ahmed, Qassim, & Manar, 2020).

The Traffic Offence Blockchain Model proposed in Malaysia is another system that uses blockchain technology for recording driving licenses. A driving license is the only document that recognizes an individual as a trained driver and ensures that he is well aware of traffic rules and would bid to them. The proposed model has five components: Application Layer, Blockchain Servers Layer, Transaction Schema, Smart Contract Model, and Decentralized Network Model (Fig. 7.4). The application layer is the user interface available to all peers in the network which includes traffic police officers and public to access all the details. Blockchain surface layer is the layer which has all blockchain nodes linked to each other with cryptographic hash codes storing all the information of the driver. Transaction Schema refers to all the details such as driver's license number, offense committed, penalty charged, and the validation time of license. Smart contract model governs the validation of license. It has the agreement which suspends the license or takes actions when any offense is committed by the licensee. This complete decentralized model has the potential to develop an effective administration for driving licenses and traffic rules across the nation (Pradana, Sing, Kumar, & Mohammed, 2018).

2.7 Land ownership records

Maintenance of land ownership records is one of the most challenging tasks for any government. It has been observed that disputes regarding even small pieces of land can lead to serious crimes. The existing system in most of the

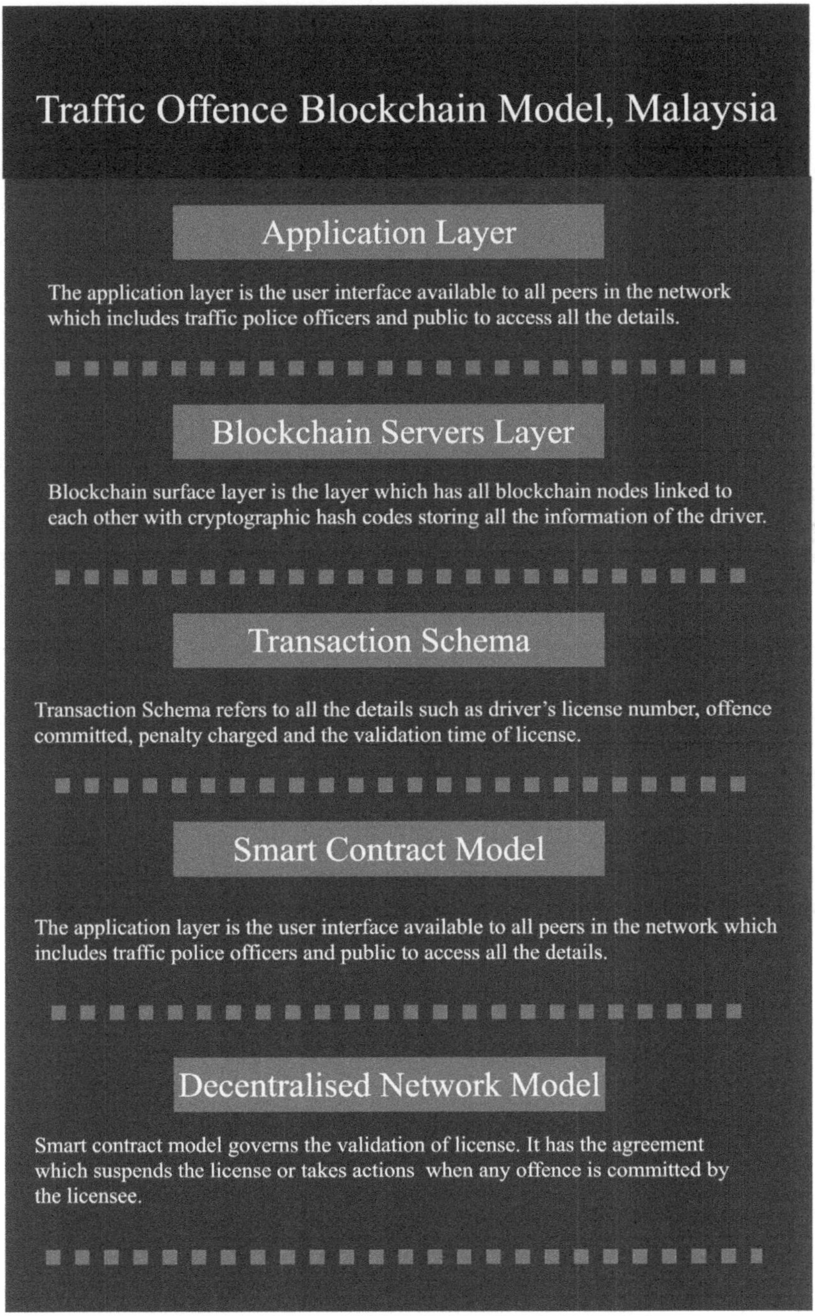

FIGURE 7.4 Traffic offense blockchain model, Malaysia.

nations is digitized to the extent of storing records on the computer hard disk but requires human intervention. The system cannot resolve the problem of corruption and forged documents. The payment of middlemen and brokers involved in the transaction between buyers and sellers increases the transaction cost for both the parties. Another problem is that the information related to the same land is managed by different authorities which makes changing the records time and effort consuming. The solution to these problems can be achieved by an immutable, secure, and decentralized land registry system using blockchain technology (Khan, Ansari, Sachdeva, & Jain, 2020; Land Registry on Blockchain, 2020).

The blockchain model contains all the information regarding a land in the immutable blocks. It can easily store the records of all the ancestral lands and ensure equal distribution among new generations with the help of smart contracts. The platform also provides an effective way for buying and selling property without the involvement of middlemen and brokers, hence reducing fraud. The three stakeholders in the system are buyer, seller, and land inspector. Firstly, the seller uploads land details on the blockchain network and then potential buyers send their requests. The land inspector checks for the validation of documents presented by the seller (Fig. 7.5). Then meetings are organized for the transaction and all details are uploaded to the blockchain database. After the deal is finalized and documents are signed, the database is updated with the new owner's information. This ensures an authentic and trustworthy transaction. Based on this model, LeewayHertz, a San Francisco-based software company, has developed an effective Blockchain Land Registry Platform using the technology of Hyperledger Sawtooth (Blockchain Land Registry Platform - Reducing Frauds And Delays, n.d.)

In January 2019, South Burlington, USA, decided to move its real estate transaction to blockchain platform with the help of Propy, a Silicon Valley Proptech Company. Furthermore, Sweden has successfully tested a blockchain

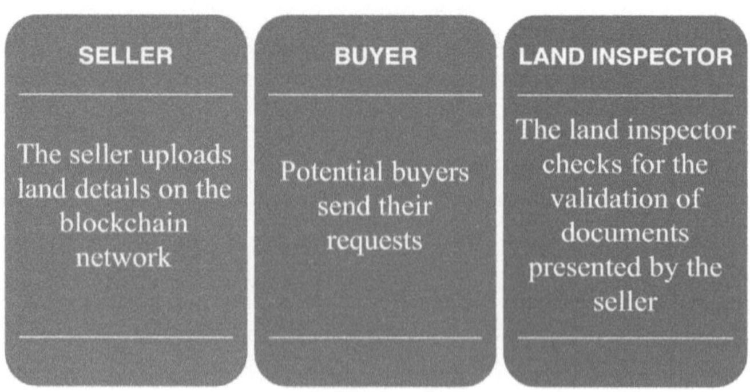

FIGURE 7.5 Actors and their roles in property transaction using smart contracts.

land registry program using ChromaWay blockchain technology. ChromaWay is a blockchain technology company providing implementation to nations across the world (Blockchain Land Registry Platform - Reducing Frauds And Delays, n.d.; ChromaWay, n.d.). In 2018, Andhra Pradesh, India, also collaborated with ChromaWay to develop a blockchain-based land registry system in the state. A debate took place in the state due to suspicion on private blockchain technology. Later it was clarified that officials from the Revenue Department, Chief Commissioner of Land Administration, and others would be monitoring the land records and whenever a change is made in the database the owner of the land would be definitely notified. Thus, the main power would reside in the hands of government officials working for the citizen welfare (Andhra Government To Adopt Blockchain Tech To End Land Record Tampering, 2019; Indian State To Implement Blockchain For Land Records - Ledger Insights - Enterprise Blockchain, 2020; Using Blockchain To Make Land Registry More Reliable In India, 2018).

2.7.1 Case study: Georgia

Land titles had been digitized in Georgia from 2006 itself. This was a part of the measures taken to combat corruption and ineffectiveness of the government. This provided a strong base for the development of blockchain platform (Lazuashvili, Norta, & Draheim, 2019). In April 2016, the National Agency of Public Registry launched a pilot project for land titling using blockchain in collaboration with BitFury. BitFury had been operating in Georgia since 2014 (Benbunan-Fich & Castellanos, 2018). After a successful completion of the pilot project, the initiative took off as a full-scale project. The service was introduced as an add-on on top of the existing system at that time. The dealing parties still need to go to the local office for registering the agreement. The local office sends the agreement for validation and digital signature. The land titles themselves are stored on a centralized database. The location and registration details are sent to a private Exonum blockchain where they are timestamped and stored. The Exonum protocol allows the hash of the documents to be stored on the Bitcoin blockchain (Lazuashvili et al., 2019). The protocol can handle 5000 transactions per second (David et al., 2019). Any citizen can check the validity of a title deed from the Bitcoin blockchain. Although this system does not remove any intermediary from the process, it offers two advantages. The confidential data of ownership records remain private. At the same time, it prevents government officials from maliciously altering any records. Over 1.5 million land titles have been registered on blockchain under this system (Qiuyun & Allison, 2019). Further there are plans to incorporate the notary service into the blockchain using smart contracts. However, the process for this is yet to be established (David et al., 2019).

2.8 Voting

Voting is an indispensable part of any kind of governance. It is the most common approach to determine the majority opinion. It facilitates collective decision-making at every level of governance. Hence, it is of utmost importance that voting is carried out as unflawed as possible. Blockchain technology has the potential to revolutionize the process of voting as it is undertaken in the present times.

A common, but age-old, method of voting is the paper ballot voting. In most scenarios, each voter is required to arrive at a specific polling station and prove their legitimacy as a valid voter using certain documents. After verification, a vote is cast by placing a ballot paper, which has been marked accordingly, inside the ballot box. This method has certain shortcomings, including lack of transparency and slow proceedings. Since the votes are recorded in a physical manner, votes may also be misrecorded, or even tampered with. The 2019 Malawi general election was entirely annulled by Malawi's constitutional court when reports of tampering of the tallying forms surfaced (Malawi Election: Court Orders New Vote After May 2019 Result Annulled, 2020).

Since blockchain is an immutable ledger, it can solve this issue of tampering of votes. Once placed in the ledger after authentication, the vote will remain encrypted which also maintains the voter's privacy.

Electronic voting or e-voting has been replacing the traditional paper ballot voting ever since the advent of the Electronic Voting Machine (EVM). In most current scenarios, an EVM consists of buttons or touch screens which the voters operate to select their choices and cast their votes. The process of verification of the voter is the same as that of the paper ballot voting. While e-voting is more secure and accurate than the previous method, it is still not devoid of shortcomings like tampering with the software or the hardware of the EVMs, or hacking the vote tallying system. One such incident happened in the Ukrainian presidential elections of 2014. Hackers had preinstalled malicious software to alter the result of the election. Further, a series of DDoS attacks were launched to delay the final tally (Clayton, 2014).

Blockchain is based on a distributed database. There is no single location which stores the entire ledger. Instead, the ledger is distributed over multiple systems across different users. Eliminating the need for a central authority eliminates the risks involved with processing votes on a centralized system. In other words, blockchain is mostly unaffected by most kinds of cyber security attacks, to which centralized systems are vulnerable, during the processing of votes.

Online voting, or Internet voting, or i-Voting allows voters to authenticate themselves and cast their vote from any place with an Internet connection. This method removes the cost and the complexity that comes along with the EVMs, but brings new technical challenges with it. In 2005, Estonia became the first country in the world to implement i-Voting in national elections. The voter is verified via a government issued identity card in the prevoting phase

(Ahmed, 2017; Nir & Jeffrey, 2018). In 2011, Norway also adopted a remote voting system but dismantled it in 2014 over privacy and security reasons (Ahmed, 2017). These are the critical issues with any i-Voting implementations. Moreover, the problem of centralization prevails with i-Voting too, making it vulnerable to cyber-attacks.

It is evident that every popular method of voting comes with its own set of challenges and limitations. Using blockchain can help control these challenges. It allows votes to be cryptographically stored such that they cannot be altered or modified. Further, being decentralized allows blockchain to be publicly accessed without compromising the privacy of the voter.

Using blockchain in voting can be exemplified by Moscow's Active Citizen Program launched in 2017. Citizens are urged to vote for municipal decisions like setting speed limits or ideas on architecture. This aims to increase the interaction of citizens with the governance and building trust with the authorities. The results are auditable on a public ledger once the voting ends. To make the system more widespread and acceptable, voters are rewarded with points with every voting they take part in (Nir & Jeffrey, 2018).

2.8.1 Case study: South Australia

In 2019, the government of South Australia collaborated with Horizon State, an Australian blockchain-based voting platform, to conduct the elections to form a Minister's Recreational Fishing Advisory Council. The elections had to be a preferential vote and there had to be provisions for reserved candidates. 42 candidates competed for five positions in the council and it was mandatory for the council to have at least one woman, one inland fisher, and one person representing the tackle industry. These requirements further increased the complexity of the process and new developments had to be made on Horizon State's blockchain-based voting solution to meet them.

The month-long voting took place from 18th February till 18th March and all citizens interested in recreational fishing were urged to visit the official website and cast their vote. Transparent results were made available right after the conclusion of the voting. The processes of voting and result calculation were then examined and approved by two separate bodies. The election was hailed as successful and historic (Palmer-Derrien, 2019; SA Government Blockchain Voting, n.d.; Voting Now Open, 2019).

Like many other use cases, blockchain has even more potential here which can be explored and utilized for a better e-governance.

3. Considerations involved in implementation

Various factors influence the feasibility of adoption of blockchain technology in the sectors discussed in Section 2. Some of those factors are briefly discussed in this section.

3.1 Transparency

The use of blockchain technology in the public sector has been hailed as a step which imparts transparency to the functioning of governments. However, it should be noted that transparency is not a result of simply incorporating a technology into the existing system. For a system to be transparent, it is necessary that it is applied in the correct way (Ølnes, Ubacht, & Janssen, 2017). Although designed to be decentralized, blockchains have given rise to a number of third parties and profitable businesses which act as intermediaries. The notion of decentralization seems to be shattered by the growing influence of enormous mining farms on the Bitcoin blockchain. Developers along with these businesses tend to occupy a central position in the ecosystem which gives them the power to decide its fate (Marcella, 2015). As this centralization shifts the power away from the common users, transparency seems to fade away.

A lot of data that the government collects are not suitable to be made available to the public. In a public blockchain, transaction pattern recognition and user meta information are considerable threats to anonymity (Marten & Kai, 2017). This makes public blockchains unsuitable for these applications. The solution to this would be to use permissioned blockchains. But that would make the tendency toward centralization even more pronounced. The data would be transparent, but only to selected officials who would anyway have access to that information. The purpose of such an application would be to enhance auditability by the virtue of its immutability (Glaser, 2017). The citizen will not view the government as being more transparent as the use of blockchain would be happening under the hood.

3.2 Trust

Blockchain is generally regarded as a trust-less or trust-free system which allows users to participate in transactions with other users whom they would no longer need to trust (Glaser, 2017). A pertinent question arises here. In what ways is trusting an algorithm different from trusting a centralized intermediary? The trust on algorithm in the present scenario appears to be blind (Lemieux, 2016). Unless the institutional systems in government are mended to be trustworthy, blockchain cannot guarantee trust. It can only aid in control and auditing.

Consensus mechanisms like proof of work (POW) provide higher levels of security against fraudsters while being uneconomical in energy usage. However, other mechanisms like proof of stake, while being energy efficient, require some degree of trust on the actors depending on the resources that they put at risk during validation (Marten & Kai, 2017). It is also required that blockchain service providers as well as smart contract developers are trustworthy (Glaser, 2017).

3.3 Privacy

Blockchain networks have significant loopholes in certain factors concerning privacy. The user on a blockchain network is not anonymous, rather, he is pseudonymous. A study by the New York Times described how sufficient amounts of pseudonymous location data can make identification of the individual trivial. The set of strings that mask the user's identity on the network is his IP address which can be traced (Snyder, 2019). In cryptocurrency like Bitcoins, pseudonymous nature allows the data to be accessed and all the transaction history to be traced to the very first node. Public blockchain ledgers have further serious privacy issues. Whenever a person makes a transaction with anyone, his public address is shared to them which gives them access to the transaction history and wallet coins. Also, some studies reveal that cryptocurrency wallet software can be forensically analyzed even without the authorized private keys (Blockchains Aren't Anonymous. But They Can Be. | Ledgerops, 2019).

The arrest of Ross Ulbricht for operating the deep website "Silk Road," a marketplace for illegal drugs and considered as anonymous, proved that identity could be traced. There are proposals for anonymous blockchain such as Bubble but privacy of an individual can still not be guaranteed (Snyder, 2019).

3.4 Security

It's true that blockchain cryptographic algorithms offer an immutable and more secure transaction platform than other existing ones. But to understand and examine its security we need to focus on some major issues. Blockchain network, being a distributed ledger, gives everyone access to transaction details which makes it vulnerable to hackers. The attacks such as denial-of-service are very common at some developed blockchain platforms (Demush, 2019). Also, blockchain has no defense against fraud or mistakes due to its immutable nature. Once a person loses their private keys or if their wallet coins have been stolen, there is no centralized security system to help them or make any reparations (Snyder, 2019).

The Decentralized Autonomous Organization hack in 2016 was a major attack on the blockchain system and led to $50m worth ether loss. The parity hack in 2017 resulted in 153,037 ETH loss which questioned one of the largest blockchain markets, Ethereum. Also, in July 2018, it was estimated that 4 million Bitcoin were lost and 2 million were stolen, when less than 18 million Bitcoin existed (Demush, 2019). Bitcoin is considered the largest implementation of blockchain technology but still cannot ensure security, hence it's hard to guarantee at other platforms.

3.5 Efficiency

Although blockchain can improve productivity and reduce the amount of unnecessary resource consumption in many use cases, it comes with certain inherent efficiency concerns. Since blockchain is an immutable ledger, it makes it an append-only data structure. This means that it will always grow in size. Also, if a node requires to append in the blockchain, it needs to have the entire copy of the blockchain locally. While this facilitates decentralization, it consumes immense storage resources. Note that while this is not an issue for some actors in the blockchain network, it affects those nodes which actually want to add blocks to the ledger, for example, miners in the Bitcoin Blockchain. The size of the entire Bitcoin blockchain is approximately 285.06 gigabytes as of the end of June 2020 (Liu, 2020).

Apart from the large storage requirements, blockchain can also be very power consuming. One of the reasons that blockchain is considered reliable and secure is its consensus algorithm. A widely used consensus algorithm is the POW algorithm. Concisely, it requires a substantial amount of computation using trial and error to make the validation of a transaction a time-consuming process. In Bitcoin blockchain, there are multiple miners competing to find a valid POW for a given transaction and only the first miner to do so is rewarded. This essentially means that the other miners wasted their resources. Every node in the network needs to be notified of the new block created so that they can update their own ledgers. Maintaining a real-time ledger ensures transparency but has a high power consumption (Golosova & Romanovs, 2018; Valentina, Fabrizio, Claudio, Chiara, & Victor, 2018).

This shows that very factors which contribute in making blockchain secure, transparent, decentralized, and reliable can also compromise the efficiency of the technology.

3.6 Cost

For the adoption of any new system or changes in the current system, it is essential that either the associated costs are reduced or any significant increase in costs is outweighed by the resulting advantages (Merlinda et al., 2019). While blockchain as a data storage technology is not cost-effective, cost benefits are expected to stem from elimination of fraud, error in payments, and transparency in government operations (Distributed Ledger Technology: Beyond Block Chain, 2016). The elimination of intermediaries in transactions also contributes significantly to the cost benefit (Waltl, Sillaber, Gallersdörfer, & Matthes, 2018).

On the other hand, the use of blockchain also has some underlying costs associated with it. Using the POW consensus mechanism for mining is a costly choice, not just from economical but also from environmental perspective.

Countries which have significant amounts of their energy resources coming from green energy sectors would find it suitable to use public blockchains for transaction cost reduction. But on the other hand, developing nations relying on coal would have to suffer due to the pollution resulting from mining activity (Strüker, Albrecht, & Reichert, 2018).

One of the most prominent blockchains which supports smart contracts is Ethereum. The transaction fees for an average sized transaction without smart contracts on Ethereum is found to be lower than that of existing payment service providers such as Paypal. However, microtransactions such as purchasing electricity from different sources several times per day are not economical. Similarly, while tokenization of land on a blockchain is cost-effective, tokenization of lesser valued assets is not (Strüker et al., 2018).

The use of blockchain in the public sector is still in its infancy and similar pilot projects are being conducted by multiple governments, each aiming to gain a strategic lead by being the first to utilize this technology. In such a situation, the reluctance to share quantitative information regarding costs is obvious (David et al., 2019).

3.7 Scalability

For any technology, its efficiency and scalability are closely linked. If a system is highly productive with minimal wastage of resources, it can be scaled, i.e., trusted with a larger amount of usage. In terms of blockchain, we discussed some issues that arise because of the same features which distinguish blockchain as a technology. Scalability is one such issue and it has come under the spotlight with increasing usage of blockchain technology.

Blockchain is known to be slower than its centralized counterparts. The primary reason for this is its transaction processing time. Decentralized networks rely on peer-to-peer networking and a consensus algorithm. We have discussed that a transaction in Bitcoin is only considered valid if it satisfies the POW, which is a resource and time-intensive process. On a large scale, this becomes problematic as more transactions keep piling up which affects the transaction confirmation latency.

Other factors which affect the performance of a blockchain on a large scale is the block time and block size. Block size refers to the size that any new block needs to be in order to be appended to the chain. In Bitcoin this is fixed to 1 MB, whereas Ethereum does not have a fixed size but most blocks are under 2 KB in size. Block time, the time taken to append a new block in the network, is proportional to the block size. It is around 10 min for Bitcoin and around 15 s for Ethereum. These limit the amount of transactions taking place in the network. In real time, Bitcoin enables only about 7 tx/s (transactions per second). Ethereum is not much better with about 15 tx/s. These numbers are very small compared to the highly scaled payment processors like Visa which can achieve more than 4000 tx/s. Decreasing the block time to increase the transactions bandwidth will compromise the consensus mechanism thus decreasing the overall security of the entire network (Qiheng, Huawei, Zibin, & Jing, 2020).

These discussions reveal an important issue known as the Scalability Trilemma or the Blockchain Trilemma. Coined by Vitalik Buterin, the cofounder of Ethereum, this states that the three important qualities of blockchain, namely decentralization, security, and scalability, cannot coexist. Scalability can be achieved if decentralization or security is compromised. The blockchain must make trade-offs in order to optimize itself to the needs of the network (Viswanathan & Shah, 2018).

4. Case studies

This section discusses the varying degrees of adoption of blockchain technology by governments around the world.

4.1 Estonia

Estonia, a Northern European nation, is one of the most technologically advanced nations of the world. Often referred to as the New Silicon Valley, it is known to have strived to find innovative IT solutions to digitize its governance. In 2000, Estonia introduced an electronic tax filing system and currently, it only takes around 3–5 min for a user to review the prefilled forms and approve the declaration form to file tax online. In 2002, Estonia introduced a digital identity card which every citizen gets on reaching the age of 15. This digital ID is highly functional as it provides every citizen with a digital signature to avail and approve the variety of e-services that Estonia has to offer. With immense focus on e-governance, Estonia has very successfully launched services like e-Law, e-Tax, e-Justice, e-Banking, and the Estonian Research Information System, to name a few (E-Estonia — We Have Built A Digital Society And We Can Show You How, n.d.b).

To enable interoperability between the various departments and services of the government, Estonia features a software-based data exchange layer known as the X-Road. It is essentially a hub for the public and the private sector information services to link and share resources. Like blockchain technology, X-Road uses cryptographic hash functions (E-Estonia, n.d.a).

In 2014, Estonia became the first nation in the world to introduce a blockchain-based e-residency which means that people all around the world can apply for an e-ID issued by the Estonian government. Developed in partnership with Bitnation, a blockchain-based Governance 2.0 initiative, this program primarily uses blockchain to authenticate identity and documents. This allows e-residents to avail a host of Estonia's e-services and start and manage their businesses remotely. Although this does not mean getting residency or citizenship in the traditional sense, it is one of the most advanced e-residency programs in the world (E-Residency, n.d.; Sullivan & Burger, 2017).

In 2011, Estonia transformed its healthcare by collaborating with Guardtime, which specializes in blockchain-based public and private solutions. Blockchain technology was used to completely decentralize and secure public health records. This made updation and retrieval of medical records by all medical stakeholders very efficient. A Keyless Signature Infrastructure blockchain is used here which was developed in Estonia itself and is used around the world (Ojo & Adebayo, 2017).

In 2017, Nasdaq Inc., an American multinational financial services corporation, collaborated with Estonia to test blockchain-enabled voting in the Tallinn Stock Exchange. The test was conducted successfully on a Nasdaq-Tallinn listed technology company LVH group. Shareholders of the company, who were Estonian citizens or Estonian e-residents, voted for corporate decisions securely using Nasdaq's blockchain technology (2017) and Ojo and Adebayo (2017).

It can be safely concluded that Estonia is one of the world's leading countries in testing and adopting blockchain technology in its e-governance fabric.

4.2 Dubai

Dubai, well known for the tallest building in the world, its numerous lavish infrastructure and promising tourism, has also accomplished milestones in the development of blockchain-based citizen e-governance. The Dubai Future Foundation and the Smart Dubai Office established the Global Blockchain Council in 2016 to explore and implement blockchain technology in governance. The Council brought forward the Dubai Blockchain Strategy to achieve the dream of becoming the world's first fully blockchain-powered city by 2021. The strategy has three key objectives: Government Efficiency, Industry Creation, and International Leadership. It means that the city aims to implement blockchain technology to make government procedures more effective and for empowering startups and attracting companies across the globe. The vision for "Smart Dubai" is to establish it as the world's leading blockchain city where all activities such as visa applications, currency, bill payments, record keeping, healthcare, and tourism are managed on a blockchain platform. The digitalization of the city is also leading the government to become paperless. On an average more than 100 million documents are generated annually and hence going paperless would save US $1.5 billion in governance. Also, the statistics show that it would reduce CO_2 emissions by 114 MTons and free up 25.1 million hours of economic activity every year (Dubai Blockchain Strategy, n.d.; Smart Dubai, n.d.; The Economic Times, 2020).

Smart Dubai, in association with various emirates, has begun government schemes to explore and develop blockchain framework. One of them is the Global Blockchain Challenge that invites blockchain startups across the world to present their innovative ideas for blockchain implementation in the city (Smart Dubai Global Blockchain Challenge 2019, n.d.). Dubai Smart City Accelerator is another government support to the companies providing blockchain and other advanced solutions for the development of smart cities and governance (Dubai Blockchain Strategy, n.d.; Dubai Smart City Accelerator, n.d.).

Smart Dubai Office collaborated with Consensys as a blockchain city advisor and IBM as the blockchain lead strategic partner for the development in different spheres (John Khaleej Times, 2017; Smart Dubai: Blockchain Case Study for Government in the UAE, n.d.). Consensys in partnership with telco Du developed BPaaS, the Blockchain Platform as a Service. BPaaS has given all the government authorities the freedom to build on a shared ledger according to their business needs. It is a cost-effective and secure platform with a pay-as-you-go model. Smart Dubai Office acknowledged BPaaS as the official tool for government prototyping (Dubai Economy Moves To Blockchain-Based Unified Business Registry Platform, 2019). Another Consensys project is Smart City University (also known as Hamdan Bin Mohammed Smart University (HBMSU)) developed with Dubai Future Foundation and inaugurated in 2018. It is a blockchain-based learning platform for the development of digital skills among citizens in the United Arab Emirates. The platform provides digital certificates, online learning material and videos, and aims to increase the percentage of digitally skilled workforce by 10% by 2020 (Dubai Inaugurates Smart University, 2018; Hamdan Bin Mohammed Smart University (HBMSU) Dubai, n.d.).

Dubai's Department of Economic Development (Dubai Economy) launched a business registry platform called Unified Business Registry Platform (UBRP) in partnership with IBM and Dubai Pulse. The platform provides smart business registration and verification. It also ensures transparent and effective administration by the government upon companies. UBRP has Dubai Silicon Oasis Authority, Electricity and Water authority, and Multi Commodities Centre as its prominent members. It is the first blockchain platform hosted in UAE that integrates hyperledger fabric and Ethereum and hence makes use of both technologies' protocols and smart contracts (Dubai Economy Moves To Blockchain-Based Unified Business Registry Platform, 2019).

Also, Dubai Economy in association with six banks that are Emirates NBD, Emirates Islamic, HSBC, RAKBANK, Abu Dhabi Commercial Bank (ADCB), and Commercial Bank of Dubai (CBD) established a blockchain consortium for sharing verified customer data between banks and licensing government authorities in UAE called as "KYC (know your customer) Blockchain Consortium." The framework allows data sharing across the country which would reduce the time consumption in verification of customers by different authorities. It also provides secure and fast banking facilities. The customer is also benefited as he would not have to present the same information repeatedly. The platform is developed by Norbloc, a leading company for blockchain KYC development. The blockchain stores a pointer to the customer's data which is shared with various parties using Norbloc P2P protocol. The platform went live in July 2020 (Dubai Economy, Emirates NBD And Norbloc Go Live With The UAE KYC Blockchain Platform | Norbloc, 2020; Dubai Economy, Emirates NBD Go Live With UAE KYC Blockchain Platform, n.d.; Global Business Outlook, n.d.).

Dubai is indeed the blockchain capital of the world with more than 100 blockchain businesses and inculcating the technology in all spheres of e-governance such as health, education, transportation, and security. The city bagged the "First Smart City on the Blockchain" award in the Barcelona Smart City Expo and got titles like the "World Capital of Blockchain."

4.3 Kenya

Kenya is an East African country which has gained recognition for being one of the most prominent hubs for technological innovation in Africa. Known as the "Silicon Savannah," Kenya has adopted new technologies in its ecosystem and attracted entrepreneurs and investors globally. It is known for its strong focus on building and expanding its ICT infrastructure. Kenya's Ministry of Information and Communication Technology has catapulted Kenya into becoming the regional leader in ICT innovations with a growth rate that outperforms all its other sectors. This can be exemplified by M-Pesa, a mobile banking service which has revolutionized the provided financial services to millions of people (Sydow, Sunny, & Coffman, 2020; Trade.Gov, 2019).

Even after such impressive advancements, Kenya faces some critical issues like many developing nations. One of these is widespread corruption which is mostly because of the middlemen present in a centralized framework. Another issue is catering to the Agriculture Industry which is the largest sector of Kenya's economy. Kenya has been keen in exploring and implementing blockchain technology as a means of resolving some of its prevalent issues and encouraging its adoption in its e-governance framework (Sydow et al., 2020). The Blockchain Association of Kenya, based in Nairobi, is dedicated toward raising awareness and providing training about blockchain technology (Blockchain Association Of Kenya | Raising Awareness Of Blockchain, n.d.). Some present implementations of blockchain in Kenyan e-governance are discussed below.

M-Akiba is a government-issued mobile-based retail bond developed using blockchain technology in 2017 with support from the World Bank. Previously, foreign and local institutional investors had most of the ownership in government bonds. The process to buy a government bond was tiresome as it included opening a Central Depository System account in the Central Bank of Kenya via some agents or the bank itself. It took about 2 days to get the ownership of a government bond. This inhibited many potential Kenyans to participate in funding for government projects and infrastructure. M-Akiba immensely sped up the process by allowing the creation of an account in less than 10 min. This not only facilitated financial inclusion for the economic development of the state but also promoted an investment culture among the masses as the government paid a considerable amount of interest for each bond. Further, users can choose to sell their bonds anytime using the underlying blockchain technology in the platform which ensures transaction immutability (About M-Akiba, n.d.).

The Kenyan government established the Distributed Ledgers Technology and Artificial Intelligence Task Force in February 2018. The aim of the taskforce is to recommend and evaluate proposals to inculcate blockchain technology in Kenya's governance framework. Further, this taskforce promotes research and development of this technology and supports startups in this field. In a report, the taskforce recommended the government about the implementations of blockchain technology in various departments like healthcare, land registry, voting, and digital currency. This depicts that the Kenyan government is keen in exploring further use cases of blockchain technology (Tokenpost, 2019).

Farmers in Kenya often face difficulty in investing in their farms due to their irregular financial conditions. Oftentimes, farmers require loans to provide for the needs of the farms. But obtaining this loan is a difficult task because of the lack of a stable credit score or collateral. Agri-wallet, a blockchain-based financing tool, has helped over 25,000 Kenyan farmers overcome this lack of a stable financial machinery. Agri-wallet enables farmers to earn revenue in terms of "tokens" which in turn can be directly used to purchase supplies from selected merchants. Further, it allows farmers to avail short-term loans when needed. Developed by Dodore Kenya Ltd., this is not a government initiative but an apt example of how blockchain can be successfully used in agricultural e-governance (Raflearning.Org, n.d.).

Kenya is home to several blockchain initiatives which are not necessarily government operated but are excellent potential use cases for inclusion in an e-governance framework. Another such example is Land LayBy which uses Ethereum-based distributed ledger to facilitate land transactions and maintain ownership records. Kenyan villagers are often troubled by the prevalent unequal land rights and the difficulty in obtaining ownership records. Land LayBy attempts to overcome these challenges by providing a transparent and incorruptible database (Bolt & Senou, 2018).

On similar lines, in July 2020, the Ministry of Lands planned to digitize land records in Nairobi. The teams that are building the system are using encryption to secure data and plan to make use of blockchain technology in the second phase of development (Capital News, 2020).

Kenya accommodates many successful implementations of blockchain which either already are made use in e-governance or have substantial potential capacity to do so. The nation is progressing toward becoming proficient in this disruptive technology and is likely to show novel advancements in this field.

5. Summary

Blockchain technology has revolutionary effects in e-governance of a nation. This chapter begins with the introduction of e-governance with the emphasis on the existing models across the globe. It is followed by some examples of e-governance without blockchain and the challenges faced in their implementation. The next section includes numerous use cases of blockchain in different fields of e-governance. This section gives the complete overview of how particular aspects of e-governance can be effectively achieved using blockchain technology. Use cases are discussed with examples of countries that have successfully implemented or proposed the use of blockchain in their e-governance framework. It also includes case studies of countries that have implemented blockchain on a major scale successfully and set a benchmark for the world. The third section presents the considerations involved in the establishment of a blockchain environment. It includes the challenges and loopholes in the blockchain system which must be resolved for justified citizen e-governance. The last section includes case studies of governments that have adopted decentralized governance with the help of blockchain technology. This chapter leaves the readers a detailed knowledge of blockchain-based governance systems existing in the world including their implementation and the challenges faced. It also broadens the minds of readers about the potential of blockchain in citizen e-governance.

References

Ableitner, L., Meeuw, A., Schopfer, S., Tiefenbeck, V., Wortmann, F., & Wörner, A. (2019). *Quartierstrom-implementation of a real world prosumer centric local energy market in Walenstadt, Switzerland.* http://arxiv.org/pdf/1905.07242.pdf.

(n.d.). *About M-Akiba.* Retrieved September 3, 2020, from: https://www.m-akiba.go.ke/index.php/about-m-akiba.

Agbo, C., Qusay, M., & Eklund, J. (2019). Blockchain technology in healthcare: A systematic review. *Healthcare, 56.* https://doi.org/10.3390/healthcare7020056.

Ahmed, B. A. (2017). A conceptual secure blockchain based electronic voting system. *International Journal of Network Security & Its Applications*, 01–09. https://doi.org/10.5121/ijnsa.2017.9301.

Ahmed, A., Qassim, N., & Manar, A. T. (2020). Novel blockchain reference model for government services: Dubai government case study. *International Journal of System Assurance Engineering and Management, 11*(6), 1170–1191. https://doi.org/10.1007/s13198-020-00971-2.

Ainsworth, R. T., & Viitasaari, V. (2017). Payroll tax & the blockchain. *Tax Notes International*, 1007–1024 (Original work published 2017) https://ssrn.com/abstract=2970699.

Alden, A. (n.d.). *Building blocks.* https://www.globalinnovationexchange.org/innovation/building-blocks.

Andhra government to adopt blockchain Tech to end land record tampering.(2019). https://www.newindianexpress.com/states/andhra-pradesh/2019/dec/15/andhra-government-to-adopt-blockchain-tech-to-end-land-record-tampering-2076359.html.

Benbunan-Fich, R., & Castellanos, A. (2018). Digitization of land records: From paper to blockchain. In *Proc. Of international conference on information systems.*

Blockcerts. (n.d.). Retrieved September 2, 2020, from https://www.blockcerts.org/.

Blockchain 101 for governments. (n.d.). *A note prepared for the committee of experts on international cooperation in tax matters.* Retrieved September 3, 2020, from: https://www.un.org/development/desa/financing/sites/www.un.org.development.desa.financing/files/2020-04/15STM_Blockchain-101.pdf.

Blockchain. (2017). *Taxation and regulatory challenges and opportunities*. https://www.wu.ac.at/fileadmin/wu/d/i/taxlaw/institute/WU_Global_Tax_Policy_Center/Tax___Technology/Backgrd_note_Blockchain_Technology_and_Taxation_03032017.pdf.

Blockchain Against Hunger. (2017). *Harnessing technology in support of Syrian refugees*. https://www.wfp.org/news/blockchain-against-hunger-harnessing-technology-support-syrian-refugees.

Blockchain land registry platform - reducing frauds and delays. (n.d.). Retrieved September 2, 2020, from: https://www.leewayhertz.com/blockchain-land-registry-platform/.

Blockchains Aren't Anonymous. (2019). *But they can be*. Ledgerops. https://ledgerops.com/blog/blockchains-arent-anonymous-but-they-can-be-05-01-2019/.

Blockchain association of kenya | raising awareness of blockchain. (n.d). Retrieved September 3, 2020, from: https://bak.or.ke/.

Bolt, J., & Senou, O. (2018). *Land layby: Using blockchain to improve landownership*. https://www.cta.int/en/digitalisation/article/land-layby-using-blockchain-to-improve-landownership-sid0de9c7e3a-501a-4455-9c0f-808e8eac7410.

Brenzikofer, A., Meeuw, A., Schopfer, S., Wörner, A., & Dürr, C. (2019). Quartierstrom: A decentralized local P2P energy market pilot on a self-governed blockchain. In *Proceedings of the 25th international conference on electricity distribution*.

Building blocks blockchain for zero hunger. (n.d.). Retrieved September 5, 2020, from: https://innovation.wfp.org/project/building-blocks.

Capital News. (2020). *Lands ministry sets up legal committees to manage digital registry transition*. Capital news. https://www.capitalfm.co.ke/news/2020/07/lands-ministry-preps-for-massive-ownership-disputes-as-transition-to-digital-lands-registry-begins/.

Chase, S. (2015). *Cyberattack deals crippling blow to Canadian government websites*. https://www.theglobeandmail.com/news/national/canadian-government-websites-appear-to-have-been-attacked/article24997399/.

ChromaWay. (n.d.). Retrieved September 2, 2020, from: https://chromaway.com/.

Clayton, M. (2014). *Ukraine election narrowly avoided "wanton destruction" from hackers*. https://www.csmonitor.com/World/Passcode/2014/0617/Ukraine-election-narrowly-avoided-wanton-destruction-from-hackers.

David, A., Maciej, S., Lorenzino, V., & Francesco, P. (2019). *Blockchain for digital government*. https://doi.org/10.2760/942739.

Demush, R. (2019). *Privacy and security on blockchains: Protection measures*. https://ledgerops.com/blog/blockchains-arent-anonymous-but-they-can-be-05-01-2019/.

Die Entstehung Vom Projekt. (2019). Quartierstrom. https://quartier-strom.ch/index.php/2019/01/17/wie-alles-began.

Distributed Ledger Technology. (2016). *Beyond block Chain*. https://www.gov.uk/government/news/distributed-ledger-technology-beyond-block-chain.

Douglas, T. (2017). *Illinois announces key partnership in birth registry blockchain pilot*. https://www.govtech.com/data/Illinois-Announces-Key-Partnership-in-Birth-Registry-Blockchain-Pilot.html.

Drift | find your power. (n.d.). Retrieved September 2, 2020, from https://www.joindrift.com.

Drift is a new startup applying peer-to-peer trading to retail electricity markets.(2017). https://www.greentechmedia.com/articles/read/drift-is-a-startup-applying-peer-to-peer-trading-to-retail-electricity.

Dubai Economy. (n.d.). *Emirates NBD go live with UAE KYC blockchain platform*. Retrieved September 2, 2020, from: https://www.khaleejtimes.com/business/local/dubai-economy-emirates-nbd-go-live-with-uae-kyc-blockchain-platform.

Dubai Economy. (2020). *Emirates NBD and Norbloc go live with the UAE KYC blockchain platform*. Norbloc. https://www.norbloc.com/news/dubai-economy-emirates-nbd-and-norbloc-go-live-with-the-uae-kyc-blockchain-platform/.

Dubai blockchain strategy. (n.d.). Retrieved September 2, 2020, from: http://www.smartdubai.ae/initiatives/blockchain.

Dubai Inaugurates Smart University.(2018). https://www.smartcitiesworld.net/news/news/dubai-inaugurates-smart-university-3421.

Dubai Economy moves to blockchain-based unified business registry platform.(2019). https://blockchain.news/news/dubai-economy-moves-to-blockchain-based-unified-business-registry-platform.

Dubai smart city accelerator. (n.d.). https://dtec.ae/dubai-smart-city-accelerator/.

E-Estonia. (n.d.a). *We have built a digital society and we can show you how*. Retrieved September 7, 2020, from: https://e-estonia.com/.

E-Estonia. (n.d.b). *What is e-residency | how to start an EU company online*. Retrieved September 8, 2020, from: https://e-estonia.com/solutions/interoperability-services/x-road/.

E-Government. (n.d.). Retrieved August 30, 2020, from: https://zerde.gov.kz/en/activity/project-management/e-government/.

E-Residency. (n.d.). *What is e-residency | how to start an EU company online*. Retrieved September 3, 2020, from: https://e-resident.gov.ee/.

Energy trading platform. Suncontract. (n.d.) Retrieved September 2, 2020, from: https://suncontract.org/.

Farah, A., & Soheib, N. (2020). Governing blocks: Building interagency consensus to coordinate humanitarian aid. *Journal of Science Policy & Governance, 16*(2). https://doi.org/10.38126/jspg160201.

Glaser. (2017). Pervasive decentralisation of digital infrastructures: A framework for blockchain enabled system and use case analysis. In *50th Hawaii international conference on system sciences (HICSS 2017)*.

Global Business Outlook. (n.d.). *Dubai'S first KYC blockchain data-sharing platform goes live*. Retrieved September 2, 2020, from: https://www.globalbusinessoutlook.com/dubais-first-kyc-blockchain-data-sharing-platform-goes-live/.

Golosova, J., & Romanovs, A. (2018). The advantages and disadvantages of the blockchain technology. In *IEEE 6Th workshop on advances in information, electronic and electrical engineering (AIEEE)*. https://doi.org/10.1109/aieee.2018.8592253.

Guang, C., Bing, X., Manli, L., & Nian-Shing, C. (2018). Exploring blockchain technology and its potential applications for education. *Smart Learning Environments, 5*, Article 1. https://doi.org/10.1186/s40561-017-0050-x.

Hamdan Bin Mohammed Smart University (HBMSU) Dubai. (n.d.). https://www.hbmsu.ac.ae/.

Health Bank (n.d.). https://www.healthbank.coop.

Indian state to implement blockchain for land records. (2020). Ledger Insights - Enterprise blockchain. https://www.ledgerinsights.com/indian-blockchain-land-records-registry-andhra-pradesh/.

Irrera, A. (2017). *Nasdaq successfully completes blockchain test in Estonia*. https://www.reuters.com/article/nasdaq-blockchain-idUSL1N1FA1XK.

John Khaleej Times, I. (2017). *Dubai picks IBM, consensys to kick off blockchain project*. https://www.khaleejtimes.com/dubai-picks-ibm-consensys-to-kick-off-blockchain-project.

Juskalian, R. (2018). *Inside the Jordan refugee camp that runs on blockchain* (Original work published 2018) https://www.technologyreview.com/2018/04/12/143410/inside-the-jordan-refugee-camp-that-runs-on-blockchain/.

Khan, R., Ansari, S., Sachdeva, S., & Jain, S. (2020). Blockchain based land registry system using ethereum blockchain. Xi'an Jianzhu Keji Daxue Xuebao. *Journal of Xi'an University of Architecture and Technology, 12*, 3640–3648.

Land registry on blockchain.(2020). https://medium.com/coreledger/land-registry-on-blockchain-a0da4dd25ea6.

Lazuashvili, N., Norta, A., & Draheim, D. (2019). Integration of blockchain technology into a land registration system for immutable traceability: A casestudy of Georgia. In *Lecture notes in business information processing* (Vol. 361, pp. 219–233). Springer Verlag. https://doi.org/10.1007/978-3-030-30429-4_15.

Lemieux, V. (2016). *Blockchain for recordkeeping; help or hype.* https://doi.org/10.13140/RG.2.2.28447.56488.

Liu, S. (2020). *Bitcoin blockchain size 2010–2020.* Statista. https://www.statista.com/statistics/647523/worldwide-bitcoin-blockchain-size/.

Malawi election: Court orders new vote after may 2019 result annulled.(2020). https://www.bbc.com/news/world-africa-51324241.

Malta begins nationwide rollout of blockcerts blockchain credentials for education and employment.(2019). https://www.newswire.com/news/malta-begins-nationwide-rollout-of-blockcerts-blockchain-credentials-20811025.

Malta is first country to put education certificates on blockchain.(2019). https://www.maltatoday.com.mt/news/national/93148/malta_is_first_country_to_put_education_certificates_on_blockchain.

Marcella, A. (2015). Blockchain technology and decentralized governance: Is the state still necessary? *SSRN Electronic Journal.* https://doi.org/10.2139/ssrn.2709713.

Marckx, C. (2014). *Dutch energy supplier BAS to accept bitcoin.* https://www.ccn.com/dutch-energy-supplier-bas-to-accept-bitcoin.

Marten, R., & Kai, S. (2017). A blockchain research framework. *Business & Information Systems Engineering*, 385–409. https://doi.org/10.1007/s12599-017-0506-0.

Merlinda, A., Valentin, R., David, F., Simone, A., Dale, G., David, J., et al. (2019). Blockchain technology in the energy sector: A systematic review of challenges and opportunities. *Renewable and Sustainable Energy Reviews*, 143–174. https://doi.org/10.1016/j.rser.2018.10.014.

Mettler, M. (2016). Blockchain technology in healthcare: The revolution starts here. In *2016 IEEE 18th international conference on E-health networking, applications and services, healthcom 2016.* https://doi.org/10.1109/HealthCom.2016.7749510.

MIT introduces new digital certification program powered by blockchain technology — observatory of educational innovation.(2017). https://observatory.tec.mx/edu-news/2017/10/20/mit-introduces-new-digital-certification-program-that-relies-on-blockchain-technology.

Muhamed, T., Marko, H., Kristjan, K., Marjan, H., & Aida, K. (2018). EduCTX: A blockchain-based higher education credit platform. *IEEE Access*, 5112–5127. https://doi.org/10.1109/access.2018.2789929.

Nir, K., & Jeffrey, V. (2018). Blockchain-enabled e-voting. *IEEE Software*, 95–99. https://doi.org/10.1109/ms.2018.2801546.

Ojo, A., & Adebayo, S. (2017). Blockchain as a next generation government information infrastructure: A review of initiatives in D5 countries. In *Public administration and information technology* (Vol. 32, pp. 283–298). Springer. https://doi.org/10.1007/978-3-319-63743-3_11.

Ølnes, S., Ubacht, J., & Janssen, M. (2017). Blockchain in government: Benefits and implications of distributed ledger technology for information sharing. *Government Information Quarterly, 34*(3), 355–364. https://doi.org/10.1016/j.giq.2017.09.007.

Palmer-Derrien, S. (2019). SA government uses startup Horizon State's democratic blockchain. In *Fisheries council election*. https://www.smartcompany.com.au/startupsmart/news/horizon-states-democratic-blockchain-fisheries-council-election/.

Pradana, A., Sing, G. O., Kumar, Y. J., & Mohammed, A. A. (2018). Blockchain traffic offence demerit points smart contracts: Proof of work. *International Journal of Advanced Computer Science and Applications, 9*(11), 375–382. https://doi.org/10.14569/ijacsa.2018.091153.

Prisco, G. (2016). *The blockchain for healthcare: Gem launches gem health network with philips blockchain lab*. https://bitcoinmagazine.com/articles/the-blockchain-for-heathcare-gem-launches-gem-health-network-with-philips-blockchain-lab-1461674938.

Qi, R., Feng, C., Liu, Z., & Mrad, N. (2017). Blockchain-powered internet of things. In *E-democracy for smart cities. Advances in 21st century human settlements*. https://doi.org/10.1007/978-981-10-4035-1_17.

Qiheng, Z., Huawei, H., Zibin, Z., & Jing, B. (2020). Solutions to scalability of blockchain: A Survey. *IEEE Access*, 16440–16455. https://doi.org/10.1109/access.2020.2967218.

Qiuyun, S., & Allison, P. (2019). A blockchain-based land titling project in the republic of Georgia: Rebuilding public trust and lessons for future pilot projects. *Innovations: Technology, Governance, Globalization*, 72–78. https://doi.org/10.1162/inov_a_00276.

Quartierstrom 1.0 - Das Projekt In Kürze. (n.d.). Retrieved September 5, 2020, from: https://quartier-strom.ch/index.php/qs1-inkuerze/.

Quartierstrom: Die Ergebnisse. Quartierstrom. (n.d.). Retrieved September 5, 2020, from: https://quartier-strom.ch/index.php/qs1-ergebnisse/.

Raflearning.Org. (n.d.). *Understanding Ag Fintechs'business models: Agri-wallet case study*. Raflearning. https://www.raflearning.org/post/understanding-ag-fintechs-business-models-agri-wallet-case-study.

Richard, H. (2001). Understanding e-governance for development. *SSRN Electronic Journal*. https://doi.org/10.2139/ssrn.3540058.

SA government blockchain voting. (n.d.) Retrieved September 2, 2020, from: https://horizonstate.com/testimonials/sa-government-blockchain-voting/.

Smart Dubai. (n.d.). *Blockchain case study for government in the UAE*. https://consensys.net/blockchain-use-cases/government-and-the-public-sector/smart-dubai/.

Smart Dubai. (n.d.). Retrieved September 2, 2020, from: https://www.smartdubai.ae/initiatives/paperless.

Smart city innovation award Für Quartierstrom 2.0. (2020). Quartierstrom. https://quartier-strom.ch/index.php/en/2020/03/31/smart-city-innovation-award-fur-quartierstrom-2-0/.

Smart Dubai global blockchain challenge 2019. (n.d.). https://dubaifutureaccelerators.com/en/blockchainchallenge/.

Snyder, S. (2019). *The privacy questions raised by blockchain*. https://www.bradley.com/insights/publications/2019/01/the-privacy-questions-raised-by-blockchain.

Strüker, J., Albrecht, S., & Reichert, S. (2018). Blockchain in the energy sector. In *Business transformation through blockchain* (Vol. II, pp. 23–51). Springer International Publishing. https://doi.org/10.1007/978-3-319-99058-3_2.

Sullivan, C. L., & Burger, E. W. (2017). E-residency and blockchain. *Computer Law & Security Report, 33*(4), 470–481. https://doi.org/10.1016/j.clsr.2017.03.016.

Sullivan, C., & Burger, E. (2018). Blockchain, digital identity, e-government. In *Business transformation through blockchain* (Vol. II, pp. 233–258). Springer International Publishing. https://doi.org/10.1007/978-3-319-99058-3_9.

Sydow, A., Sunny, S. A., & Coffman, C. D. (2020). Leveraging blockchain's potential — the paradox of centrally legitimate, decentralized solutions to institutional challenges in Kenya. *Journal of Business Venturing Insights*, e00170. https://doi.org/10.1016/j.jbvi.2020.e00170.

TCS backs out of Kochi corporation's e-governance system.(2020) (Original work published 2020) https://www.thehindu.com/news/cities/Kochi/tcs-backs-out-of-corporations-e-governance-system/article31817247.ece.

The Economic Times. (2020). *Dubai- the world capital of blockchain development*. https://economictimes.indiatimes.com/tech/ites/dubai-the-world-capital-of-blockchain-development/articleshow/74224682.cms.

Theresia, T., & Novelo, D. M. F. (2019). Leveraging blockchain technology in humanitarian settings — opportunities and risks for women and girls. *Gender and Development*, 317—336. https://doi.org/10.1080/13552074.2019.1627778.

Thomas, M., Raymond, C. K.-K., Zhechao, L. C., & Debiao, H. (2019). Blockchain in healthcare applications: Research challenges and opportunities. *Journal of Network and Computer Applications*, 62—75. https://doi.org/10.1016/j.jnca.2019.02.027.

Thompson, A. R., & Andrew, S. (2016). Blockchain (distributed ledger technology) solves VAT fraud. *SSRN Electronic Journal*. https://doi.org/10.2139/ssrn.2853428.

Tokenpost. (2019). *Kenyan blockchain and AI taskforce recommends CBK to consider creating digital currency*. tokenpost. https://tokenpost.com/Kenyan-Blockchain-and-AI-taskforce-recommends-CBK-to-consider-creating-digital-currency-2785.

Trade.Gov. (2019). *Information, communications and technology (ICT)*. International trade administration. https://www.trade.gov/knowledge-product/kenya-information-communications-and-technology-ict.

Using blockchain to make land registry more reliable in India.(2018). https://blogs.lse.ac.uk/businessreview/2018/04/13/using-blockchain-to-make-land-registry-more-reliable-in-india/.

U.S. Social Security Administration. (n.d.). Retrieved September 2, 2020, from: https://www.benefits.gov/agencies/U.S.%20Social%20Security%20Administration .

Valentina, G., Fabrizio, L., Claudio, D., Chiara, P., & Victor, S. (2018). To blockchain or not to blockchain. *That Is the Question. IT Professional*, 62—74. https://doi.org/10.1109/mitp.2018.021921652.

Viswanathan, S., & Shah, A. (2018). *The scalability Trilemma in blockchain*. https://medium.com/@aakash_13214/the-scalability-trilemma-in-blockchain-75fb57f646df.

Voting now open.(2019). https://yoursay.sa.gov.au/blog/minister-s-recreational-fishing-advisory-council-voting-now-open.

Waltl, B., Sillaber, C., Gallersdörfer, U., & Matthes, F. (2018). Blockchains and smart contracts: A threat for the legal industry?. In *Business transformation through blockchain* (Vol. II, pp. 287—315) Springer International Publishing. https://doi.org/10.1007/978-3-319-99058-3_11.

What is the Open Source University at OS.UNIVERSITY and why does it matter?.(2017). https://medium.com/the-os-university-blog/what-is-the-open-source-university-at-www-os-university-and-why-does-it-matter-e84fc36f83f0.

Williams, J. (2014). *USPS loses database and backup in hardware failure*. https://www.fedscoop.com/usps-loses-database-backup-hardware-failure/.

Zambrano, R., Young, A., & Verhulst, S. (2018). *Case Study : Connecting refugees to aid through blockchain enabled ID management: World Food programme's building blocks*. https://blockchan.ge/blockchange-resource-provision.pdf.

Chapter 8

Cloud/edge computing for smart cities

J. Nulyn Punitha Markavathi[1], D. Kesavaraja[2]

[1]Department of CSE, PSN College of Engineering, Tirunelveli, Tamil Nadu, India; [2]Department of CSE, Dr. Sivanthi Aditanar College of Engineering, Tiruchendur, Tamil Nadu, India

Chapter outline

1. Introduction

The term "edge computing" includes some technologies with a wide range and also including peer-to-peer, fog computing, and blockchain. Edge computing provides the ability to compute and store data closer to end devices. It is managed by the operator, which may extend beyond the network edge and user edge (Nakamoto, 2008). It enables the applications by providing computation and storage with low latency. It is also the process of offloading computing and storage from the centralized cloud to the network's logical extremes. It means processing a few processes in the cloud and moving to the local edge server. It reduces the number of long-distance communication between the client and the server. Data produced by the Internet of Things (IoT) devices are processed at the place of data created by taking away the routes to data centers with the help of edge computing. It is mainly based on the use cases. Edge computing improves the speed and performance of transporting data through the applications on the edge. It increases the efficient usage of bandwidth by analyzing the data at edges. The cloud, which requires data transfer from the IoT requires large amounts of bandwidth. It is mainly used in a remote location with

Blockchain for Smart Cities. https://doi.org/10.1016/B978-0-12-824446-3.00011-9

minimum cost. It can process data without putting it on the public cloud, ensuring full security. Edge computation provides a limitation of cloud usage by extending computing closer to the data source. Edge computing provides greater business method and enables latency through better control and faster insights, lower operating expenses, and gives network bandwidth support. It is essentially the process of decentralizing computer services and moving them closer to the source of data. It can reduce the latency because it can reduce the volume of data moved and distance it travels. Distributed edge computing is analogous to, and can be regarded as an extension of the evolution of content distribution over the last decades (Lin et al., 2007). To improve the performance and efficiency for delivering content, especially the video content with dominating broadband traffic, broadband service providers also typically deploy distributed storage caches. To enable applications with edge computing extends both computing and storage. Enabling applications to be localized in edge computing near end-users helps improve network transit latency. Data takes time to travel across the network, and the resulting delay can occur. In real time at the edge point only the network process can be processed in edge computing. It is also necessary for the localization of data and efficient data processing. The local processing of information to reduce the volume of traffic over transport resources. Edge computing serves as the decentralized extension of the cellular network, campus networks, and data center networks. Also, it reduces the backhaul traffic by keeping the right content at the edge. It means the traffic is between a remote site and the internet via a centralized backbone. The computation resources can be distributed geographically in a variety of location types based on the use case requirements. Edge computing is the structure of enhanced security and lower latency that enables faster analysis and decision making, and utilization of capacity by sending a small amount of data to the centralized processing of data centers. Edge technologies deliver a real-time, fast experience to customers and provide the flexibility to meet the trade requirements with centralized data storage. Edge technology makes the device as a supercomputer. Edge computing has security intimations to the network by moving data processing. It is possible to develop a multilayered approach to security that takes the communication layer and cloud security simultaneously. Let's consider a building secured with dozens of IoT video cameras. The cameras give a raw video signal as output and continuously stream that signal to a cloud server. In the cloud server, the video output from all the cameras is detecting motion through application to ensure that only clips activities are saved into the server's database. Simultaneously, the video footages are processed to the cloud with a heavy load. Every camera uses its internal computer to run the motion-detecting application. The cloud server is only responsible for storing the important footage; it means the server cloud should communicate with a higher number of cameras without overloading. Edge computing allows capturing and storing videos at the edge and directly in the camera. Edge computing technologies help to achieve better image quality, even in challenging conditions. Smart cameras will analyze data in real-time, taking advantage of advanced capabilities and give context to all that data around us and this device gives us the precise amount of information that user needs (Fig. 8.1).

2. Characteristics

To support strong and good environment, the edge computing uses the physical structure and artificial intelligence. To enable network slicing in smart cities it is imperative to tackle the challenges such as end-end management and orchestration. Engage effectively with local people in local governance and decision by e-participation. **Geographically distributed**: It is a cutting-edge computing paradigm characterized by a geodistributed system. It migrates computing resources like that computing power, applications, and data from the remote cloud to the next edge and enables a large number of real-time smart city services. **Orchestration intelligence**: Smart cities initiate community based problems and solving the collaborative process of the entire system. Orchestration is very complex because the decentralized process makes the security scheme must be autonomous of edge computing. **Empowerment intelligence**: It provides an open platform and gives experimental facilities and smart city infrastructure to cluster innovation. Resource allocation resizing the same node among a cluster node. It enables the self-learning solution by transforming edge computing capabilities. It can be enabled by advanced network capabilities and machine learning. **Instrumentation intelligence**: Real-time data collections are made by the city infrastructure. A massive amount of data will be generated by smart devices and real-time applications that require instant processing. **Heterogeneous**: Edge computing enables the massive number of heterogeneous devices to run

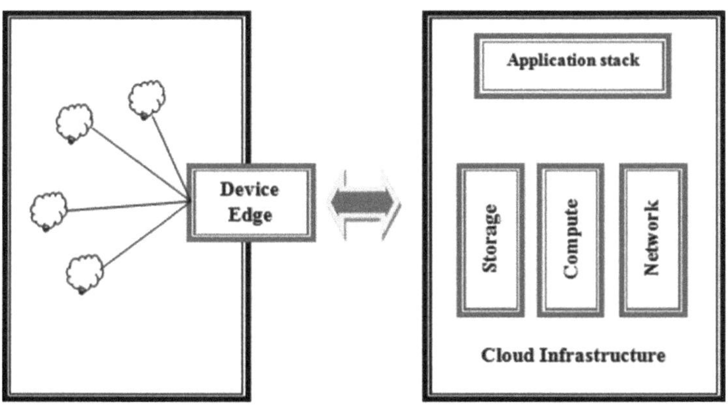

FIGURE 8.1 Structure of edge computing.

the different protocols. Interoperability can be used to support seamless operation in edge computing. It is also used to implement smart cities by utilizing interoperable interfaces. **Real-time communication**: A large amount of data will be generated by the smart devices and smart city applications that require instant processing and it used a small part of the data. For example, to enable real-time autonomous cars, edge computing enables the road side units is a variable solution (Fig. 8.2).

3. Frameworks

It is a platform for developing applications using edge computing in smart cities. Edge computing allows us to bring the resources closer to where they are needed. The frameworks can be divided into five main parts of smart city development. **Technology framework:** To maintain interaction between human and technology system to form smart city technologies with variety of levels. Digital service-oriented infrastructure is used to connect devices and smart cities (Schneider et al., 2017). Smart city always provides a large quality of data by multiplicity devices. **Human framework:** The main knowledge of human framework is based on network and human support system. It is the main indicator to achieve smart city using it is mainly based on the knowledge of networks and the human support system. It is an important indicator of a smart city using edge computing. The mobility is used to building capable human resources to develop smart city. The development of economic understanding is central to a smart city. **Institutional framework:** It mainly refers to formal provisions that assign primary responsibility and also authority to an agency for the collection, processing, and statistics. It also includes arrangements for data sharing and coordination. Interoperability can be implemented in smart cities by interoperable interfaces and open source frameworks. **Energy framework:** The smart city provides sustainability through the use of data and technology. Sustainability in edge computing can achieve by using renewable energy resources. Sustainability refers to energy

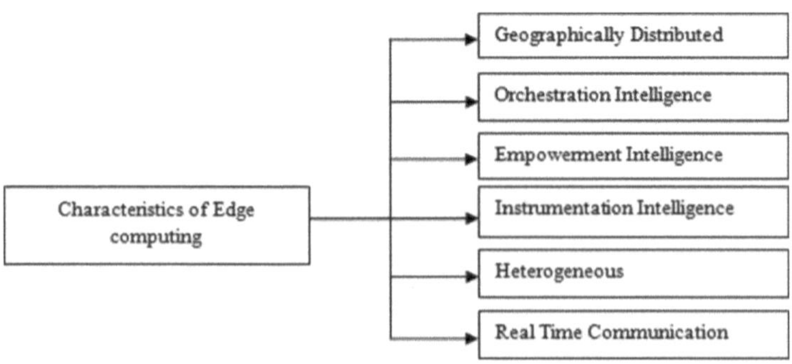

FIGURE 8.2 Characteristics of edge computing.

gathering that takes energy from environmental sources. So sustainable developments are required in edge computing. **Data management framework:** This technique is used to ensuring the reuse-ability and consistency of smart IoT device data. The users requires some data like validity, storing and protecting, it is the main administrative process to ensure the accessibility and reliability.

4. Existing system

Edge computing is obtaining more and more popularity in the IoT domain. In parallel, it has given massive amounts of data and need to optimize computational resources and also increasing tendency to send data to the cloud. The origin of edge computing is the Content Delivery Network (CDN). It introduces nodes at locations geographically closer to the end-user for delivery. But in this method, there are some problems such as cost, complexity, and deployment procedures. Pervasive computing allows decentralized distributed applications used as a different peer-to-peer overlay networks. It avoids the long-distance links between peers and decreases the network load and latency. It cannot maintain the entire system which is not secure and also a very expensive cost to operate. Cloud computing expects the availability of computer system resources, especially for data storage. The major leverage of cloud computing is to backup and restore the data, and maintenance cost is very low (Han et al., 2019). It also provides unlimited storage capacity and data security. The cloud server provides the best security but you will be aware of sending all your organization's sensitive information to a third party that is a cloud computing service provider. There may be a chance that your information is hacked by hackers, so this model moves on to cloudlet computing, where mobility is enhanced in small scale cloud data center (Edge Computing, 2017). The major benefits of the cloudlet are elasticity and scalability. Another one is fog computing, which is also supporting mobility, low latency, and improvement of the quality of services, but it does not verify the user is authorized or not. Encryption has failed in securing data from the attackers (Qiang and Nirwan, 2018). Now the trend is to use edge computing; this is data processing that happens off the network on the edge of the network and closer to the source of the information. Edge computing provides low latency, enhancing the security and improving reliability because the information's are processed and sent to the cloud (Giang et al., 2018) (Table 8.1).

5. Infrastructure architecture

A networked computerized world is associated with things and individuals through actual areas in the edge. The circulated registering geography, where data is handled, is found near the edge. Things and individuals burn through that data. Edge registering is geography since distributed computing is a way

TABLE 8.1 Existing system table.

S.No	Author name	Year of publications	Title of the paper	Description
1	Bettina Kemme Marta Patino	2007	Enhancing edge computing with database replication	Wide area data replication protocol fast response time fault tolerance properties to reduce data replication
2	Bukhary Ikhwan Ismal Eshan Motajeran ·	2015	Evaluation of Docker as edge computing Platform	Decentralized system latency is high distance between client and server calculated by multiple hops
3	Blesson Varghese Nan Wang	2016	Challenges and opportunities in edge computing	Quality of service offloading to properly find out the edge node
4	Jiao Zhang Zhaolong ning	2017	Energy latency trade off for energy –aware offloading in mobile edge computing networks	To reduce total cost to get optimal solution but its increased network load and latency
5	Michael Schneider Jason Rambach	2017	Augmented reality based on edge computing using the example of remote live support	Offloading end-to-end latency is elevated deployment is the drawbacks of the system
6	Nam Ky Jiang Rodger Lia	2018	Fog at the edge: Experiences building an edge computing platform	Centralized cloud computing to solve the problem of distributing application across edge device reusability scalability
7	Jaiwen Kang Rong Yu	2018	Block chain for secure and efficient data sharing in vehicular edge computing and networks	P2P data sharing centralized system secure data and sharing high quality of data sharing
8	Seyed Akbar Mostafavi Mohammed Asif	2019	Edge computing for IoT: Challenges and solutions	Security scalability latency

of computing. There are five main partners in edge computing. Actual area implies area is a site or territory inside a site where something will be found, and the presence of things is unquestionable in business geography. IoT is loaded with things. It frames basic temperature sensors in the rush hour gridlock framework. Without these things, the framework can't gather this information to perform edge registering. Individuals use cases and prerequisites are distinctive for individuals and things. Individuals are frequently working like things that interface or separate. Edge registering requires an association with a more extensive computerized world. There are numerous edges, where various types of things and individuals associated with various types of things. The center edge doesn't need a hyper-scale center. It is fundamentally a specialist organization or cloud supplier. Some key segments can be utilized to build up the foundation of edge registering, like distributed computing topology, which is a self-sufficient, shut circle framework that never interfaces with a center yet it can't straightforwardly associate with edge computing, so it alludes to the neighborhood processing. It requires some data transmission and low-power availability. Bluetooth is the most well-known low-power availability choice, however, the downside is an absence of consistency. WiFi offers the fairest data transmission, however, the short-range denies its consistency and the force utilization is a lot higher than Bluetooth (Lim et al., 2018). Data preparing requires edge computing, a blend of registering, stockpiling, organizing, information examination, information the board, and performing data handling. The intricacy of data handling with the cutting-edge investigation and AI, like self-ruling driving, implies the vehicle is equipped for detecting its current circumstance and working without human inclusion (Maheshwari et al., 2018). Found near the edge processing prerequisites are that it could require installed knowledge, registering inside a couple of feet, in a similar structure, nearby, or in the metro region or area. Edge figuring expects sensors to screen the drivers at the dashboard (Chiang et al., 2017), Figs. 8.1, 8.2, 8.3 and 8.4.

6. Smart city challenges and concerns

Edge means the growing importance of computing capacity in smart cities, smart cars, or hand-held devices used by technicians and service centers. The local devices in the field are getting more intelligent with every quarter that interacts with edge applications. Each and every industry has one server as

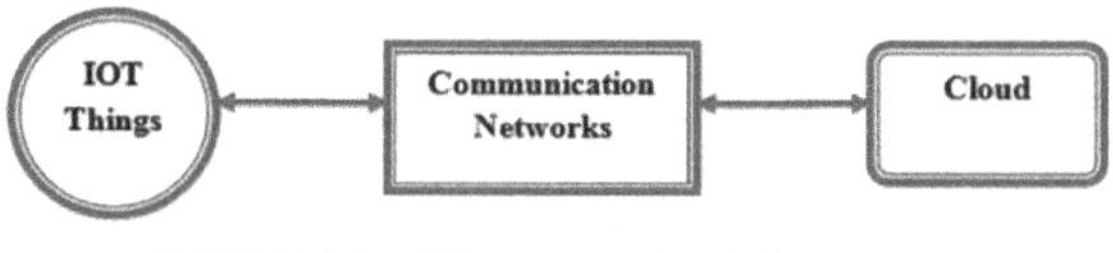

FIGURE 8.3 Edge computing infrastructure.

running host applications and they are used as a new type of distributed computing of IoT of edge computing. Processing at the edge brings data accumulation closer to computation but it provides a new set of challenges to the network. **Network bandwidth** is when data are stored at the edge, and as more computation happens at the edge, the network bandwidth will shift. It allocates higher bandwidth to data centers and lower bandwidth to the end points (Porambage et al., 2018). It requires less data transmission bandwidth. So the challenge for edge computing is the need to balance more bandwidth toward the network. **Distributed computing** is a distributed cloud, which refers to used cloud technology, local, or wide area networks to connect those distributed resources. Edge computing is a part of a distributed cloud system. It includes routers, servers, WiFi, or local data centers that are connected by the cloud network. It behaves as a single system; it is efficient, flexible, and maximizes the performance (Hou et al., 2018). The computing models are dissipating; computing now needs to include networking as a key element of the network. Edge computing can be almost anywhere, and anyone uses the connected device. It provides the same services and can be activated when needed. **Latency** is the delay between users and web applications response to that action. It is often referred to in networking terms as the total round trip time it takes for a data packet to travel (Bi and Zhang, 2018). Application latency is reduced based on decision making latency. To compute, it is located at both the core and the edge data and traverses the network in each direction, sharing data and dealing with access rights. It means data transfer is no longer a simple one-way process. **Security** is important especially when edge computing struggles with the same network models and security for remote servers. The centralized data center is applied with computing resources and applications. It is possible to build around the resources for easier security. But edge computing forces businesses to struggle, which applies the same network security models and physical security for more remote servers. The traffic patterns are the example or challenge of the security in edge computing (Zhang et al., 2018b). **Backup** edge computing collects a large amount of data and from different locations. Data protection is a strategy for all data needs. Network bandwidth requires critical storage media when deciding how to protect assets because backup over the network may not make sense. **Data accumulation** is a key asset, and collecting at the edge brings new challenges and can create liabilities if it is handled with existing data handling rules. There are two options for accumulating data by table or index. Accumulating with a table appends all of the source data to the elastic tube table and by its index. Data storage and access are critical, both of which will need to surround the networks as part of the data lifecycle (Li et al., 2019).

7. Why do we need smart cities?

Cloud computing promises scalability and security for data systems. But edge computing's prime feature is comprised of closer proximity between data storage and processing. Edge computing supports lots of benefits compared to another computing process to enhancing the data management and processing of smart cities. Smart cities feed on a mass stream of data to operate, as connected devices and services grow, such as the growing risk of network congestion, which possibly affects performance (Shamim et al., 2018). To achieve delegate system task and streamline operations by using decentralized system of computing and processing, edge computing will be outstanding for data-laden projects such as traffic management and the development of autonomous vehicles. The processing of data with minimal disruption is a key to consolidate mobility solutions and edge computing is a natural solution. Most of the smart cities using edge computing to generate real-time applications to traffic network like cameras, streetlights, and traffic lights. **Good infrastructure** included the main aim of constructing a smart city, which is to provide good infrastructure to the residents with water and sanitation services. The analytics are compiled into small grids and fed into computers. The process of edge computing is mainly focused on making the city efficient (Stojkoska and Trivodaliev, 2017). **Reduced communication cost** is incorporating the essential ideas to decrease correspondence costs that can't be any regular correspondence between the squares at every single time. The expense of the information assortment network is utilized for the worldwide assortment of information from trades and gives the switches/scaffolds and conveyance of information from replicators. **Autonomy** has the capacity to make an informed, uncoerced decision. It distributes application data and services where they can optimize outcomes its best in a growing set of connected assets. This includes edge analytics software. **Enhanced quality of data** refers to the quantitative pieces of information. The data are generally considered high quality of data to use in operations, decision making, and planning. The number of data sources increases and the internal data consistency become significant (He et al., 2018). **Smart Solutions:** Smart cities are required and it provides a smart solution to maintain the water quality, treatment of wastewater, delivering electronic service, and providing public data. **Promotes development**: Smart cities enhance the development activities of the region. The activities are building schools, shopping malls, hospitals, and so forth. **Employment:** A smart city is an economy of agglomeration. It provides various opportunities and advantages to residents. The construction of a smart city requires a lot of labor and it provides employment for many people (Fig. 8.4).

8. Applications of edge computing

Edge computing is mainly used for data, applications, and services to move away from centralized pivots to logical networks. Edge computing covers a scope of technologies, for example, remote sensing systems, circulated information, and augmented reality. Some of the innovative applications include

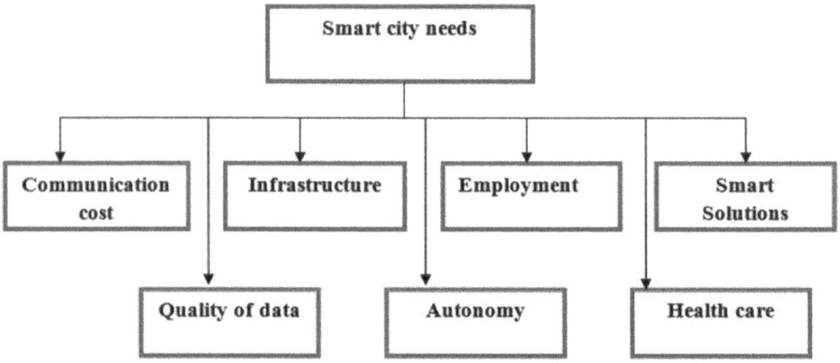

FIGURE 8.4 Smart city needs.

smart cities; the architecture of edge computing makes the utilization of controlling devices and administrations to react to condition changes in real-time. Smart cities can change how individuals live and use the benefits of urbanization (Zhang et al., 2018a). Edge computing depends upon an instrument for collecting information and fundamental processing tasks. It is more advanced and increasingly reliant on computers capable of humanlike reactions and the ability to make decisions on a device which will become more prevalent in the successful smart city. Edge computing to generate real-time insights into traffic patterns through devices such as cameras, streetlights, and traffic lights to create rerouting services (Khan et al., 2020). **Health care**: With IoT gadgets fit for delivering measures of patient-generated health information, health care suppliers might access essential data about their patient's continuously moderate and fragmented databases. Medical devices could be made to assemble and process information throughout treatment. Sharing and exposure of data make any edge solution challenging to execute other safety efforts. For example, blockchain innovation could give better approaches to address. **Augmented reality (AR) devices** contain a maximum number of realities, and a minimum number of virtual glasses and headsets at times used to make this effect. Most of the clients have experienced augmented methods through their cell phone displays. Snapchat or Instagram has utilized AR. It expects devices to process visual information and visual components continuously (Erol-Kantarci and Sukhmani, 2018). Visual information is conveyed back to the server without edge computing because the digital components are added before sending it back to the gadget. It avoids some actions to bring critical latency (Iosup et al., 2011). **Security solutions** are when edge computing technology gives some benefits of security and surveillance systems to respond to threats within a particular time. The security system can identify threats and alert users about unusual activity in real-time. Edge computing is the major part to maintain in a unified manner to ensure all devices and proper security protocols are followed. Encryption, patching, and the use of artificial intelligence to monitor and detect threats are essential. **Autonomous vehicles** make a decision immediately, for example, to stop a pedestrian crossing in front of autonomous vehicles. To handle this situation or to make an immediate decision, a remote server is not reasonable. A vehicle that utilizes edge technology can interact more efficiently because they can communicate with each other first as opposed to send data on accidents and traffic to remote servers (Ismal and Goortani, 2015). The autonomous cars, using more than 100 vehicle sensors will generate approximately 40 TB of data for eight hours of driving. Wheel-edge computing plays a dominant role in sending all the data to the cloud, which is also unsafe and unsuitable. The car should immediately respond to the events that have valuable data when coupled with the performance of other cars of its class. **Fleet management** includes, for example, truck companies, and the main goal of the company is to combine and send data from multiple operational data points like batteries, brakes, etc to the cloud. Every key element in operational components is analyzed by the cloud (Varghese et al., 2016). And also, edge computing saves fleet costs by managing, avoiding breakdowns, collisions and it has the potential to save fleet cost on ongoing data transmission and storage. **Video conferencing** sets the server-side of video conferencing software closer to participants, and quality problems can be reduced. Video conferencing is a modern communication tool for increasing the decision-making process. This technology allows users in different locations to hold meetings without having to move to a single location. Some slow links to the cloud can cause many video conferencing frustrations (Jiao et al., 2018).

9. Features of edge computing

Edge computing works on a decentralized data-processing approach, like a cloud where large quantities of data can be turned into actionable insights quickly. The percentage of real-time data processed outside the cloud has been increasing due to quicker reaction times to maintain an advantage. Data processing occurs at or near the source of data generation (Quwaider et al., 2016). **Latency** is the performance of your network based on latency, bandwidth, and throughput. Reducing latency in edge computing is the most important benefit to increase the performance of the network. IoT edge computing process data locally or nearby edge data centers. The information they collect does not have to travel much far in traditional cloud architecture. The data computation closer to the source can reduce the distance; edge computing can greatly reduce latency. So the results of high speed at end-users with latency are measured in microseconds than milliseconds. Latency is the measurement of a data packet that travels from the original point to its destination. By locating key processing tasks closer to end-users, edge computing can deliver faster and responsive services. IoT can greatly improve performance speed (Ianuale et al., 2016). Each node using dedicated networks to reduce streamline network path and also it provides direct communication between nodes.

Stack path provides private networks that allow bypassing the public internet. The private networks reduce latency by providing more efficient paths for data packets to travel. Also, the CDN allows the client to reach their users in the most efficient manner possible (Kang et al., 2019). These delivery channels greatly reduce the network lag between the user and its applications. To make your web applications with less latency using the CDN and also private network to data transformation (Jiao et al., 2018). **Security** helps to increase attacks of networks using IoT edge computing devices; it also provides some important security advantages. The centralized cloud architecture makes it vulnerable to denial service attacks. Edge computing distributes processing, storage, and applications across data centers. The architecture incorporates specialized edge data centers, it provides additional security measures to guard against attacks. In real-time the edge data center offers the variety of tools and quality of data to client can use to secure and monitor the networks. By using the smart contract methods unwanted attacks from third parties can be avoided. Without third party people it allows the credible transformation (Atzei et al., 2017). A blockchain using a smart contract is visible to all users. It is a decentralized system between all authenticated parties with no need to pay middleman and it saves you time and conflict. It is a transaction protocol that is intended to automatically execute. Smart contracts reduce of the need for trusted intermediates and fraud losses. It is the line of code that are stored on a blockchain and automatically execute when predetermined terms and conditions are met. Blockchain transaction records are encrypted, and that makes it very hard to hack for attackers; because each record is directly connected with previous records. So the whole chain would need to be altered to change a single record (Zyskind et al., 2015). **Decision making:** In a smart city, the emergence of the cloud has made building ecosystems possible, but additional time and connectivity are required to inform fact-based decisions after analysis. Decision making using context-awareness concept helps to take immediate action. It is mainly used to define the location. It deals with the location of smart devices and their surrounding environment (Mostafavi & Asif, 2019). It improves the performance of the emergency response system. It enables automated services to immediate action. The prediction of content popularity using deep learning algorithms to enable intelligent cache decision making. It needs less manual data aggregation and document reviewing and it takes some time for processing and analyzing upon the data. It analyzes the vast data sets faster and more accurate results. **Scalability**: In smart cities, it will increase the number of actuators, smart sensors, and other smart devices. To enable the resource-intensive and latency smart city applications, edge computing provides a solution. Consider reliable performance for an increasing number of end-users and diverse networks (Hou et al., 2018). It reflects the ability of a system to expand its services and to increase the demand of the users (Hou et al., 2018). Edge computing allows the smart city to expand its computing capacity through IoT devices and edge data centers. **Versatility:** Edge computing also makes smart cities incredibly versatile. Edge data centers allow them to service end-users efficiently with latency. It is valuable for content providers to deliver uninterrupted streaming services. It also empowers new IoT devices to edge network architecture. Smart cities can offer new services to IT infrastructure. It is a huge benefit for smart cities to expand network reach to regions with limited connectivity. **Reliability:** To enhance the reliability, performing the tasks near network edge and also use the robust hardware, fault-tolerant networking protocols. Edge computing reduces the amount of data flowing to the primary network. Edge networks ensure a faster, seamless experience. Edge computing devices and data centers are connected to the network; it is more difficult to analyze the failure to shut down service entirely. Data can be changed through multiple pathways to ensure users retain products and information that they need. IoT edge computing devices and data move into edge architecture, and it provides reliability (Satyanarayanan et al., 2009).

10. How edge computing making cities smarter and better

Smart cities are connected to the internet through cloud connections. The issue of road safety, infrastructure, and smart cities are implementing edge computing strategies. It keeps the data at the location or much closer to the location of the data source. The roles of storage and processing transition from cloud servers to the computers at network edge possible to the devices. **Smart public services** in edge computing allow for more immediate responding to situations when smart cities use this technology to improve the quality of data by performing analytics at the site where the data against the location. Edge computing also allows reduced network latency that occurs only using the cloud and also it improves your speed. CCTV allows high-speed network connections to edge computing and also controls the high resolution video and quality of data to make the smart cities facilities (Xe et al., 2019). Smart cities introduce the principles of fog and edge computing to transfer the intelligence to centralize to local location of sensors and gateway. **Sensor monitoring** for edge computing is only transmitting data to the central hub when something unusual happens and it is monitored by the use of the sensor. **Worker messaging:** This platform can support secure high-bandwidth messaging when the absence of a web connection. With IoT sensors produced all types of endpoints, from traffic lights to trash. So processing at the edge becomes increasingly attractive to contain cost. **Security Camera:** Backhaul cost savings are substantially related to security cameras. The system can store the camera footage internally and only access by the edge, allowing you to launch affordable IP cameras. CCTV cameras are the keys to modern techniques. **Streetlights:** In smart cities they choose lower-power LED lights to physical fitting of new street lights will be substantial cost. **Smart Private services**: To implement the proper functioning of the network of automated taxis and delivery services in smart cities, edge computing is a natural solution. For self-driving cars, the expense of bandwidth and getting latency within reasonable limits is pushing companies to process some data as close as possible to its origin (Kesavaraja and Shenbagavalli, 2017). It requires software capable of controlling operations of edge devices and data centers within the network infrastructure to improve their operational feasibility (Ahmed et al., 2016). **Quality of services (QoS):** The main problem of allocating resources to the application based on the level by the performance, availability, and reliability but the main challenge is providing quality of service. These problems can be overcome by the edge computing process. It increases the number of services or applications of resources and system performance is maintained (Kesavaraja and Shenbagavalli, 2017). The quality of service coordinating between end-to-end networks to identify and marking techniques. It is performed within a single network. QoS policy management functions to control and administer end-to-end traffic across a network. **Sensors in the smart city:** A huge amount of data being generated by the IoT sensors, and it enhances the embedded system from bins to traffic lights. By using edge computing, sensors can be monitored locally with data only being communicated centrally when there is a reading that is out of the ordinary or a fault that needs reporting. **Dynamic infrastructure:** To maintain balanced security of building and roads in smart cities by improving infrastructure, it uses the earth detectors, which can monitor areas and structures that are more vulnerable to tremors. By using structural sensors the user can monitor the vibrations and collections of the materials in buildings, and the waste sensor is used to detect the garbage levels and maximizes waste collection routes (Jila & Han, 2018). **Environment harmony:** To increase awareness of sustainability and global warming, smart cities use IoT to lower their carbon footprint. For example, weather prediction monitors are used to sense atmospheric conditions and unstable activity to improve response planning (Higashino et al., 2017). And also smart cites using air pollution sensors used to monitor the CO_2 emissions and car pollution. **Connected parking meters** are the edge systems that provide a strong network for monitoring parking meters. The data are transmitted to the central hub only when it is required to improve smart parking meter availability, to show parking spots to drivers, and to reduce idling time as well as provide data about available parking spaces.

11. Conclusion

In this chapter, this work represents about edge computing paradigm to enrich smart cities with computing and storage resources. Cloud computing offers a centralized remote location to the edge of the network. While centralized data, processing, and storage have their limitations, to enable real-time applications the use of edge computing is required. It ensures shorter response time, security, decision making, and better reliability. This model analyzed the IoT based on architecture, its challenges, and current state-of-the-art solutions. It also describes what a smart city needs and its purpose in real-time. Edge computing is enhancing security and reducing latency time between networks.

References

Ahmed, E., Yaqoob, I., Gani, A., Imran, M., & Guizani, M. (2016). Internet-of-things-based smart environments: State of the art, taxonomy, and open research challenges. *IEEE Wireless Communications, 23*(5), 10–16. https://doi.org/10.1109/MWC.2016.7721736.

Atzei, N., Bartoletti, M., & Cimoli, T. (2017). A survey of attacks on ethereum smart contracts (SoK). In *Lecture notes in computer science (including subseries lecture notes in artificial intelligence and lecture notes in bioinformatics)* (Vol. 10204, pp. 164–186). Springer Verlag. https://doi.org/10.1007/978-3-662-54455-6_8.

Bi, S., & Zhang, Y. J. (2018). Computation rate maximization for wireless powered mobile-edge computing with binary computation offloading. *IEEE Transactions on Wireless Communications, 17*(6), 4177–4190. https://doi.org/10.1109/TWC.2018.2821664.

Chiang, M., Ha, S., Chih-Lin, I., Risso, F., & Zhang, T. (2017). Clarifying fog computing and networking: 10 questions and answers. *IEEE Communications Magazine, 55*(4), 18–20. https://doi.org/10.1109/MCOM.2017.7901470.

In comparison of edge computing implementations: Fog computing, cloudlet and mobile edge computing (pp. 1–6). (2017).

Erol-Kantarci, M., & Sukhmani, S. (2018). Caching and computing at the edge for mobile augmented reality and virtual reality (AR/VR) in 5G. In *Lecture notes of the institute for computer sciences, social-informatics and telecommunications engineering, LNICST* (Vol. 223, pp. 169–177). Springer Verlag. https://doi.org/10.1007/978-3-319-74439-1_15.

Giang, N., Lea, R., & Blackstock, M. (2018). Fog at the edge: Experiences building an edge computing platform. In *IEEE edge computing conference*.

Han, B., Wong, S., Mannweiler, C., Crippa, M. R., & Schotten, H. D. (2019). Context-awareness enhances 5G multi-access edge computing reliability. *IEEE Access, 7*, 21290–21299. https://doi.org/10.1109/ACCESS.2019.2898316.

He, D., Chan, S., & Guizani, M. (2018). Security in the internet of things supported by mobile edge computing. *IEEE Communications Magazine, 56*(8), 56–61. https://doi.org/10.1109/MCOM.2018.1701132.

Higashino, T., Yamaguchi, H., Hiromori, A., Uchiyama, A., & Yasumoto, K. (2017). Edge computing and IoT based research for building safe smart cities resistant to disasters. In *Proceedings - international conference on distributed computing systems* (pp. 1729–1737). Institute of Electrical and Electronics Engineers Inc. https://doi.org/10.1109/ICDCS.2017.160.

Hou, W., Ning, Z., & Guo, L. (2018). Green survivable collaborative edge computing in smart cities. *IEEE Transactions on Industrial Informatics, 14*(4), 1594–1605. https://doi.org/10.1109/TII.2018.2797922.

Ianuale, N., Schiavon, D., & Capobianco, E. (2016). Smart cities, big data, and communities: Reasoning from the viewpoint of attractors. *IEEE Access, 4*, 41–47. https://doi.org/10.1109/ACCESS.2015.2500733.

Iosup, A., Ostermann, S., Yigitbasi, M. N., Prodan, R., Fahringer, T., & Epema, D. (2011). Performance analysis of cloud computing services for many-tasks scientific computing. *IEEE Transactions on Parallel and Distributed Systems*, 931–945. https://doi.org/10.1109/TPDS.2011.66.

Ismal, B. I., & Goortani, E. M. (2015). Evolution of docker as edge computing platform. In *IEEE international conference on open systems*.

Jiao, Z., Xiping, H., Zhaolong, N., Ngai, E. C. H., Li, Z., Jibo, W., et al. (2018). Energy-latency tradeoff for energy-aware offloading in mobile edge computing networks. *IEEE Internet of Things Journal*, 2633–2645. https://doi.org/10.1109/jiot.2017.2786343.

Jila, G., & Han, G. (2018). Edge computing based-intelligent manhole cover management system for smart cities. *IEEE Internet of things Journal, 5*(3), 1648–1656, 1-1.

Kang, J., Yu, R., Huang, X., Wu, M., Maharjan, S., Xie, S., et al. (2019). Blockchain for secure and efficient data sharing in vehicular edge computing and networks. *IEEE Internet of Things Journal, 6*(3), 4660–4670. https://doi.org/10.1109/JIOT.2018.2875542.

Kesavaraja, D., & Shenbagavalli, A. (2017). Scalable robust implementation of reliable streaming (SRIRS) with tristate aware quality of service. *Wireless Personal Communications, 96*(2), 2797–2820. https://doi.org/10.1007/s11277-017-4325-x.

Khan, L. U., Yaqoob, I., Tran, N. H., Ahsan, K. S. M., Nguyen, D. T., & Seon, H. C. (2020). Edge computing enabled smart cities: A comprehensive survey. *IEEE Internet of Things Journal*. https://doi.org/10.1109/jiot.2020.2987070, 1–1.

Lim, C., Kim, K. J., & Maglio, P. P. (2018). Smart cities with big data: Reference models, challenges, and considerations. *Cities, 82*, 86–99. https://doi.org/10.1016/j.cities.2018.04.011.

Lin, Y., Kemme, B., Patiño-Martínez, M., & Jiménez-Peris, R. (2007). Enhancing edge computing with database replication. In *Proceedings of the IEEE symposium on reliable distributed systems* (pp. 45–54). https://doi.org/10.1109/SRDS.2007.4365683.

Li, F., Shi, Y., Shinde, A., Ye, J., & Song, W. (2019). Enhanced cyber-physical security in internet of things through energy auditing. *IEEE Internet of Things Journal, 6*(3), 5224–5231. https://doi.org/10.1109/JIOT.2019.2899492.

Mostafavi, S. A., & Asif, M. (2019). Edge computing for IoT: challenges and solutions. *Journal of Communications Technology, 26*, 5–8. ISSN:2457-905X.

Maheshwari, S., Choudhury, S., Seskar, I., & Raychaudhuri, D. (2018). Traffic-aware dynamic container migration for real-time support in mobile edge clouds. In *International symposium on advanced networks and telecommunication systems, ANTS* (Vol. 2018) IEEE Computer Society. https://doi.org/10.1109/ANTS.2018.8710163.

Nakamoto, S. (2008). *Bitcoin: A peer-to-peer electronic cash system.*

Porambage, P., Okwuibe, J., Liyanage, M., Ylianttila, M., & Taleb, T. (2018). Survey on multi-access edge computing for internet of things realization. *IEEE Communications Surveys and Tutorials, 20*(4), 2961–2991. https://doi.org/10.1109/COMST.2018.2849509.

Qiang, F., & Nirwan, A. (2018). Towards workload balancing in fog computing empowered IoT. *IEEE Transactions on Network Science and Engineering*. https://doi.org/10.1109/TNSE.2018.2852762, 1–1.

Quwaider, M., Al-Alyyoub, M., & Jararweh, Y. (2016). Cloud support data management infrastructure for upcoming smart cities. In *Procedia computer science* (Vol. 83, pp. 1232–1237). Elsevier. https://doi.org/10.1016/j.procs.2016.04.257..

Satyanarayanan, M., Bahl, P., Caceres, R., & Davies, N. (2009). The case for VM-based cloudlets in mobile computing. *IEEE Pervasive Computing*, 14–23. https://doi.org/10.1109/MPRV.2009.82.

Schneider, M., Rambach, J., & Stricker, D. (2017). Augmented reality based on edge computing using the example of remote live support. In *Proceedings of the IEEE international conference on industrial technology* (pp. 1277–1282). Institute of Electrical and Electronics Engineers Inc. https://doi.org/10.1109/ICIT.2017.7915547.

Shamim, H. M., Ghulam, M., & Umar, A. S. (2018). Improving consumer satisfaction in smart cities using edge computing and caching: A case study of date fruits classification. *Future Generation Computer Systems*, 333–341. https://doi.org/10.1016/j.future.2018.05.050.

Stojkoska, B. L. R., & Trivodaliev, K. V. (2017). A review of Internet of Things for smart home: Challenges and solutions. *Journal of Cleaner Production*, 1454–1464. https://doi.org/10.1016/j.jclepro.2016.10.006.

Varghese, B., Wang, N., Barbhuiya, S., Kilpatrick, P., & Nikolopoulos, D. S. (2016). Challenges and opportunities in edge computing. In *Proceedings - 2016 IEEE international conference on smart cloud, SmartCloud 2016* (pp. 20–26). Institute of Electrical and Electronics Engineers Inc. https://doi.org/10.1109/SmartCloud.2016.18.

Xe, J., Tang, H., Huang, T., Yu, F. R., Xie, R., & Liu, J. (2019). A Survey of blockchain technology applied to smart cities: Research issue and Challenges. *IEEE Communication Surveys Tutorials, 21*(3), 2794–2830.

Zhang, J., Chen, B., Zhao, Y., Cheng, X., & Hu, F. (2018a). Data security and privacy-preserving in edge computing paradigm: Survey and open issues. *IEEE Access, 6*, 18209–18237. https://doi.org/10.1109/ACCESS.2018.2820162.

Zhang, P. Y., Zhou, M. C., & Fortino, G. (2018b). Security and trust issues in fog computing: A survey. *Future Generation Computer Systems, 88*, 16–27. https://doi.org/10.1016/j.future.2018.05.008.

Zyskind, G., Nathan, O., & Pentland, A. S. (2015). Decentralizing privacy: Using blockchain to protect personal data. In *Proceedings - 2015 IEEE security and privacy workshops, SPW 2015* (pp. 180–184). Institute of Electrical and Electronics Engineers Inc. https://doi.org/10.1109/SPW.2015.27.

Chapter 9

Waste management in smart cities using blockchaining technology

D. Jeyabharathi[1], Angelin Merling Thava[2], S. Jeba Prasanna Idas[3], T. Sangeetha[1]

[1]*Department of Information Technology, Sri Krishna College of Technology, Coimbatore, Tamil Nadu, India;* [2]*Department of Information Technology, Jayaraj Annapackiam CSI College of Engineering, Nazareth, Tamil Nadu, India;* [3]*Department of CSE, College of Engineering and Technology, International University of Technology Twintech, Sanaa, Yemen*

Chapter outline

1. Introduction

1.1 Blockchaining and IoT-based waste management system

The world population is growing tremendously. The cities that accommodate more people are facing astounding urban living pressure. While smart cities are evolving throughout the world, there are still more issues in the efficient

Blockchain for Smart Cities. https://doi.org/10.1016/B978-0-12-824446-3.00014-4

disposal of waste. With the increased rate of population and massive amount of waste produced by the people has increased is the biggest challenge. Accumulation of waste and disposal are the main issue. Internet of things (IoT) can play an important role to make the cities greener, safer, and more efficient. IoT technologies are used to handle waste containers. Also blockchaining technology is used make safe transactions, decentralizing databases and growing network adoption in the system. Improvement in safety and quality of life can be achieved by connecting devices, vehicles, and infrastructure all around in a city. We present a waste collection management system which uses an IoT prototype with sensors and blockchain technology to provide intelligence to waste bins or containers.

1.2 Blockchaining technology

Blockchain is a digital, distributed ledger. It keeps records of all transactions in a peer to peer network. Blockchaining has the advantage of immutable shared record ledger, block can be shared all the nodes of network, but no one can change the record information, because blockchaining methodology provides trustworthy. By using blockchaining smart contract in working, it triggered the action automatically. Blockchaining technology managed the distributed network. Normally, in a network independent computers are called nodes. If someone requests a transaction in the network, Network nodes validate the transaction using optimized algorithm. Once the transaction is verified, combine the new block of data to the ledger. Then new block is combined with the existing blockchain. Finally, the updated blockchain is distributed into all the nodes in the network. So, no one can change the information, records must be transparent and maintain in a secure manner. Blockchain technology has enormous potential when it comes technology has shaping the enhanced Smart Communities of the future in ways that are more efficient and provide better quality of living. For technology to deliver on its promise it must be allowed to change the status quo. Blockchain technology can provide a solution to a wide variety of problems which are generally faced by the smart cities. But the implementation of this technology in practice will depend on the administration and preferences of the community.

1.3 Internet of things

IoT is a sort of universal global neural network that links various devices in the cloud. IoT is a smartly connected device and system consisting of smart machines that interact and communicate with other machines, environments, objects, and infrastructure and the Radio Frequency Identification (RFID) and sensor network technologies will meet this new challenge. IoT-based technology placed a vital role to build smart cities. Due to faster population growth in cities, waste management is an essential one. Two important challenges are in waste management. First one, automatic estimation of the level of boxes. Second one, transparency is needed to track the path in waste management. First one is achieved through IoT-based technology. Some solution has introduced the use of IoT sensors to find the route of the waste track which have large amount of wastage. It is used to avoid high fuel costs and reduce the work amount, by applying machine learning algorithm. In some papers, the proposed smart solid waste bins were implemented for monitoring and collecting both wet and dry garbage stationed at different locations. This system involved four significant phases such as the hardware and software system in the cloud, mobile application, and routing information. All these components are adopted in our system plus desktop management system, SMS on the bin status, and geolocation information. IoT-based waste management systems were proposed. For each case, an implemented smart bin that uses a microcontroller with a sensor for the detection of the garbage level and a module for Internet connectivity was presented. Both systems had no means of monitoring the garbage collectors on duty, but this would be addressed in our system. The relevance of addressing waste management problems in the smart city keeps receiving attention in recent works. These works have a primary goal of proposing a framework for addressing waste management. This article presents a novel platform for smart trash control at the university, which is able to keep the university clean at low cost, requiring low labor resources. A novel microcontroller system is designed with a sensor module for measuring the filling height of garbage using ultrasound and geolocation of collected data based on LoRa technology. Furthermore, the paper presents a new method for predicting the probability of the filling level of each trash bin by applying LR in ML; furthermore, a graph theory–based optimization solution is proposed to compute the paths of waste collection with different time schedules in order to minimize the environmental and socioeconomic impacts, as well as supporting the workers of the university (Abu-Mahfouz & Hancke, 2018; Andrea, Nicola, Angelo, Lorenzo, & Michele, 2014; Anuga, Silva, Hancke, & Abu-Mahfouz, 2018; Ogbodo, Dorrell, & Abu-Mahfouz, 2017; Penzes, 2018; Sharma, Moon, & Park, 2017; Sun, Yan, & Zhang, 2016; Tasatanattakool & Techapanupreeda, 2018; Watanabe et al., 2016).

2. Objective

- Safeguarding biodiversity and saving energy through recycling using IoT smart waste management system
  - To decrease the various dreadful disease due to unclean and dirty surroundings all over the cities
    - To minimize dustbin overspill using volume sensor fixed in the lid of the garbage bins to avoid harm to the atmosphere.

- Classification of waste with automatic waste sorting algorithm.
- Developing safe transactions, decentralizing databases, and growing network adoption using blockchain technology.

2.1 Contribution of this chapter

The following are the contribution of this chapter.

Infrared sensor is used to detect the entry of the waste dumped in the waste container. Volume sensor is used to denote the fullness of the waste container. Inductive proximity sensor, capacitive sensor, and resistive plate sensor are used to identify the categorization of waste. Automatic waste sorting algorithm is used to classify the waste. Blockchain waste management platform could be used for proper sanitation.

2.2 Outline of the chapter

Section 1 describes introduction about this chapter. Section 2 describes the previous work related to waste management system, problems, issues, and solution. Section 3 discusses the working process, architecture, and methodology of the proposed research. Section 4 presents the implementation results and depicts the conclusion and references for further research.

2.3 Related work

A detailed survey on previous works on blockchaining technology, IoT-based sensors on waste management in smart cities is made in this chapter, and the problems associated with these approaches are highlighted.

2.4 Blockchaining-based waste management in smart cities

Blockchaining-based waste management is tremendous technology. Peters and Panayi (2016), have proposed the combination of BIM and blockchaining technology for sharing ability to serve as a single source of truth in industries. Here blockchain technology is used for design decisions, source of data to initiate action such as payments and orders. Pieroni, Scarpato, Di Nunzio, Fallucchi, and Raso (2018) have proposed smart city applications and services to overcome the challenges. Here blockchaining granting ledger method provides the solution for joining grid, exchanging information, buy or sell energy between nodes. Biswas & Muthukkumarasamy (2017), have discussed the integration of blockchaining with devices in a smart city will produce a common platform where all devices securely communicate information in a distributed environment.

2.5 IoT sensor–based waste management in smart cities

Pritchard et al., 2017, have proposed a garbage storing up framework for data gathering using GSM module. This framework utilizes the camera at each position of a waste holder. The camera takes entry of the reject holder, while the stack cell sensor takes the weight to get whether full or not. Lee and Wu, 2014, have proposed the waste association structure. This framework is used to utilize the GPRS to send the sensor information to a flexible application over the cloud.

2.6 Motivation

From the observation, immediate collection and sanitation when dust box filled with waste is very essential factor. For these contributions are

- Using IoT sensors such as ultrasonic sensors for detecting garbage level, inductive proximity sensors for detecting metals, capacitive sensors for detecting plastic, wood other raw materials, flame sensors for detecting fire, and weight sensors for detecting weight of collected waste.
  - Using blockchaining technology developing safe transaction, distributed ledger, secure network in management.

After reviewing the related literature of RFID sensor managing and other related managing process, it is observed that proper waste management is a critical task. From the literature survey the main issues are

- Rectify improper waste management
  - Only 20%–28% waste only is processed and treated
    - RFID is used for waste management, its usage is limited
    - Lack of accountability because wastes in city streets are very high. This proposed system is developed to address all these issues.

3. Proposed work

Blockchaining and IoT sensor–based waste management system has two main steps: (i) Working process for IoT sensors in waste management and (ii) Working process for blockchaining in waste management.

3.1 Working process for IoT sensors in waste management

Fig. 9.1 shows the overall architecture of the proposed system. The proposed framework is working dependent on IoT sensors. It finds the amount of the trash and promptly report the status to the driver who gather the trash through

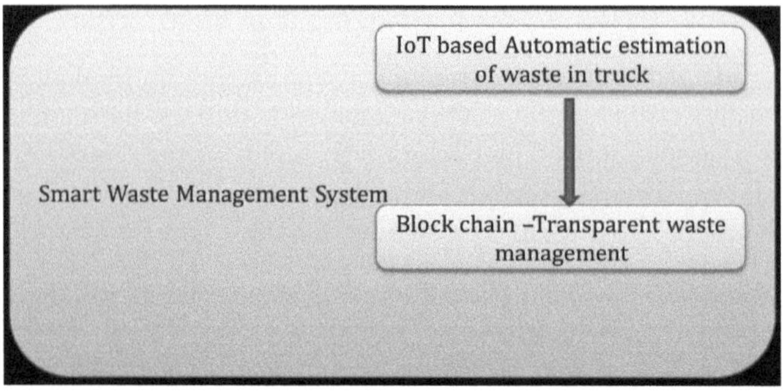

FIGURE 9.1 Overall architecture of the proposed system.

versatile application. The sensors are fixed in the trash box. The sensors like separation sensor and a light sensor to distinguish the launch of the case, a weight sensor to recognize that another thing is dropped inside and to compute the heaviness of the things inside the case. Moreover, one more sensor is additionally positioned in the container amount sensor that can be utilized to discover the amount of the trash. Fig. 9.2 shows the IoT-based sensor technology for waste management. The IoT trash observing framework is an imaginative framework which will assist with keeping the urban areas clean. This framework screens the trash canisters and advises about the degree of trash gathered in trash receptacles by means of page. For this the framework utilizes ultrasonic sensor put over the containers to recognize the trash level and contrast it and the trash receptacles profundity. This framework utilizes Wi—Fi modem for sending information. A website page is worked to show the status to the client checking it.

To collect the dataset of waste, we observed number of resident buildings around the waste container. This helps to predict the time taken to fill the

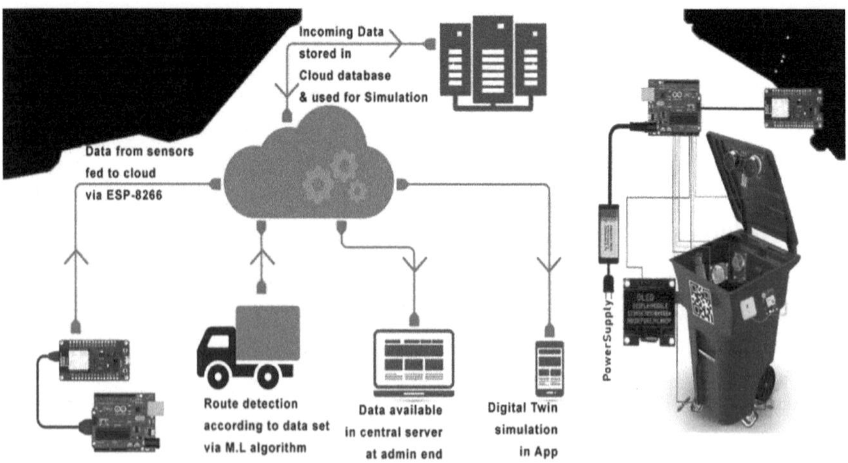

FIGURE 9.2 IoT-based smart waste management.

container and the size of the container also considered. Moreover, shortest path should be proceeded to collect and dispose the waste. The sensing module of the proposed system consists of four components that altogether act as a system. Raspberry Pi, sensor, network interface, and power are the four components used in this system.

3.2 Raspberry Pi

The information collected by sensor will be processed by microcontroller. Raspberry Pi 3 model could be used. The allocated RAM will be of 1 GB. It will primarily use Raspbian, a Debian-based Linux operating system. The Raspberry Pi is a low-cost, credit card–sized computer that plugs into a computer monitor and uses a standard keyboard and mouse. Raspberry Pi has faster clock speed and it is good at software applications.

Sensor is a device, module, or machine whose purpose is to detect events or changes in its environment and send the information to other electronics. In this chapter, we have used five different types of sensor to detect various things in waste collecting bins. They are ultrasonic sensor, proximity sensor, capacitive sensor, flame sensor, weight sensor, and temperature sensor.

3.3 Ultrasonic sensor

Ultrasonic sensor will be used to detect the level of garbage filled in the waste bin. The level of garbage will be depicted in terms of distance between the sensor and garbage in dustbin. This module has 4 pins. They are VCC (5v), Trig, echo, and GND. Trig has, to be used to send out an ultrasonic high level pulse. Echo pin will then automatically detect the returning pulse. Furthermore, sensor will determine the time interval between sent signal and received signal by echo. In this study we have used ultrasonic sensor to measure the waste level based on the distance inside the bins. It determines the status of waste inside the bins.

3.4 Inductive proximity sensor

An inductive proximity sensor is a device that can only detect metal targets. They do not detect nonmetal objects. It uses the principle of electromagnetic induction to detect or measure objects. An inductor develops a magnetic field when current flows through it; alternatively, a current will flow through a coil containing an inductor when the magnetic field changed. The operating principle is based on a coil and oscillator that creates an electromagnetic field in the close surroundings of the sensing surface. The presence of a metallic object in the operating area causes a dampening of the oscillation amplitude. The rise or fall of such oscillation is identified by a threshold, circuit that changes the output of the sensor. The operating distance of the sensor depends on the size, shape, and strictly linked to the nature of the material. In this study, inductive proximity sensor is used to detect metal-related objects from the waste containers.

3.5 Capacitive sensor

Capacitive sensors are capable of detecting plastic, wood, and other raw materials including metal. A common application is the detection of liquids, plastics, and grains. A capacitive sensor acts like a simple capacitor. A metal plate in the sensing face of the sensor is electrically connected to an internal oscillator circuit and the target to be sensed acts as the second plate of the capacitor. The external capacitance between the target and the internal sensor plate forms a part of the feedback capacitance in the oscillation circuit. As the target approaches the sensors face the oscillation increase until they reach a threshold level and activate the output. In this chapter, we use capacitance sensor to detect plastic-related materials which is thrown by the public to the waste bins.

3.6 Flame sensor

Flame sensor detects presence of fire or flames inside the bins. We configured it on 760 −1100 nm wavelength and detection distance 100 −200 cm. Flame sensors works to minimize the risk associated with fire. There are different types of flame sensors available. In this waste management system ultraviolet flame detector sensor is used to detect fire in the waste containers. In addition, automatic fire suppression system is also attached in the dustbin or the waste container. The main advantage of UV flame detectors is they are capable of detecting explosions and fires within 4 ms with the time delay of 3−4 s.

3.7 Temperature sensor

This type of sensor is used to detect the level of temperature and humidity of the bins that helps in forecasting fire accidents.

3.8 Working process for blockchaining in waste management

In smart city, waste management operator of the corporation can place the smart dustbin or garbage collector into the city street at particular distance gap. Smart dustbin has the specialty to separate the waste, when people put the waste into the dustbin. Here, automatic waste sorting algorithm can be used for classifying the waste and send into the corresponding dust box inside the dustbin because smart dustbin contains one open and several dust box inside its open. The smart dustbin lid contains several IoT sensors. Each sensor is used for different purposes such as separating the waste type of plastic, metal and organic materials. When the dustbin is filled with the waste, lid ultrasonic sensors response quickly and informed as filling alert transaction block to the waste management operator, who is working for the corporation through blockchaining environment. So, all the nodes in the network also view filling alert transaction block. Filling alert transaction block contains dust box, dustbin, waste type, street name, city name, hash value, previous hash value, date, and timestamp. Then waste management operator validates alert message and also analyzes the request information and responds to the action immediately. After that, waste management operator allots optimized root map, waste type, and driver to the Transport Officer. When transport officer has viewed allotment for the collection process, he generates new collection transaction block to all the nodes in the network. If any new transaction is initiated and formed as block that will be updated in all the nodes of the network using blockchaining technology. Updated record ledger is distributed into all nodes. The transport officer allots particular truck for the collection process. The collection truck has the information such as collection waste type of smart dustbin, optimized root map to reach the smart dustbin, and truck weightage. When the truck reached target site smart dustbin, it collects the particular waste type from the bottom open of the bin. Based on the type of waste, truck will automatically dispose the waste into the corresponding waste buyer such as E-waste collector or input to plastic bank otherwise dumpster into the landfill area of the corporation. Waste collector collects the waste and provides the waste as input for the recycle factories. Variety of waste collector would have collect the waste from the truck they give money to the corporation, send the response transaction message to the nodes of the network. When transaction is validated, response transaction block is combined into the blockchaining ledger. That record is distributed into all the nodes of the network. Suppose the waste may be a dumpster waste, it disposed on landfill area. By using blockchaining technology and IoT sensors waste collected and disposed recycle into the proper and secure way. Figs. 9.3 and 9.4 show the result of smart waste management when the garbage has reached a particular level. The results of the experiment are shown above.

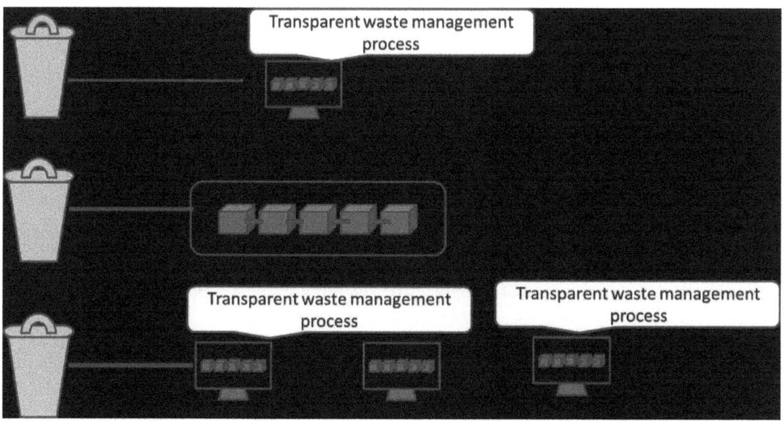

FIGURE 9.3 Transparent waste management using blockchain.

FIGURE 9.4 Experimental result.

4. Conclusion

Our proposed system is used to manage the waste with the help of IoT sensors and blockchaining technology. Ultrasonic sensors detect garbage level in the waste box. Then inductive proximity sensor which can be used to detect metal items and capacitive sensors are used to identify plastic and wood. Furthermore, blockchaining technology is used for proper sanitation. This chapter also gives the full idea about waste management in smart cities. Novel approaches like IoT sensors and blockchaining technology are used to detect waste fullness and sanitation of waste in a secure manner.

References

Abu-Mahfouz, A. M., & Hancke, G. P. (2018). Localised information fusion techniques for location discovery in wireless sensor networks. *International Journal of Sensor Networks, 26*(1), 12–25. https://doi.org/10.1504/IJSNET.2017.10007406.

Andrea, Z., Nicola, B., Angelo, C., Lorenzo, V., & Michele, Z. (2014). Internet of things for smart cities. *IEEE Internet of Things Journal*, 22–32. https://doi.org/10.1109/jiot.2014.2306328.

Anuga, A. G., Silva, J. B., Hancke, G. P., & Abu-Mahfouz, A. M. (2018). A survey on 5G networks for the internet of things: Communication technologies and challenges. *IEEE Access*, 3619–3647. https://doi.org/10.1109/ACCESS.2017.2779844.

Biswas, K., & Muthukkumarasamy, V. (2017). Securing smart cities using blockchain technology. In *Proceedings – 18th IEEE international conference on high performance computing and communications, 14th IEEE international conference on smart city and 2nd IEEE international conference on data science and systems, HPCC/SmartCity/DSS 2016* (pp. 1392–1393). Institute of Electrical and Electronics Engineers Inc. https://doi.org/10.1109/HPCC-SmartCity-DSS.2016.0198.

Lee, C. K. M., & Wu, T. (2014). Design and development waste management system in Hong Kong. In , *Internet of things', IEEE international conference on smart technologies and management for computing, communication, controls, energy and materials (ICSTM), 20172014 IEEE international conference on industrial engineering and engineering management, bandar sunway* (pp. 798–802). https://doi.org/10.1109/IEEM.2014.7058748.

Ogbodo, E. U., Dorrell, D., & Abu-Mahfouz, A. M. (2017). Cognitive Radio based sensor network in smart grid: Architectures, applications and communication technologies. *IEEE Access, 5*. https://doi.org/10.1109/ACCESS.2017.2749415, 19084–19098.

Penzes, B. (2018). *Blockchain technology in the construction Industry: Digital transformation for high productivity.* London: Institution of Civil Engineers (ICE). https://www.ice.org.uk/ICEDevelopmentWebPortal/media/Documents/.

Peters, G. W., & Panayi, E. (2016). Understanding modern banking ledgers through blockchain technologies: Future of transaction processing and smart contracts on the internet of money. *New Economic Windows*, 239–278. https://doi.org/10.1007/978-3-319-42448-4_13.

Pieroni, A., Scarpato, N., Di Nunzio, L., Fallucchi, F., & Raso, M. (2018). Smarter City: Smart energy grid based on Blockchain technology. *International Journal of Advanced Science, Engineering and Information Technology, 8*(1), 298–306. https://doi.org/10.18517/ijaseit.8.1.4954.

Pritchard, S. W., Hancke, G. P., & Abu-Mahfouz, A. M. (2017). Security in software-defined wireless sensor networks: Threats, challenges and potential solutions. In *Proceedings – 2017 IEEE 15th international conference on industrial informatics* (pp. 168–173). Institute of Electrical and Electronics Engineers Inc. https://doi.org/10.1109/INDIN.2017.8104765. INDIN 2017.

Sharma, P. K., Moon, S. Y., & Park, J. H. (2017). Block-VN: A distributed blockchain based vehicular network architecture in smart city. *Journal of Information Processing Systems, 13*(1), 184–195. https://doi.org/10.3745/JIPS.03.0065.

Sun, J., Yan, J., & Zhang, K. Z. K. (2016). Blockchain-based sharing services: What blockchain technology can contribute to smart cities. *Financial Innovation, 2*(1). https://doi.org/10.1186/s40854-016-0040-y.

Tasatanattakool, P., & Techapanupreeda, C. (2018). Blockchain: Challenges and applications. In *International conference on information networking* (Vol. 2018, pp. 473–475). IEEE Computer Society. https://doi.org/10.1109/ICOIN.2018.8343163.

Watanabe, H., Fujimura, S., Nakadaira, A., Miyazaki, Y., Akutsu, A., & Kishigami, J. (2016). Blockchain contract: Securing a blockchain applied to smart contracts. In *2016 IEEE international conference on consumer electronics* (pp. 467–468). Institute of Electrical and Electronics Engineers Inc. https://doi.org/10.1109/ICCE.2016.7430693. ICCE 2016.

Chapter 10

Introduction to blockchain and distributed systems—fundamental theories and concepts

Neha Sharma[1], Madhavi Shamkuwar[2], Sakthi Kumaresh[3], Inderjit Singh[4], Amol Goje[5]

[1]*Analytics and Insights, Tata Consultancy Services, Pune, India;* [2]*Zeal Institute of Business Administration, Computer Application and Research, Savitribai Phule Pune University, Ganesh Khind, Pune, Maharashtra, India;* [3]*M.O. P. Vaishnav College for Women, Chennai, Tamil Nadu, India;* [4]*Vara Technology Pvt Ltd, Pune, India;* [5]*Society for Data Science, Pune, India*

Chapter outline

Blockchain for Smart Cities. https://doi.org/10.1016/B978-0-12-824446-3.00002-8

1. Introduction

The speed with which the new technologies are being created due to fourth industrial revolution cannot be compared with any period in the history and is certainly exponential. The previous revolutions can be said to have evolved at a linear rate, with the first industrial revolution starting in 1765, followed by second in 1870 and third in 1969. The fourth industrial revolution happening right now is boosted by the emergence of Internet and digitization. Internet came in 1990s and unlike previous revolutions, within just 30 years a host of technologies like Internet of Things (IoT), Cloud Computing, etc., have emerged out of Internet itself. Hence, we can see more and more technologies supporting each other and being the reason for the emergence of some other technology. Technological revolution has placed the world at the threshold that will change the way the world is progressing. The change will be such that its scale, sophistication, and transformation will be unlike the world has ever seen before, blurring the lines between digital, physical, and biological spheres. Mankind can only vision what developments this technological revolution will unravel, but to leverage it to the maximum the global leaders in technology, governments, and business should join hands to come to a consensus.

Technology does miracles toward the business growth and societal development, but at the same time poses major risks associated with data flow, security, privacy, and integrity. In traditional system, generally system administrator is the centralized single trusted authority responsible for providing access rights to different users across the network. To make changes in the system is easy in this case as a single person is responsible for the same and changes done are irreversible. System controlled by a single person is thus susceptible to changes and vulnerable to risks in terms of privacy, integrity, and security. When a system is person oriented there may be lot of data adversaries and the system is more prone to single-point-of-failure. To overcome this, there is a need of technology which will eliminate the requirement of human intervention and related threat. Blockchain is a solution to the aforesaid problems where dependency on the centralized human authority is not required thus providing security from single-point-of-failure, along with other benefits like data immutability, reliability, and verifiability (Casado-Vara, Prieto, De La Prieta, & Corchado, 2018a; Dabbagh, Sookhak, & Safa, 2019; Reyna, Martín, Chen, Soler, & Díaz, 2018). The transactions performed with blockchain are executed, stored, and monitored on decentralized network infrastructure (Aste et al., 2017). The emergence of the blockchain technology initially was for financial transactions, but later it emerged as the most promising technology in multiple domains such as electronic voting, healthcare, education smart system, and certificate authorities (Henry et al., 2018).

The blockchain is the generic name for the database and transaction processing capabilities, which does ordinary things in amazing ways. Blockchain transactions are transparent, secure, and free, which encourages multiple applications to be built for financial and nonfinancial domain. Transferring the money without the need of third party is the most popular application, e.g., Bitcoin (Dwyer, 2015). Implementing the contracts in neutral and unbiased way for achieving ethical practices in business is possible through another blockchain use case called Smart Contract. Blockchain can be used to store the identity and accomplishments of the individuals in secured way so that it cannot be altered or stolen. It can also be used to store vast information in encrypted manner on distributed network and save on charges for cloud storage, which is heavy as we pay to the intermediaries, i.e., cloud owners for using the services. Blockchain also keeps the proof of provenance which helps in tracking the origin of the record and presents the credible record of the whole process (Becker et al., 2013; Peters & Panayi, 2016) (Hawlitschek, Florian, Benedikt, & Timm, 2018; Sara, Mahtab, & Joseph, 2018). Blockchains can become a very useful base to provide communication within a certain company or system, by creating a platform called ADEPT (Autonomous Decentralized Peer-to-Peer Telemetry), which implements blockchains to form a base for cheap autonomous communication between all machines involved. In the other words, it is a successful attempt to create a blockchain-based Internet of Things (a system of electronic devices, which are connected to the Internet, and interconnected among themselves) (Utilization of Blockchain Technology to Overthrow the Challenges in Healthcare Industry et al., 2015).

In this chapter, the endeavor is to systematically present all the concepts that have been adopted in a blockchain to make it a disruptive technology of this era. The next section focuses on the current research status with state-of-art review, highlighting systematic literature review and evolution of blockchain. Section 3 appraises the blockchain concepts like operations, networks, features, and consensus mechanisms, whereas Section 4 discusses types of blockchain and various platforms. Sections 5 and 6 present various interesting application and challenges, respectively. Finally, this chapter is concluded with probable future of the technology.

2. Literature review

This section is further divided into two subsections. First subsection presents a systematic literature survey, whereas second subsection presents the journey of evolution of blockchain technology.

2.1 Systematic literature review

The first step toward literature study is to identify related articles and carry out a systematic literature review. The authors of this chapter were keen to know the origin of blockchain concepts and its reflections in various articles types. From the study of research database IEEE explore, Science Direct and J-Gate, it is evident that the actual publication of the concept has emerged in the year 2013 (IEEE, nd). The concept has gained popularity over the years and same is depicted in the increased number of (Fig. 10.1) publications with time as depicted in Fig. 10.1.

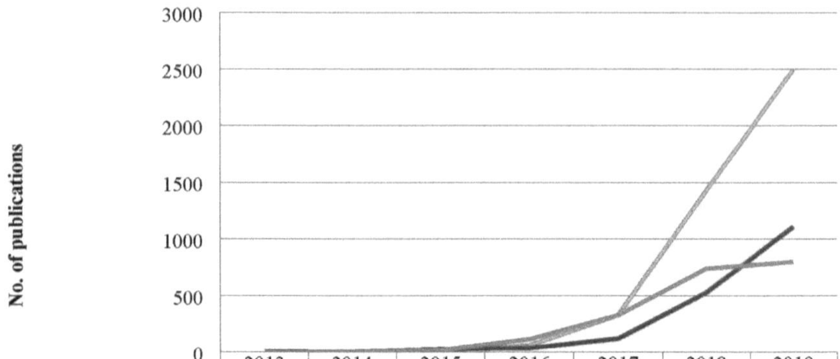

	2013	2014	2015	2016	2017	2018	2019
IEEE explore	2	1	14	61	330	1429	2495
ScienceDirect	5	4	27	36	122	528	1107
J-Gate	0	0	19	116	328	739	799

FIGURE 10.1 Year-wise publications in research database IEEE explore, Science Direct and J-Gate (IEEE, nd; ScienceDirect.com | Science, health and medical journals; J-Gate – Largest e-Journal Gateway | Journal Finder | Journal).

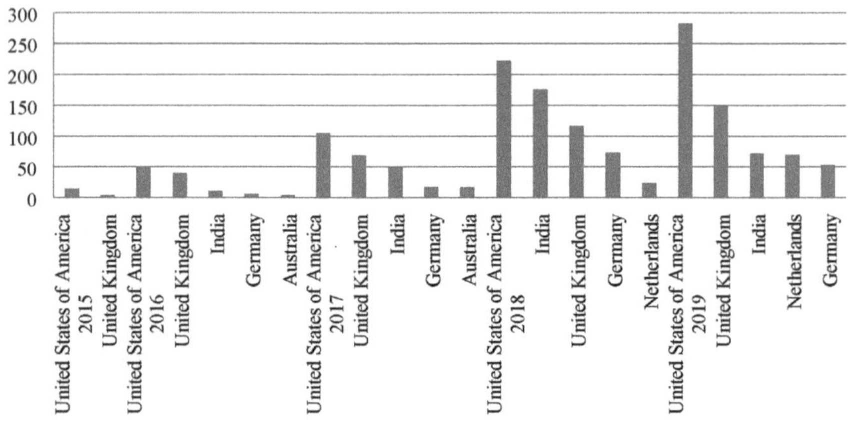

FIGURE 10.2 Country-wise publications in research databases J-Gate (J-Gate – Largest e-Journal Gateway | Journal Finder | Journal).

The authors were also curious to know the country-wise publications related the "blockchain" concept and hence research databases J-Gate and IEEE explore were explored as shown in Figs. 10.2 and 10.3.

The Gartner hype cycle shown in Fig. 10.4 was published in the year 2019 (Gartner: Global Research and Advisory Company). It clearly suggests that blockchain technology is emerging and will reach its maturity not before next decade. So, it holds lots of opportunity to experiment and innovate.

Finally, to know the research interest of community toward blockchain technology, Google worldwide trends were studied for last 17 years, i.e., from year 2004 till current date (Google Trends). The trend shows rise in the search of the term "blockchain" over a period of time as clearly evident in Fig. 10.5.

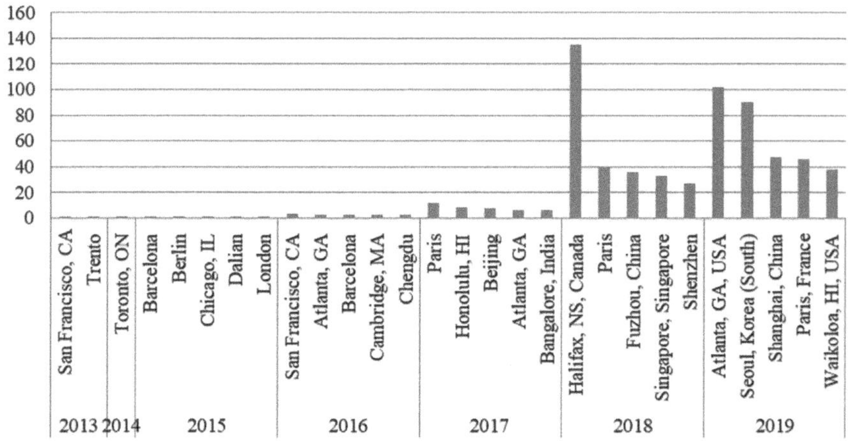

FIGURE 10.3 Country-wise publications in research database IEEE explore (IEEE, nd).

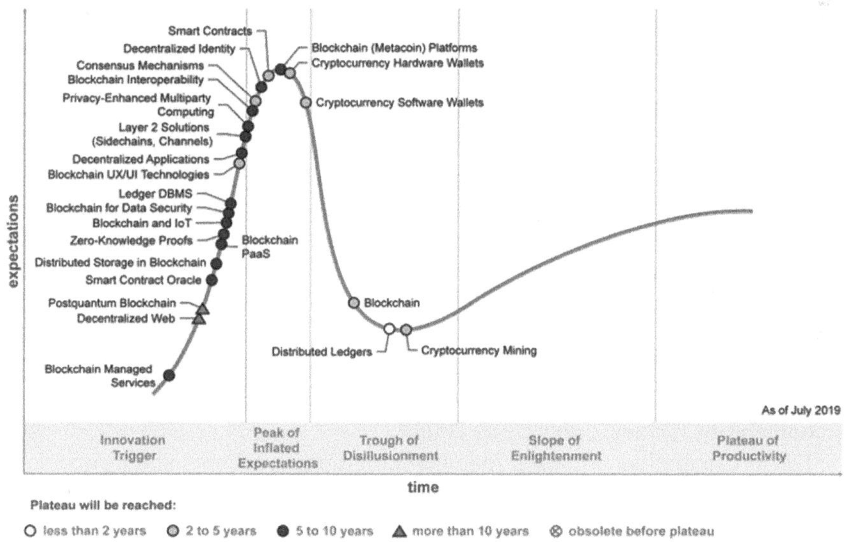

FIGURE 10.4 Blockchain and Gartner Hype cycle (Gartner: Global Research and Advisory Company).

2.2 Evolution of blockchain

Blockchain technology has evolved immaculately since early 1990s and is being successfully applied from financial to nonfinancial applications like manufacturing, sales, education, etc. In this section, the evolution of blockchain is presented systematically.

FIGURE 10.5 Google Worldwide trends for search term "blockchain" from 2004 till July 2020 (Google Trends).

It is very important to know from where you have come, in order to appreciate where you are and where do you want to go? The term "blockchain" came into light for the first time when a paper titled "Bitcoin: A peer-to-peer electronic cash system" was published online in 2008 by an anonymous person or group of people known as Satoshi Nakamoto (Nakamoto, 2008) and then with the implementation of Bitcoin in 2009. The paper revolutionized the way data are stored and transactions are done. The data are packaged in the block and block is added to the chain after approval. Once the block is there on the network, it cannot be altered. The change can be reflected only by adding new block to the chain.

Bitcoin or blockchain did not start in 2008 or 2009, but is a part of larger movement. This is a technological solution to a social and political problem, which will continue to evolve in a greatest possible way. However, we cannot deny their origins. The philosophies from various electronic cash or digital cash paved the way for cryptocurrencies in general and bitcoin in particular. The movement from cash to digital started around 1990s. The fundamental requirement from any e-cash system is that it should address anonymity and accountability (Danezis and Diaz, 2008).

David Chaum has proposed many cryptographic concepts and protocols in his contributions from 1981 to 1990, like blind signature where the content is blinded before signature, anonymous communication network which was also called mix network where message and sender are encrypted that can be decrypted by the receiver for whom the message is meant, and secret sharing which is about detecting the use of same e-cash token twice (double spending), that built up the foundation for blockchain (D. Chaum, 1985; D.L. Chaum, 1981; Danezis & Diaz, 2008; Nakamoto, 2008). He is the originator of digital cash system, which took care of two fundamental requirements from any e-cash system and addressed anonymity and accountability. David Chaum found a first electronic cash company known as "Digicash" in 1990 and the first virtual currency was created in 1995 (Greenberg, 2012). The US-based bank located at Missouri, Mercantile bank, which acquired the Mark Twain

Bank, was the first and only US bank to support DigiCash systems. Deutsche Bank, based in Germany, was the second backing bank of DigiCash systems (DigiCash's eCash™ to be Issued by Deutsche Bank - David). Digicash was ahead of its time and focuses on making transactions anonymous; however, the problem for solution offered was not evident as online trades and e-commerce were not popular. Chaum stated in an interview in 1999 that the DigiCash project, and its technology system, entered the market before e-commerce was fully integrated within the Internet [26]. In 1998, DigiCash filed a Chapter 11 bankruptcy and in 2002 the company was sold for assets to eCash Technologies, another digital currency company, which was acquired by InfoSpace on Feb. 19, 2002. (Pitta, 1999) (A Cypherpunk's Manifesto - Activism.net).

Eric Hughes a mathematician from University of California, Berkeley; John Gilmore a computer scientist who was Sunmicrosystems' fifth employee, Tim May, a retired businessman who worked for Intel and many others, came together in late 1992 to form a group called "Cypherpunks." The group published a "Cypherpunk Manifesto" which was written by Eric Hughes in 1993 highlighted the basic idea of the movement. The manifesto says that "Privacy is not Secrecy" and to achieve privacy the society requires anonymous transaction systems and cryptography as well as it should be part of social contract. Privacy is about revealing the identity only when it is desired (Back, 2002).

The next attempt was made in 1997 by Dr. Adam Back, who developed Hashcash system (AALWA: Ask any LessWronger anything - LessWrong). The system is also known as proof of work system which uses hash cryptography and allows anonymous communication basically to control email spams and denial-of-service attack. Every email is added with a hashcash stamp as a header to prove that substantial amount of CPU time and computational power is used to calculate the stamp. The idea is to deter and make sending spam mails difficult and expensive. The header is prepared by the sender who initializes a random number and appends a counter to the header and computes 160-bit SHA-1 hash of the header. The header is acceptable if first 20 bits, i.e., 05 most significant digits are all 0s, else the counter is increased and hash algorithm is applied again. Only 2^{140} hash values satisfy the criteria out of 2^{160} possible hash values, therefore probability of a random selected header to have a hash with 20 leading 0s is one in 2^{20} or approximately one in million and approximately one second of time is needed to compute the hash. A legitimate email sender would not mind spending this much time and computational power to generate a hash; however, this is significantly uneconomical for a spammer who has to send large number of mails. The computation at receiver's end is pretty simple and has to just calculate 160-bit SHA-1 hash of entire string received and takes 2 μs in 1 GHz machine. The recipient computer checks date, email, and hash string to validate the mail received. Hashcash system is an efficient brute force approach to find a valid header and is a basis for mining algorithm used in blockchain.

Wei Dai, who is a computer engineer and a cypherpunk, published an article on B-Money—an anonymous, distributed electronic cash system, in 1998, which ignited interest in cryptocurrencies (Ridley, 2017; Decoding the Enigma of Satoshi Nakamoto and the Birth of ...; Bitcoin: The Cryptoanarchists' Answer to Cash - IEEE Spectrum). His paper on b-money suggested many core concepts that are used in cryptocurrencies, and more specifically bitcoin like

1. Proof of Work for every transaction, i.e., amount of time and computational power needed.
2. The transaction is verified by a group of people/community who update a combined ledger book, and they are incentivized for their efforts and being honest. This is called as proof of stake (PoS), where the group of people invests their money in a special bank and will lose all their stakes if they attempt to process any fraudulent transaction.
3. All the transactions and fund exchanges are maintained in two ways, i.e., every participant maintains a separate ledger regarding the money that belongs to them and a collective bookkeeping is maintained by group of people. The transactions and fund exchanges are authenticated with cryptographic hashes.
4. Contracts are enforced through the broadcast and signing of transactions with digital signatures (i.e., public key cryptography).

In the same year, i.e., 1998, Nick Szabo, a computer scientist, cryptographer, and a legal expert, designed a concept of decentralized digital currency called "bit gold," which was never implemented but played a vital role in the architecture of a bitcoin (O'Leary, 2015; Peck, 2012; Tschorsch & Scheuermann, 2015). Bit Gold introduces the concept of timestamping and creation of money using hashcash. Computer power is dedicated to solve cryptographic puzzles by the participants and the solved puzzles are sent to byzantine fault-tolerant public registry and a public key of solver is assigned to it. Each solution would become part of the next challenge, creating a growing chain of new property. This aspect of the system provided a way for the network to verify and timestamp new coins, because unless a majority of the parties agreed to accept new solutions, they could not start on the next puzzle (Szabo, 1998). However, there were certain features which were missing in bit gold like incentives to keep nodes honest, a way to keep tokens fungible (no agreed way to set difficulty, one token might be made with significantly more difficulty than the other).

As can be seen, various people from across the world have been working tirelessly on the blockchain technology and cryptocurrencies since the 1990s and there have been multiple attempts to solve the complex issues surrounding cryptocurrency, by arguably some of the most brilliant minds in this space.

3. Blockchain concepts

A blockchain is a list of transactions or a record that is kept up by a network of clients, instead of a focal specialist. It is referred to as blockchain in light of the very fact that new exchanges are prepacked into "blocks" of information and composed onto the finish of a "chain" of prevailing blocks depicting every earlier exchange. The blocks are linked by joining the hash of the earlier block to the recent block, the recent block to the subsequent block, and so on. Successive nested blocks undertake transactions in a serial order, thus a transaction cannot be changed antedated without changing its block and all the following blocks. A subchain of the block is regarded only if it is larger than a competitor chain. Honest nodes should produce fresh blocks quicker than an opponent's attempt to do so. The scheme delivers this efficiency by implementing a proof-of-work system, which testifies that a node has invested the required funds to do a computationally challenging job.

A standard blockchain is an outstanding structure of information that stores historical statements and transactions. All nodes in the framework are in agreement with the operations and their order. In blockchain, the records are distributed in different blocks and each block is connected by a cryptographic pointer to its predecessor, and way back to its initial block. The basic structure of the blockchain is shown in Fig. 10.6, where each block has two main sections: block header and block body. Each block header has a unique identifier in the form of hash which is generated using hash algorithm. Apart from hash, the other components of block header are index, previous hash, number of transactions in the block, timestamp, and nonce as shown in Table 10.1. The body of the block is a collection of all transactions stored in various arrays and implemented using a data structure called Merkle tree. Hence, every block in the network is connected altogether, forming a chain of integrity. Transactions in an open blockchain are publicly visible, while the view is restricted in the case of personal blockchain or a blockchain consortium.

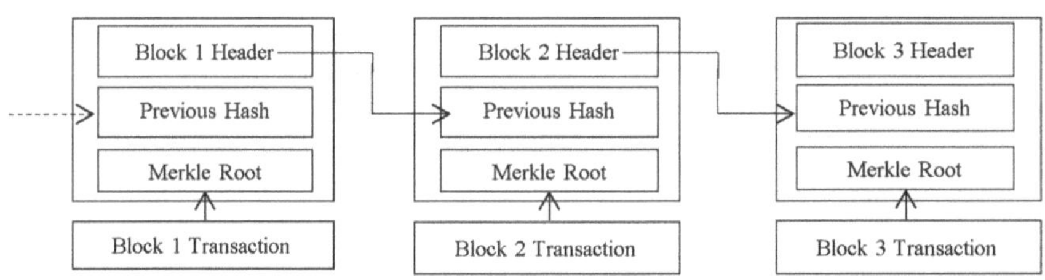

FIGURE 10.6 The structure of a blockchain.

TABLE 10.1 The component of the block.

Name of the component	Description
Index	It is a unique sequential number that informs about the block position in the blockchain. The index of the first block is "0."
Hash	It is a cryptographically generated unique number, mainly used for quick identification of data.
Previous hash	It is the hash of the preceding block.
Time stamp	It stores the time of block creation.
NumTx	Number of transactions embedded in the block.
Nonce	It is the number which is used only once in the mining process.
Transaction	It is a data record that has to be saved on the blockchain.

3.1 Blockchain operations

As discussed earlier, each block consists of some value, like the hash of the block and also of the previous block. The value that is kept within a block is based on the kind of blockchain. For instance, Bitcoin blockchain stores the information of transactions, like the sender, receiver, and quantity of coins. The mining process ensures the validity of the transaction (Gatteschi et al., 2018). A hash of the block is like a fingerprint that uniquely identifies the block and all its content. The hash is computed, once the block is formed. Altering anything in the block would result in changed hash, which means it is no longer the same block. Now, this block with new hash value needs to be updated to the local copy of blockchain of every node in the network. So, that blockchain architecture maintains a strong consistency among nodes in the network.

A block is a container data structure with average size of 1–8 MB, and is large enough to handle huge number of transactions. The structure of block consists of two components such as block header and list of transactions. Block header consists of metadata about block such as previous block hash, mining statistics, and merkle tree root. Previous block hash is used to create the new block hash as represented in Fig. 10.7 (How does a blockchain work - Simply Explained? — Steemit). If all the transactions in a block are arranged in a tree form, then it is called as merkle tree. In a merkle tree, every leaf node consists of the hash of the transactions, and the intermediate nodes consist of the combined hash values. The structure of merkle tree is presented in Fig. 10.8. At the root level, the merkle tree contains the combined hash of

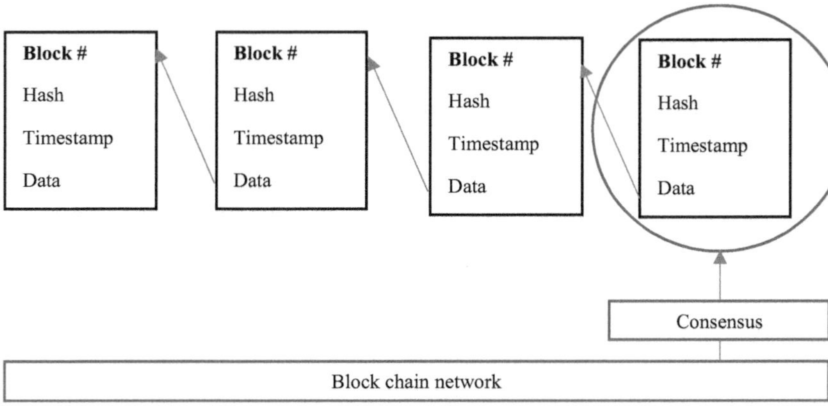

FIGURE 10.7 Blockchain operations.

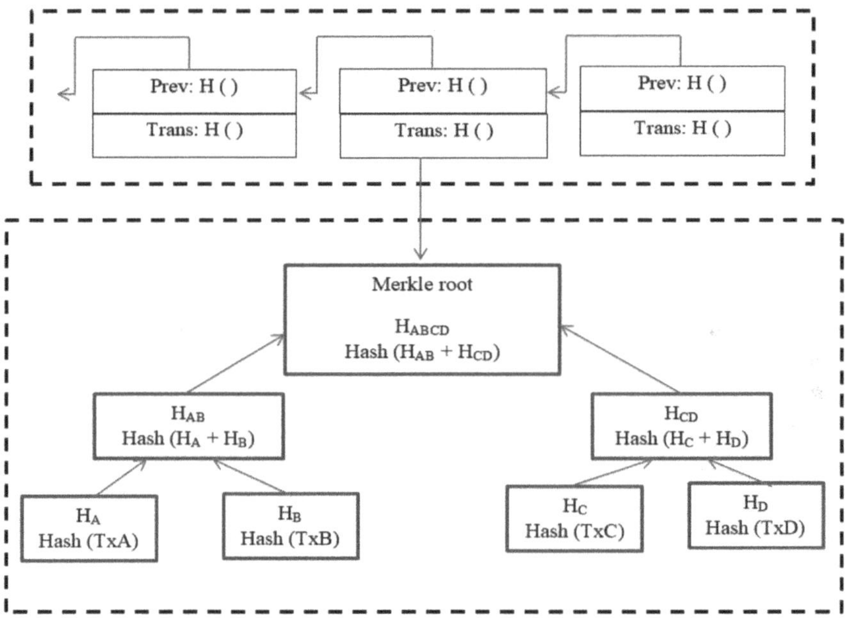

FIGURE 10.8 Merkle tree implementation of a blockchain.

the right tree and left tree. If anyone wants to change the transaction, the hash will get changed and at the same time the changes will be reflected in the root hash as well. If the root hash changes, the hash of the linked blocks will also change, which will lead to a tamper-proof architecture (Computer Science and Engineering - NOC … - NPTEL).

Figure seven depicts an example of a blockchain which has a series of four blocks and every block features a hash along with the hash of the previous block, which binds them in a chain. The first block is known as the genesis block as it does not direct to any previous blocks. Now, if the second block gets tampered, the hash value of the block will get changed. Therefore, the hash value of the second block will not match with the previous hash value of the third block. This will result in making block three and every other block invalid as the blocks do not contain a legitimate hash of the previous block. So, altering one block can make the subsequent blocks invalid (conda).

However, utilizing hash values is not enough to stop tampering, as computers are powerful and very fast to compute tons of hash values per second. In case of any tampering, it can compute all the hash values of other blocks to form the blockchain valid once again. So, in order to eliminate tampering, a concept of proof of work (PoW) is used, which is a procedure that slows down the formation of latest blocks. In case of Bitcoins, it generally takes 8 to 10 minutes to compute the desired PoW and add a brand-new block to the chain. This procedure makes alteration with the blocks, extremely difficult. As a result, if a block is altered, the PoW will have to be recomputed for all the subsequent blocks.

There is another method which makes blockchains secure and that is by being distributed. Whenever any user joins the network, he will get the complete copy of the blockchain. The node will utilize this to check that everything continues to be in the sequence. Now, as a new block is generated, each and every node in the network receives a copy of it. The nodes then check the block to ascertain that it is not been altered or modified, and then include this block to their own network and produce an agreement regarding valid and invalid blocks.

3.2 Blockchain network

The "blockchain" is the core principle behind Bitcoin digital currency (Hayden & Choi, 2019). It is decentralized, distributed technology that collects a digital record of any event and store it in a distributed database that is shared among all the users connected with it. Every transaction in a blockchain is verified by the majority of the connected nodes of the system. Once information is inserted it cannot be erased or altered. Bitcoin which is the decentralized peer-to-peer (P2P) digital currency medium is among the list of majorly accepted and favored implementations of blockchain technology.

To understand decentralized, distributed P2P network which is the backbone of blockchain, there is a need to grasp the concept of various networking systems, its functioning as well as its pros and cons. The network system is like a hierarchy in government administration or client—server architecture in database management systems. The basic difference in various network systems is the concept of authority. There are some common definitions of authority, like

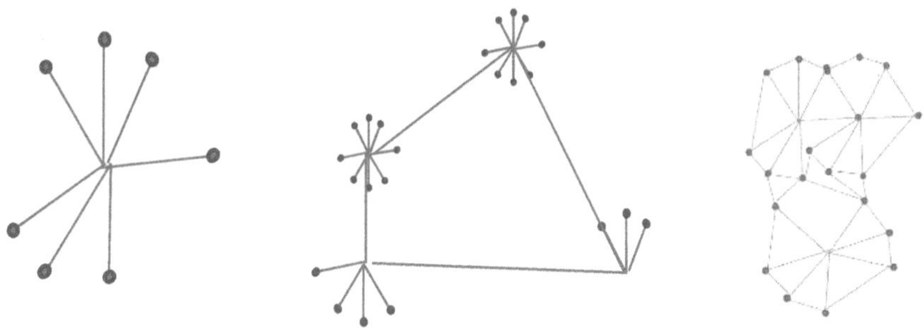

FIGURE 10.9 Types of network systems—(A) centralized, (B) decentralized, (C) distributed.

the right to give commands, be in-charge, make decisions, and enforce obedience in the organization or person, having control in a typically administrative sphere and the power to influence others. Therefore, on the basis of center of authorities, the network systems are divided into three types: centralized, decentralized, and distributed network systems as shown in Fig. 10.9.

3.2.1 Centralized system

Centralized systems have just one point of authority and this is where all the control is concentrated. All the processes and decisions are carried out at the single location, which makes the system extremely susceptible to collapses, meaning any disruption at central server can collapse the whole system.

Advantages of Centralized System:

1. It is very easy to implement and helps in quick decision-making. Since single authority takes the decision, there is no consensus required.
2. With just one single authority, there is no chance of multiple authorities performing the same function and this reduces the economics of scale.

Drawbacks of Centralized System:

1. There is dependence on one single authority for data and security. The system is unsafe as any attack on the sole authority shall destabilize the system and might lose primary source of information.
2. It's bureaucratic in nature and thus it adds many layers and hierarchies.
3. It is not transparent and thus prone to fraud. The lack of transparency will induce irresponsibility at the top level since that is where the power is concentrated. Some examples of centralized network systems are a banking system, food franchises like McDonald's or Silversea.

3.2.2 Decentralized systems

The decentralized systems have several points of authority and thus power is more diversified. This makes the system less susceptible to downfalls as the fall of a single point of authority would not bring down the entire system. The hierarchy is thus flat compared to a centralized system.

Advantages of Decentralized System:

1. The decisions are now made much closer to the consumer. Thus, the decision-making authorities have more information about the end users.
2. Second the system is less susceptible to collapses because now there are multiple points of authority. The downfall of one point is not going to bring about a destabilization of the system as we saw in centralized systems.

Drawbacks of Decentralized System:

1. There are still the economies of scale as with multiple authorities the duplication of task problem persists.
2. Although they are safer than centralized system, they are still susceptible to collapses. Thus, they are not completely safe.

A few examples of these are supply chains like Johnson and Johnson's.

3.2.3 Distributed systems

In distributed systems, everyone has an equal authority which makes the system virtually unsusceptible to collapses. However, this does not guarantee that the system is safe from being hacked unless and until at least 50% authority is with the system. So, a distributed network has a flat hierarchy.

Advantages of Distributed System:

1. The biggest benefit that distributed systems introduce is that it removes the need for intermediaries.
2. The network is economical and unviable to bring down, which makes it extremely secure and safe system.
3. The complete transparency in the system prohibits fraud.

Drawbacks of Distributed System:

1. These systems are still new and the technology is still nascent. So, it will take a lot of time to stabilize.
2. There will be economies of scale with time; however, initial implementation of the system is very expensive.

A few examples of these are cryptocurrencies and blockchain networks.

3.3 Features of blockchain

A blockchain technology has many features and some of them are listed below:

(a) **Decentralized:** Each transaction in the centralized systems requires to be validated through the central trusted agency (e.g., the central bank), inevitably resulting in the cost and the performance bottlenecks at the central servers. Hence, instead of a single source, multiple participants are considered who agrees upon the transaction with the help of consensus algorithm (Croman et al., 2016). Besides, a transaction in the blockchain network can be conducted between any two peers (P2P) without the need of authentication by the central agency. In this manner, blockchain can significantly reduce the server costs (including the development cost and the operation cost) and mitigate the performance bottlenecks at the central server (Zheng et al., 2018).

(b) **Security:** The transactions in blockchain are encrypted and linked with the previous transaction. Therefore, the criminal activities like hack, fraud, or unauthorized access of critical information can be prevented (Top five blockchain benefits transforming your industry - IBM).

(c) **Immutable:** A blockchain follows the append-only ledger principle which makes it an immutable technology. If information in one block is changed, it will have a ripple factor, i.e., it will demand change in subsequent blocks making the overall working infeasible (Eyal & Sirer, 2014; Nakamoto, 2008). In this immutable ledgers, the consensus of the participants is verified, thus the transactions log created cannot be altered or replaced (Lemieux, 2016).

(d) **Distributed Ledger:** The blockchain shares undefined copies of all data. Participants approve data independently without an authority being integrated. The remaining nodes continue to work, even if one node fails, ensuring no disturbance (Lavanya, 2018).

(e) **Scalability:** The block rate consists of throughput and information propagation time. However, it depends on the consensus algorithm and the number of participants involved in the transaction. The large number of participants will lead to limit the desired high throughput (Croman et al., 2016). In blockchain every node holds the information of other blocks, so scalability becomes crucial when the data size is large (THE EUROPEAN UNION BLOCKCHAIN OBSERVATORY).

(f) **Persistent:** Every transaction expanding across the network has to be confirmed and registered in blocks shared within the network, which makes it infeasible to tamper. In addition to this, every broadcasted block will be verified by other nodes and help in identifying the spam, if any.

(g) **Anonymity:** All users can make communication with the blockchain network through a generated address. However, a user could also create more than one addresses to mitigate the exposure of identities. There is no centralized entity storing user's personal data. This characteristic preserves the privacy of the transactions. But to be noted, blockchain cannot ensure the 100% authorization due to the underlying limitations.

(h) **Auditable:** Since every transaction in the blockchain is attested and registered along with a timestamp, users can check and discover the past records with so much of ease. Users can do so through accessing any node in the distributed network. In Bitcoin, all the transactions within the network could be discovered to earlier transactions repeatedly. This increases the transparency of the data kept inside the blockchain.

(i) **Transparency:** Blockchain is a distributed ledger in which all the participants in the network share the same information. Hence, transactions become more transparent.

(j) **Digital:** Entire information in blockchain is digitized disposing of requirement for common documentation (Blockchain – revolutionary change or not? Cross sector use ...).

3.4 Different blockchain consensus mechanisms

Consensus mechanisms are protocols that make sure that all the nodes are synchronized with each other and agree on transactions which are legitimate and are added to the blockchain. These consensus mechanisms are crucial for a blockchain in order to function correctly and to minimize the risk of different types of attacks. They make sure everyone uses the same blockchain. Everyone can submit the data to be added to the blockchain; therefore, it is essential to check the transactions constantly and to perform continuous audit of the blockchain by all nodes. There are many ways to reach consensus, so this section discusses the most popular ones.

3.4.1 Proof of work

Proof of Work (PoW) is the first blockchain consensus mechanism and was first used by Bitcoin. However, this has been adopted by many cryptocurrencies successfully. The PoW creates a new block with the help of mining process which is carried out by the few special nodes in the network, which are known as miners. Each miner competes with other miners to solve complex mathematical puzzles and get rewarded. However, it is important to note that solving the puzzle requires enormous amount of computational power as there is no way to find the solution except for continuously guessing it till the time we get the right answer, which might take lots of time. More computational power is needed to solve the puzzle in faster way, making the process further expensive. Nevertheless, verifying the correctness of the solution is extremely simple, which makes PoW process very popular.

3.4.2 Proof of stake

Proof of Stake (PoS) works on a philosophy that the miner who owns most coins in the blockchain network holds the authority to validate or mine the new transaction of a block. In a system that uses PoS, a randomized process is used to determine the miner who gets to produce the next block. Users can stake their tokens to become a validator (someone who can produce blocks), which means they lock their tokens up for a certain time. After doing so they are eligible to produce blocks. The factors affecting the process of deciding the next miner to produce the next block depends upon the design of the blockchain, maximum number of coins, and also the duration of coins being staked.

PoS is much more energy efficient than the PoW system. In a PoW system, a miner may not own any coins they are mining, which means they only seek to maximize their profits without actually improving the network, whereas in a PoS system, validators have a much bigger incentive to maintain the network as they actually hold the coins of the blockchain which they are validating.

3.4.3 Delegated Proof of Stake

Delegated Proof of Stake (DPoS) is one of the fastest consensus protocols where stake-weighted voting system mechanism is implemented for digital democracy. In a DPoS-based blockchain system, miner (participants/stakeholders) can vote for a certain number of delegates by staking their coins and the weight of their vote depends on their stake. Hence, the stakeholders delegate or outsource their responsibility to a third party. For instance, if two miners stake 10 and 20 coins on two different delegates, respectively, then the vote of second miner weighs double than the vote of first miner. So the authority to produce the block goes to the delegates that receive the highest votes and they are rewarded for the same. The number of delegates that gets a permission to create blocks depends on the blockchain design.

3.4.4 Proof of Capacity

The next consensus mechanism is Proof of Capacity (PoC) that makes use of a process called plotting, where solutions are prestored. In PoW process, massive computational power is needed by the miners to guess the correct solution, whereas in PoC process, the digital storages like hard disks are used to prestore the solutions, and this process is called plotting. Once the storage has been fully plotted with solutions, then the particular miner can take part in the process of block creation. However, the miner who solves the puzzle first, gets the opportunity to create the new block. In PoC process, the emphasis is on storage capacity, as bigger storage allows stocking more solutions and thus increasing the chance for creating a block.

3.4.5 Proof of Elapsed Time

The last consensus mechanism is Proof of Elapsed Time (PoET), which is a random process to decide the miner who will create the next block on the basis of waiting time. In this process, a random wait time is assigned to each node and the opportunity to create the new block is given to the node which completes its wait time first. PoET can work effectively if there exist a system to authenticate that multiple nodes are not run simultaneously and assignment of wait time is genuinely random.

4. Blockchain taxomony

In this section, different types of blockchains and various platforms have been discussed.

4.1 Type of blockchains

Depending on the requirements of applications, the blockchain can be broadly categorized into four types as shown in Fig. 10.10:

1. Public Blockchain
2. Private Blockchain
3. Consortium or Federated Blockchain
4. Hybrid Blockchain

4.1.1 Public blockchain

Public blockchains are publicly accessible and has no restriction on who will take part, share, or be a validator. In this blockchain, nobody has the full

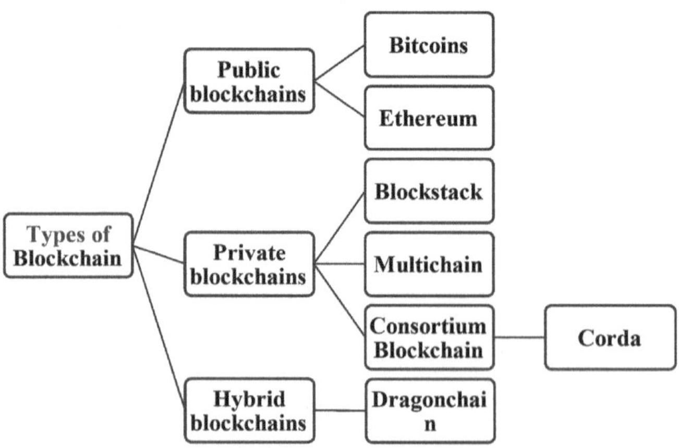

FIGURE 10.10 Types of blockchain.

control over the network but everybody can read or write. This characteristic ensures the security of data and helps to achieve immutability, as it becomes impossible for single individual to deceive this type of blockchain. The control on the blockchain is equally distributed among each node in the network and hence is known as thoroughly distributed and fully decentralized network. They are majorly used for creating cryptocurrencies like Bitcoin, Litecoin, Ethereum by blockchain companies like Factom and Blockstream.

4.1.2 Private blockchain

Private blockchains are also referred as permissioned blockchain. It has certain limitations on who can access it and participate in the transactions. Only selective entities are permitted to access it, rather it can be said that only owner being a centralized authority can access it or grant permission to make changes in it. The roles and access rights are defined at the time of building the blockchain application; however, network administrator can take care of assigning the rights to new user or to revoke it from an already created user. These blockchains are majorly implemented in private companies to efficiently manage the confidential data that should be accessible or available solely to the authorized person or people within the organization. Due to the very fact that private blockchain is nothing but a closed blockchain, the information is restricted within the organization's sphere and it is completely inaccessible or unattainable to any unknown entities. Eris Industries, Blockstack, and Multichain are few example of private type of blockchain [53].

4.1.3 Consortium blockchain

Consortium blockchain is a type of private blockchain and is governed by a group rather that a single entity. It is a collaborative model that brings group of organizations that work and compete with each other. This type of blockchain can be more efficient as some rights can be kept restricted to individual like data access rights in order to keep it secure from public access, whereas group can collaborate on some common aspects of the business. There can be two types of users of such blockchain:

i. The users who have full charge over the blockchain and who can select who should have the rights to access the blockchain.
ii. The users who will simply access the blockchain, like participants from bank, government, etc.

4.1.4 Hybrid blockchain

A hybrid blockchain draws its feature from public as well as private blockchain. The permission-based structure of a private blockchain system is combined with the security and transparency of a public permissionless system. This makes the hybrid blockchain very flexible for the users, who can effortlessly join a multiple public blockchains from private blockchain. Usually, in a hybrid blockchain, the transaction of a private network is verified within that network. However, it can also be verified in the public blockchain by increasing the hashing function with the help of more nodes to improve the transparency and security of the blockchain network. A popular example is Dragonchain.

4.2 Major blockchain platforms

The abovementioned types of permissioned and permissionless blockchains can be used to develop different interesting applications in different platforms discussed in this subsection.

4.2.1 Hyperledger

Hyperledger is an open-source project that aims to create a suite of tools for enterprises to deploy blockchain technologies quickly and effectively. The protocol is commonly used in blockchain software solutions because it comes with its libraries that help to speed up development. The Linux Foundation is a strong supporter of Hyperledger, and it has supplied significant expertise to accelerate the creation of the protocol. Hyperledger is also highly compatible with Linux, so it is designed to work effectively on the same servers that are widely used in today's business world.

4.2.2 Ethereum

Ethereum was proposed by a Russian–Canadian coder, Vitalk Buterin in 2013. It is an open-source and distributed computing platform, mainly used for executing smart contracts. An Ethereum platform has a run-time environment, Ethereum Virtual Machine (EVM), which needs to run on each node. It is essentially a permisssionless platform that works on Proof of Work which makes it very slow. Ethereum ecosystem is fueled with the help of native cryptocurrency called Ether. Anyone who wishes to uses the platform for running an application and executing the transactions has to pay in Ethers.

4.2.3 Ripple

Ripple platform was discovered in 2012 to enable global payments by connecting banks, digital asset management company, payment providers, corporates, etc., using RippleNet. It is an open-source distributed consensus ledger that uses the native currency called Ripple or XRP that achieves the consensus between nodes using probabilistic voting. Ripple is faster and scalable in comparison to other blockchains as it is built on advanced technologies. It has three prominent applications like xVia to send payments across multiple networks, xRapid to reduce liquidity costs, and xCurrent for cross-border payments. Many industry giants are testing Ripple to integrate it to their existing payment system.

4.2.4 IOTA

Internet of Things Application (IOTA) was founded by David Sønstebø, Sergey Ivancheglo, Dominik Schiener, and Dr. Serguei Popov of Germany in 2015. It is a decentralized currency which is developed on a futuristic technology known as Internet of Things and is not based on blockchain technology. Cryptocurrency based on blockchain technology is generated by miners by solving complex puzzle, whereas IOTA eliminates this transaction cost as there is no associated fee. An IOTA transaction is executed if the user verifies other two random transactions in this decentralized environment ensuring a faster network. IOTA is different than existing blockchains as it uses a blockless distributed ledger called Tangle and incorporates direct acyclic graph (DAG) feature. IOTA coin cannot be purchased with the traditional cash and can only be bought in the exchange of other cryptocurrencies like Ethereum and bitcoin. It's a global trend to maximize the sharing of resources like tools, appliance, drones, storage devices, processors, eBikes, wifi bandwidth, etc., instead of keeping it idle. IOTA coins help to lease and share such resources and also can be used profusely for applications toward e-voting, e-governance, masked messaging, etc.

4.2.5 Chain core

Chain core is a blockchain infrastructure which uses a chain protocol to implement a permissioned network. It is mainly used for issuance and transfer of assets, especially related to finance like derivatives, gift cards, securities, loyalty points, and coins. The developer edition of this blockchain is free; however, the commercial edition is available for the organizations which want to run it in production environment. The process of asset creation, control, and transfer among the participants is decentralized, but the network is mainly governed by designated set of entities known as a federation.

4.2.6 Corda

Corda was developed in 2015 by an association of world's leading financial institutions called R3. It is a blockchain platform that uses a smart contract protocol to record, supervise, and synchronize the financial agreements between regulated financial institutions by removing costly frictions in business transactions. There is no built-in token or cryptocurrency in corda and is an open-source permissioned blockchain, where only approved participants are allowed to access the data, which certainly boosts privacy and offers fine-grained access control to digital records. It was initially designed for the financial sector by the consortium of financial industry; however, now it has multiple applications in trade, supply chain, healthcare, government sector, etc. As per the record, more than 60 firms including front runners use Corda blockchain platform and very heavy investments around 107 million USD into Corda.

5. Applications of blockchain

There are many applications that are built using blockchain technology for transforming society.

(a) Smart Contracts: It is a computer protocol or a program which controls the transactions of digital money among different parties with specific conditions. These contracts are stored on blockchain technology, which intends to facilitate, check, and enforce the performance of contracts digitally. In real-life scenario, an arbitrator makes sure that everybody follows the terms. The blockchain not solely waives the necessity for third person or parties, but additionally make sure that all ledger participants have the fair knowledge of the contract details and that agreement terms implement mechanically once conditions are met. Smart contracts are often used for various things, like monetary services, insurance policies, property laws, legal documents related to crowd funding, etc.

i. One such application of smart contracts is drought insurance systems where the users are insurers and famers, who are engaged in the entire insurance process where the farmers need weather-based insurance programs. The applications are developed using NEO blockchain platform, executed on NeoVM and use Oracle as a Service to fetch the external data using Application Program Interface (API); the system calculates the compensation values for the farmers who become victim of extreme weather conditions (Nguyen et al., 2019).

ii. The next application discusses supply chain operations between Consumer and Provider of service in supply chain operations by creating a Document of Understanding (DOU). To build Blockchain network, Hyperledger Fabric was installed on Intel-based servers running Red Hat Enterprise Linux 7 (Montes et al., 2019).

(b) Tokenization: For authenticating a singular physical item, the things are combined up with a corresponding digital token or stamp. This implies tokens are accustomed to link the physical and digital worlds. These digital tokens are very helpful in terms of providing chain management, anticounterfeiting properties, and scam detections.

i. The research paper is based on real estate (RE) with having RE tokenization integrates the RE investors, tenants and sellers in the process. These tokens are "unit of account" with properties countable, fungible, and divisible. With such token properties, people tend to use them to denote value of property assets they possess, these T-tokens hence become a "medium of exchange" (Latifi et al., 2019).

ii. Steemit is a Multi-Token Economy with three different economic activities: Steem, Steem Power, and Steem Dollars, the beneficiaries of these systems get benefitted by Steem and Steem Dollars. Both of these activities are economic activities and are responsible for external exchanges. This public content platform is entirely laid on blockchain and has much complicated token economy design (Kang et al., 2019).

(c) Interorganizational data management: Blockchain technology brought a revolutionary change in the way information is gathered and collected. It is less regarding maintaining information and more about managing a system of records. Hence, this helps in management of interorganizational data with so much of ease.

(d) For governments: The components of blockchain technology can be implemented to maintain all the activities of government. Blockchain technology proves to be a possible way to enhance government services and stimulate additional clear government—citizen relations. The distributed technologies can work to dramatically optimize various business procedures through more effective and secure sharing of data.

i. Blockchain is used for "reward checking system" for Thai Government where customer and lottery administrator play their role in the lottery wining process. The customer owns Ethereum address with details such as first-time lottery purchase, consequent purchase, award announcement, and create Ethereum wallet for customers (Saichua et al., 2019).

ii. South Korean government proposes five phase—based framework for National Digital ID based on the blockchain (NIDBC) for identity and authentication management. The procedure of NIDBC is explained with respect to national health insurance service of the National Health Security Office (NHSO) has been presented (Chakraborty, Aich, & Kim, 2019).

(e) Healthcare: Blockchain features a vast scope in healthcare industries. The blockchain technology facilitates the efficient and secure transfer of medical records of patients, manages the health supply chain, and helps the healthcare researchers or analyst to unlock the genetic code. Various companies use the concept of blockchain to improvise the way the medical data are shared and used. The integration of Internet of Things (IoT), Blockchain, and Machine Learning framework is responsible to detect anomaly in the patient's health such as heart beats, pulse rates, calories burns, sleep time monitoring using two networks, i.e., Personal Health Care (PHC) BlockChain and the External Record Management (ERM) Blockchain (Lee et al., 2019).

i. SHAREChain is a Smart healthcare data sharing framework which has five components Patient Identity Source, Consumer, Data Source, Repository, and Blockchain registry responsible for providing reliable and interoperable data to solve problems which arise due to data sharing process.

ii. Gem, a United States—based startup, prepares an ecosystem in healthcare using Ethereum Blockchain technology termed as Gem

Health Network. The system uses patient-centric approach where transparency of data and process exists among multiple stakeholders (Mettler, 2016).

(f) Financial Service: Blockchain financial services are redefining the prevailing rails of recent financial markets' infrastructure. This sector experiences vital activities that range from backend clearing and settlement to the worldwide capital markets' design. In a number of these cases, the distributed ledger systems do not require being completely decentralized, and various monetary establishments are gazing to build their own "private blockchains."

Besides, there are many application of blockchain technology like Abra which allows the money transaction without the need of bank account or paying transaction fee, whereas Guardtime ensures industrial cyber security for Estonian Project. Augur performs market prediction using blockchain technology, which can also be used for betting on sports, stocks, natural disasters, elections, etc. Ujo is an application that implemented smart contracts to bring transparency in music industry, whereas Ubitquity implements for ensuring provenance and avoid middle man in real estate industry. Storj is cloud-based decentralized storage which allows people to save their data in an economical way with more security. OpenBazaar employs smart contracts to establish connections between buyers and sellers.

6. Challenges of blockchain

Blockchain is still in nascent stage and there are many challenges which are listed in this section:

(a) Blockchain Regulation—It is highly important to define the legal scope of blockchain as companies in various sectors are waiting to get concrete direction on regulation and legislation so that it can be implemented without any legal risks. There are areas such as Smart Contracts where there is a looming uncertainty on regulation and existing regulations do not cover Smart Contracts.

(b) Blockchain Complexity—Blockchain uses complex technical jargon which could be difficult to understand by nontechnical users and hence efforts are taken to create glossaries and indexes to make it understandable. Even though companies and organizations are using distributed ledger technology in a great deal, there is a shortage of skilled blockchain developers.

(c) Blockchain Security Issues—51% attacks are one of the most known security vulnerabilities where blockchain's hash rate is controlled by malicious entities and these entities perform double spends by reversing the transactions and blocks the miners to confirm the blocks. Poor security practices also open security loopholes and hackers can easily attack on the cryptocurrency exchanges. Blockchain has been attacked by various types of malware such as crypto jacking where the malware consumes computer's resources to mine cryptocurrency.

(d) Blockchain Scalability—Blockchain's transaction history keeps growing on daily basis and thereby there is a constant risk of slowness in the system. Blockchain's scalability depends on various factors such as block size, response time, fees, etc. At the beginning, the capacity of a block in blockchain was 1 MB and each block could hold almost 2020 transactions. Over the time, the number of transactions has started to grow causing a scalability problem. Since every transaction goes through a validation process, the number of transactions need to stay up for an extended time in the queue to get validated causing slower response time and scalability concern.

(e) Cost and Efficiency—The Blockchain technology is quite competent in cost reduction. But it still faces specific challenges while implementing the legacy systems. Setting up the initial blockchain infrastructure is expensive. Small financial companies or banks would not prefer investing in something that does not hold a promising future. As we discussed, like scalability, many other factors contribute to high maintenance cost. With the increase in popularity of blockchain, validation process has become more expensive due to complication and increasing of mining resources. This has started to cause problem as the transactions waiting for validation need to wait for long time to finish validation unless a higher processing fee is paid for quick validation.

7. Conclusions

This chapter shares the systematic literature review which showcases the sudden hype in the blockchain-related searches and thus foresees the future of blockchain and its applications in various sectors. An overview on various dimensions like types, taxonomy, features of blockchain, the exciting blockchain applications, and major challenges has been presented in detail. Blockchain is a revolutionary idea which brings the transparency among different users and has become very popular in no time. Blockchain technologies and its applications are constantly evolving with time and its invention culminates many people's interest. In almost no time, others realized that this technology might be applied for several other things conjointly, like storing and sharing the logs of patients in medical sectors, generating a digital registrar or collecting taxes in financial sectors, etc. Blockchain's ability to support different business application makes blockchain one of the revolutionary technologies which would bring change in society, economics, business, and technology.

References

Aste, T., Tasca, P., & Di Matteo, T. (2017). Blockchain technologies: The foreseeable impact on society and industry. *Computer, 50*(9), 18–28. https://doi.org/10.1109/MC.2017.3571064.

Back, A. (2002). Hashcash- A Denial of Service Counter-Measure. *Technical report*. Submitted for publication.

Becker, J., Breuker, D., Heide, T., Holler, J., Rauer, H. P., & Böhme, R. (2013). Can we afford integrity by proof-of-work? Scenarios inspired by the bitcoin currency. In *The economics of information security and privacy* (pp. 135–156). Springer Berlin Heidelberg. https://doi.org/10.1007/978-3-642-39498-0_7.

Casado-Vara, R., Prieto, J., De La Prieta, F., & Corchado, J. M. (2018a). Blockchain framework for IoT data quality via edge computing. In *BlockSys 2018 - proceedings of the 1st blockchain-enabled networked sensor systems, part of SenSys 2018* (pp. 19–24). Association for Computing Machinery, Inc. https://doi.org/10.1145/3282278.3282282.

Chakraborty, S., Aich, S., & Kim, H. (2019). A Secure Healthcare System Design Framework using Blockchain Technology. *21st International Conference on Advanced Communication Technology (ICACT), PyeongChang Kwangwoon_Do, Korea (South)*, 260–264. https://doi.org/10.23919/ICACT.2019.8701983.

Chaum, D. L. (1981). Untraceable electronic mail, return addresses, and digital ™pseudonyms. *Communications of the ACM, 24*(2), 84–90. https://doi.org/10.1145/358549.358563.

Chaum, D. (1985). Security without identification: Transaction systems to make big brother obsolete. *Communications of the ACM, 28*(10), 1030–1044. https://doi.org/10.1145/4372.4373.

Croman, K., Decker, C., Eyal, I., Gencer, A. E., Juels, A., Kosba, A., et al. (2016). On scaling decentralized blockchains (A position paper). In *Lecture notes in computer science (including subseries lecture notes in artificial intelligence and lecture notes in bioinformatics)* (Vol. 9604, pp. 106–125). Springer Verlag. https://doi.org/10.1007/978-3-662-53357-4_8.

Dabbagh, M., Sookhak, M., & Safa, N. S. (2019). The evolution of blockchain: A bibliometric study. *IEEE Access, 7*, 19212–19221. https://doi.org/10.1109/ACCESS.2019.2895646.

Danezis, G., & Diaz, C. (2008). *Survey of anonymous communication channels*.

Dwyer, G. P. (2015). The economics of Bitcoin and similar private digital currencies. *Journal of Financial Stability*, 81–91. https://doi.org/10.1016/j.jfs.2014.11.006.

Eyal, I., & Sirer, E. G. (2014). *Financial Cryptography and Data Security*. In Christin, N., & Safavi-Naini, R. (Eds.), *Majority Is Not Enough: Bitcoin Mining Is Vulnerable* (pp. 436–454). Springer, Berlin Heidelberg.

Gatteschi, V., Lamberti, F., Demartini, C., Pranteda, C., & Santamaria, V. (2018). To blockchain or not to blockchain: That is the question. *IT Professional, 20*(2), 62–74. https://doi.org/10.1109/MITP.2018.021921652.

Greenberg, A. (2012). *This machine kills secrets: How wikiLeakers, cypherpunks, and hacktivists aim to free the world's information* (Vol. 0525953205). Dutton Adult.

Hawlitschek, Florian, Benedikt, Notheisen, & Timm, Teubner (2018). The limits of trust-free systems: A literature review on block-chain technology and trust in the sharing economy. *Electronic Commerce Research and Applications, 29*, 50–63. Submitted for publication.

Hayden, C., & Choi, Y. (2019). Blockchain and Bitcoin: Concept, Functionality, and Security. *International Journal of Cyber Research and Education*, 27–37. Submitted for publication.

Henry, R., Herzberg, A., & Kate, A. (2018). Blockchain access privacy: Challenges and directions. *IEEE Security and Privacy, 16*(4), 38–45. https://doi.org/10.1109/MSP.2018.3111245.

https://ico.conda.online/the-crypto-guide-for-beginners-%E2%80%93-what-is-blockchain/. (Accessed 9 April 2021).

https://jgateplus.com. (Accessed 10 April 2021).

https://nptel.ac.in/courses/106105184/3/lec3.pdf. (Accessed 6 April 2021).

https://spectrum.ieee.org/computing/software/bitcoin-the-cryptoanarchists-answer-to-cash. (Accessed 10 April 2021).

https://steemit.com/blockchain/@thesumitbanik/how-does-a-blockchain-work-simply-explained. (Accessed 11 April 2021).

https://trends.google.com/. (Accessed 10 April 2021).

https://www2.deloitte.com/content/dam/Deloitte/in/Documents/strategy/in-strategy-innovation-blockchain-revolutionary-change-noexp.pdf. (Accessed 10 April 2021).

https://www.activism.net/cypherpunk/manifesto.html. (Accessed 7 April 2021).

https://www.chaum.com/ecash/articles/1996/05-07-96%20-%20DigiCash_s%20Ecash™%20to%20be%20Issued%20by%20Deutsche%20Bank.pdf. (Accessed 7 April 2021).

https://www.eublockchainforum.eu/sites/default/files/reports/report_security_v1.0.pdf. (Accessed 10 April 2021).

https://www.gartner.com. (Accessed 10 April 2021).

https://www.ibm.com/blogs/blockchain/2018/02/top-five-blockchain-benefits-transformingyour-industry/. (Accessed 8 April 2021).

https://www.lesswrong.com/posts/YdfpDyRpNyypivgdu/aalwa-ask-any-lesswronger-anything. (Accessed 10 April 2021).

https://www.nytimes.com/2015/05/17/business/decoding-the-enigma-of-satoshi-nakamoto-and-the-birth-of-bitcoin.html. (Accessed 10 April 2021).

https://www.sciencedirect.com. (Accessed 10 April 2021).

IEEE. (n.d). ieeexplore.ieee.org.

Kang, S., Cho, K., & Park, K. (2019). On the effectiveness of multi-token economies. In *ICBC 2019 - IEEE international conference on blockchain and cryptocurrency* (pp. 180–184). Institute of Electrical and Electronics Engineers Inc. https://doi.org/10.1109/BLOC.2019.8751242.

Latifi, S., Zhang, Y., & Cheng, L. C. (2019). Blockchain-based real estate market: One method for applying blockchain technology in commercial real estate market. In *Proceedings - 2019 2nd IEEE international conference on blockchain, blockchain 2019* (pp. 528–535). Institute of Electrical and Electronics Engineers Inc. https://doi.org/10.1109/Blockchain.2019.00002.

Lavanya, B. (2018). *Blockchain technology beyond bitcoin: An overview.*

Lee, A. R., Kim, M. G., & Kim, I. K. (2019). SHAREChain: Healthcare data sharing framework using blockchain-registry and FHIR. In *Proceedings - 2019 IEEE international conference on bioinformatics and biomedicine, BIBM 2019* (pp. 1087–1090). Institute of Electrical and Electronics Engineers Inc. https://doi.org/10.1109/BIBM47256.2019.8983415.

Lemieux, V. L. (2016). University of British columbia. In *Blockchain technology for record keeping: Help or hype?* (Vol. 1).

Mettler, M. (2016). Blockchain technology in healthcare: The revolution starts here. In *2016 IEEE 18th international conference on e-health networking, applications and services, healthcom 2016*. Institute of Electrical and Electronics Engineers Inc. https://doi.org/10.1109/HealthCom.2016.7749510.

Montes, J. M., Ramirez, C. E., Gutierrez, M. C., & Larios, V. M. (2019). Smart contracts for supply chain applicable to smart cities daily operations. In *5th IEEE international smart cities conference, ISC2 2019* (pp. 565–570). Institute of Electrical and Electronics Engineers Inc. https://doi.org/10.1109/ISC246665.2019.9071650.

Nakamoto, S. (2008). *Bitcoin: A peer-to-peer electronic cash system.*

Nguyen, T. Q., Das, A. K., & Tran, L. T. (2019). NEO smart contract for drought-based insurance. In *2019 IEEE Canadian conference of electrical and computer engineering, CCECE 2019.* Institute of Electrical and Electronics Engineers Inc. https://doi.org/10.1109/CCECE.2019.8861573.

O'Leary, M. (2015). *The mysterious disappearance of Satoshi Nakamoto, founder & creator of bitcoin.* The Huffington Post.

Peck, M. (2012). Bitcoin: The cryptoanarchists' answer to cash. *IEEE Spectrum, 49*(6).

Peters, G. W., & Panayi, E. (2016). Understanding modern banking ledgers through blockchain technologies: Future of transaction processing and smart contracts on the internet of money. *New Economic Windows*, 239–278. https://doi.org/10.1007/978-3-319-42448-4_13.

Pitta, J. (1999). *Requiem for a bright idea.* Forbes. N.

Reyna, A., Martín, C., Chen, J., Soler, E., & Díaz, M. (2018). On blockchain and its integration with IoT. Challenges and opportunities. *Future Generation Computer Systems, 88*, 173–190. https://doi.org/10.1016/j.future.2018.05.046.

Ridley, M. (2017). The Bitcoin revolution is only just beginning. The Times. Retrieved 23 December 2017. In *The legal scholar and computer scientist Nick Szabo.*

Saichua, P., Khunthi, S., & Chomsiri, T. (2019). Design of blockchain lottery for Thai government. In *Ecti DAMT-NCON 2019 - 4th international conference on digital arts, media and technology and 2nd ECTI northern section conference on electrical, electronics, computer and telecommunications engineering* (pp. 9–12). Institute of Electrical and Electronics Engineers Inc. https://doi.org/10.1109/ECTI-NCON.2019.8692241.

Sara, S., Mahtab, K., & Joseph, S. (2018). Blockchain technology: A panacea or pariah for resources conservation and recycling? *Resources, Conservation and Recycling*, 80–81. https://doi.org/10.1016/j.resconrec.2017.11.020.

Szabo, N. (1998). *Secure property titles with owner authority.*

Tschorsch, F., & Scheuermann, B. (2015). *Bitcoin and beyond: A technical survey of decentralized digital currencies.*

Utilization of Blockchain Technology to Overthrow the Challenges in Healthcare Industry, Hemalatha, K., Hema, K., & Deepika, V. (2015). *Advances in intelligent systems and computing 1054, emerging research in data engineering systems and computer communications proceedings of CCODE* (pp. 199–208).

Zheng, Z., Xie, S., Dai, H. N., Chen, X., & Wang, H. (2018). Blockchain challenges and opportunities: A survey. *International Journal of Web and Grid Services, 14*(4), 352–375. https://doi.org/10.1504/IJWGS.2018.095647.

Chapter 11

Blockchain for green smart cities

S. Porkodi[1], **D. Kesavaraja**[2]

[1]*Dr. Sivanthi Aditanar College of Engineering, Tiruchendur, Tamil Nadu, India;* [2]*Department of CSE, Dr. Sivanthi Aditanar College of Engineering, Tiruchendur, Tamil Nadu, India*

Chapter outline

1. Introduction

1.1 IoT in digital era

The Internet of Things (IoT) sensors and devices are increasing tremendously all around the world. In 2025, there may be more than 70 billion sensors and IoT devices connected to the Internet across the world. Any sensor, actuator, and devices that acquire data, compute, transmit, and receive data and

Blockchain for Smart Cities. https://doi.org/10.1016/B978-0-12-824446-3.00004-1

instruction over Internet that is used in everyday life comes under IoT. The IoT devices can also be interconnected to each other to form a network. From smart brush, camera to satellite, anything can be a part of IoT and are connected to the Internet (Ocampos, 2020). The IoT is evolving continuously over years that comprises of a very vast network consisting, sensors, smart gadgets, electronics, vehicles, software, home applicants, etc. The IoT is mainly used to enhance profitability, maximize productivity, and minimize risk factor due to human errors. The data acquired from the IoT devices are used to improve communication among the people and the IoT processes within the organization, used for sharing data and information among various entities, to develop green smart city applications, smarter technologies, and various infrastructures. As various countries are racing to develop 5G, the introduction of 5G will boost the development of IoT and green smart city. Since in IoT there are huge volumes of acquired data that need to be transferred via Internet on real-time basis and smart city also deals with larger amount of data and also the data are processed instantly and the required actions are taken by the smart devices (Spec India, 2019).

1.2 Green smart city

In the recent past, green smart cities are being developed in both developed and in developing nations. IoT devices are mainly used in green smart cities to acquire data from the citizens, devices, and environment (Sarkar, 2020). The acquired data are used for improving the quality of resident's life, improve urban system functionalities, and enhance various industries and businesses with the help of modern technology for the efficient performance, overall cost minimization, and to reduce resource deterioration in the green smart city. The data analytics and Artificial Intelligence (AI) are mainly used in the green smart city to efficiently use the collected data for better use of physical infrastructure, for building a healthy and strong economy of the country, for cultural and social development of the citizens, and for enhancing the communication between people and the government. Learning and adapting new innovative technologies and ideas in the green smart city are for maximizing the intelligence of the services provided for the citizens to solve major problems and respond more effectively and efficiently according to the emergency situation that arises in the city. Some green smart city projects including smart street lighting (Yusaku, Noriaki, Akihir, & Naoya, 2013), smart energy meter (Chaudhari, Rathod, Shaikh, Vora, & Ahir, 2018), smart traffic management (Javaid, Sufian, Pervaiz, & Tanveer, 2018), smart irrigation system (Gummalla et al., n.d.), smart parking (Barriga et al., n.d.), smart energy management (Pieroni, Scarpato, Di Nunzio, Fallucchi, & Raso, 2018), driverless vehicles (Leprince-Ringuet, 2020; Sharma, Moon, & Park, 2017), smart garbage collection and recycling (Mustafa & Ku Azir, 2017), smart healthcare (Pramanik, Pareek, & Nayyar, 2019), smart water management (Kenda et al., 2018), e-services (Alotaibi, 2019), etc., are been implemented in various parts of the world. As the IoT and green smart city deals with enormous amount of citizen data that include private and sensitive information, privacy and security is the most important feature to be developed along with all the green smart city application (Zhang et al., 2017). There are various technology and cryptographic techniques to protect the data and systems from hackers and intruders, one such technology is blockchain.

1.3 Need of blockchain in green smart city

Blockchain is one of the emergent technologies that is gaining popularity and interest among people of various industries in the past few years. Blockchain is a distributed ledger on a decentralized network, it plays a major role in handling large volume of data, highly tamper resistant against various attacks, immutable, simplify verification and auditing requirements, highly cost-efficient, and provides security to the sensitive data. The blockchain technology is necessary in the development of green smart city to solve privacy and security issues in various fields of smart city, to protect IoT devices, and to effectively manage data (Fernández-Caramés & Fraga-Lamas, 2018). The green smart city deals with citizen's location coordinates, medical history, credit card numbers, travel history, etc., which are sensitive and vulnerable to attacks and thus needs protection. Any unauthorized data or intruders may cause serious harm to the citizens. Such as an intruder accessing GPS data of a citizen can use them to rob the home when user is not around or a Hacker can gain control over the smart system demanding ransom or an unauthorized person in the smart transportation or traffic control systems can manipulate vehicles that may even lead to accidents and the whole transportation system can collapse creating lots of disturbance to the citizens. Thus, to handle these many data and to avoid any intruders and hackers blockchain is necessary in the green smart city. Blockchain also helps in interoperability, thus data from various fields can be integrated and can be efficiently used in the fields of government, finance, healthcare, etc. (2019). Also the traditional cyber security protection to the smart devices is not enough due to its dynamic, heterogeneity, and scalability characteristics. Thus, blockchain can be used as a solution to handle various challenges in the green smart city development.

There are various study and surveys conducted based on the concepts and components of smart city and security of IoT ecosystem in the smart city. In 2019, Clauson, Breeden, Devidson, and Mackey (2018) performed a case study focused mainly of the sustainability of the smart city in the urban development by identifying the real-world smart city concepts but the smart city characteristics need advancement in the availability of economic and social infrastructures. In 2017, Gharaibeh et al. (2017) proposed achievements of smart city along with its present issues in security and privacy from data-centric approach perspective also presents cryptography, machine learning, and blockchain to secure the smart city. In 2017, Karamitsos (2017) explain about the blockchain technology and present the related platforms of blockchain and blockchain applications in various domains mainly on the education, various governmental, and global initiatives. In 2020, Guimaraes (De Guimarães et al., 2020) explains about the e-government on urban area and the quality of citizen's life in the smart city. In 2020, Sun (Sun & Zhang, 2020) researched on various blockchain applications that deal with big data mainly focusing on creating a green environment and minimized carbon emission smart city (Clauson et al., 2018).

In this chapter, current overview of green smart cities and concepts of blockchain are studied to find the challenges in the green smart city and how various blockchain-based applications could solve the issues in a green smart city is discussed, creating a way to analyze the better usage of blockchain in the research and development of future green smart cities.

2. Green smart city

2.1 Architecture

The green smart cities are in the developmental phase and there is a continuous development to these cities at various part of the world. Thus, there are many different architecture of the green smart city, which are evolving every day. Perception layer, network layer, support layer, and application layer are basic layers of the green smart city shown in Fig. 11.1.

Perception Layer: The main function of the perception layer is to collect or acquire data from the users and the environment; this layer can be also termed

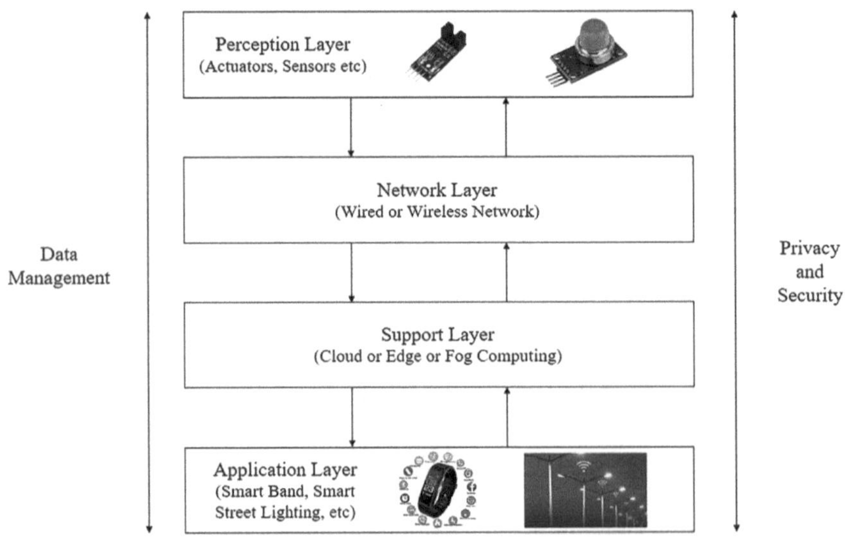

FIGURE 11.1 The architecture of the green smart city.

as data sensing or data recognition layer. The function of this layer is to gather data from all connected actuators, sensors, and IoT devices of the network. The acquired data are transferred to the back-end via network layer.

Network Layer: The network layer carries the acquired data from the perception layer, which may be wireless network including Wireless Sensor Network, communication network, Internet, or wired network. The gathered data are sent either to smart things to act immediately on any emergency or server to store all the collected data.

Support Layer: The purpose of the support layer is to give support to the application layer by providing all the necessary requirements, intelligent computing techniques that include cloud, edge, or fog computing.

Application Layer: The main functionality of application layer is to provide intelligence for decision-making depending upon customized user requirements and provides quick services based on real-time data.

2.2 Characteristics

The characteristics of the smart city include physical infrastructure, connectivity, access control, authentication, availability, interoperability, heterogeneity, scalability, handling big data, resident involvement, and mobility as shown in Fig. 11.2.

Physical Infrastructure and Connectivity: One of the basic necessities for developing smart city is physical infrastructure such as assets, buildings, roads, etc. These infrastructures of the city can be used efficiently along with the AI and data analytics for developing smart city applications and services. For example, solar panels can be attached at the top of the buildings and smart

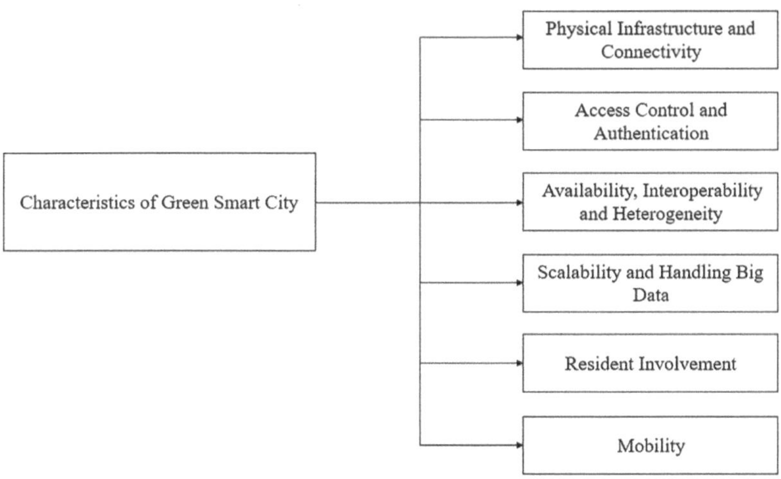

FIGURE 11.2 The characteristics of the green smart city.

home to produce green energy, bus stops can be built along with chargeable ports to support electric vehicles, roads can be built along with the Wi-Fi to support Vehicular ad hoc network (VANET), etc.

Access Control and Authentication: IoT devices generate an enormous amount of big data, which should be managed and controlled properly. Also access control should be set, determining who can access the data and protected from unauthorized access. The responsibility of the green smart city is to protect the citizen's data by preventing intruders and unauthorized access. So a strong user authentication and secure communication network should be developed in green smart city. Cryptographic algorithm, access control mechanisms, and biometrics are used for user authentication in the smart city for the protection, privacy, and security of user data. But much stronger technology is needed since hackers are also developing and intruding into the systems.

Availability, Interoperability, and Heterogeneity: Green smart city applications and services mainly rely on data generated by the IoT devices on real time. Thus, the green smart city cannot function properly with the downtime of any smart device. So, data availability is much needed for green smart city to function actively at the times of emergency and critical situations. The acquired data from various sources should be shared among different sectors for the better usage of data. Thus, interoperability should be maintained within the smart city. The green smart city systems are independent, distributed, and are accessed by various level of users in the smart city for different needs. So, heterogeneity is maintained in green smart city for handling diverse technologies, different IoT devices, various platforms, communication protocols, applications, etc.

Scalability and Handling Big Data: The data acquired and generated by IoT devices are increasing every day, which should be managed, and to handle the network traffic due to huge volume of increasing data the system must adapt to the situation and scalability is important to be maintained throughout green smart city. The enormous quantity of data is not only generated from various sources but also with different velocity, thus requiring an efficient technology for handling and managing the big data. Huge data are gathered, created, transferred, stored, and processed at green smart city. The knowledge and information which are extracted from all these data can be extremely utilized when used for managing asserts and for better decision-making in the applications of green smart city.

Resident Involvement: The quality of the resident's life is being improved by the development of the green smart city and so the resident involvement is much needed for building new applications and for improving the smart applications based on user requirements. In order to involve residents in the development of smart city, creativity, education, and capability to quickly learn and adapt to advanced new technologies are much needed. Identifying resident's needs and understanding the user better can be used for efficiently creating a smart application or smart service.

Mobility: Monitoring traffic based on real-time data and controlling the vehicle movement within the city comes under mobility. If there arises any emergency situation or any problem, then quick solution or instant reaction should be provided. For example, if there is an accident or a traffic jam in the city, then intimation of the problem occurred and route diversion must be given automatically and quickly to all the vehicles in that route. The effectiveness of the mobility can be achieved with the development of physical infrastructure (smart road) and communication infrastructure.

2.3 Challenges in green smart city

The main objectives of green smart cities are to improve the quality of resident's life, maximize productivity, minimize environmental impacts such as pollution, global warming, degradation of natural resources, etc, promote local economies, and improve the health of the residents. But green smart city consists of some other challenges like cyber-attacks, hackers, privacy, security, infrastructure, connectivity, education, and socialization as shown in Fig. 11.3.

Cyber-Attacks and Hackers: The IoT devices and smart applications handle variety of data regarding the residents. These data are highly vulnerable to cyber-attacks such as identity attack, eavesdropping, background knowledge attack, Sybil attack, collision, Denial of Service (DoS) attack, forgeries, and spamming. The cyber-attacks can be caused by injecting botnets into the resident's system. For example, in 2020, cyber criminals used Emotet botnet for sending malicious E-mails to various user for infecting the system with

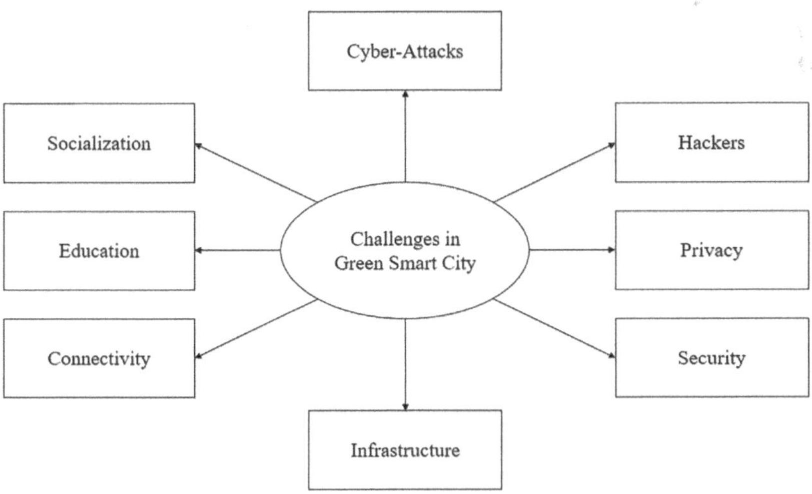

FIGURE 11.3 The challenges in the green smart city.

ransomware and Distribute Denial of Service (DDos) attacks (Cimpanu, 2020). In the smart city, the IoT devices are collaborated with various other technology and multiple devices are interconnected. If a Hacker manages to get into the system, then the Hacker can encrypt the whole system and demand ransom from the user. For example, in 2018, cyber security breach took place at one of the online marketplaces called Taskrabbit and 145 million users were affected and their data such as Social Security Number (SSN) and bank account details were compromised (Jignasa, 2018). In 2015 and 2020, power grid system of Ukraine and Israel's smart city water structure was attacked by the intruders, respectively, creating disturbance within the cities (Brumfield, 2020; Greenberg, 2017). Thus, a strong technology is needed to prevent hackers and intruders from entering the system and stop them from committing cyber-attacks. (Jignasa, 2018a).

Privacy and Security: The green smart city deals with huge amount of data including resident's sensitive data such as current coordinates, location history, medical history, digital identity information, etc. Any harmful change of details in the smart healthcare system will impact the patient's health and may even cost the life of the patient. If the current coordinates and the location history can be accessed on hacking the smart car, then an attacker might cause a burglary at the resident's home or even an assault on the resident (Jill, 2018). So, privacy and security of the user is much important and a strong technology must be established in the green smart city for the protection of the residents.

Infrastructure and Connectivity: At present, mainly in the developing countries, infrastructure is not enough to develop a smart city. So first sensors can be implemented for small smart city applications and gradually physical infrastructure can be built for developing smart city more easily. The smart city also operates on vast amount of big data, where the collected data are to be sent via Internet on the real-time basis. The 4G network technology can only provide up to 100 Mbps of maximum download data rate and 12 Mbps of maximum upload data rate. 4G cannot support when every city in the country starts to works as a smart city. The utilization of 5G technology theoretically can provide up to 20 Gbps of maximum download data rate and 10 Gbps of maximum upload data rate. The connectivity in the smart city can be boosted with the utilization of 5G so that billions of IoT devices can transfer data on real-time basis without any delay.

Education and Socialization: The green smart city needs smart citizens to actively participate in development of green smart city which needs education and awareness among people to keep updates on new technologies. Socialization is also important, since lots and lots of data are collected from the user to make a better personalized solution at the times of decision-making and so the user must be willing to share the data with various IoT devices and applications.

3. Blockchain technology

3.1 Blockchain and its components

Blockchain is decentralized distributed ledger where all data are stored as records that are sequentially chained together in chronological order on a peer-to-peer network. The data are stored across many servers globally and any user of the network can view all the data on the real time as soon as a block has been created by which data transparency is achieved. These records are tamper resistant, immutable, verifiable, and permanent on the blockchain. Thus, the blockchain provides a huge security boost to the system. The blockchain can be used in any field where there is a need to maintain multiple copies of data regularly so that even if the data in one node are somehow changed by a Hacker, the original data can be verified and retrieved from other nodes of the network and replaced in the place of hacked node easily. Cryptographic algorithms are also used in the blocks to protect the data from hackers and to make data more secure. A block mainly consists of timestamp which is the exact time when a block is created, previous hash is the cryptographic hash of previous block, nonce is a unique random number used by blockchain miners to solve the complex computation and transaction root is the Merkle tree root hash. The structure of the blockchain is shown in Fig. 11.4. Once a transaction joins the blockchain, it can never be altered and any modification will only create a new transaction without altering the old transaction thus making the blockchain immutable.

The components of the blockchain include peer-to-peer network, ledger, cryptography, rules for validity, and consensus mechanism. In blockchain data sharing is done on a peer-to-peer network. Transaction or data are listed and linked cryptographically in the ledger as blocks. There are various cryptographic technique such as public key infrastructure, Merkle trees, and one way

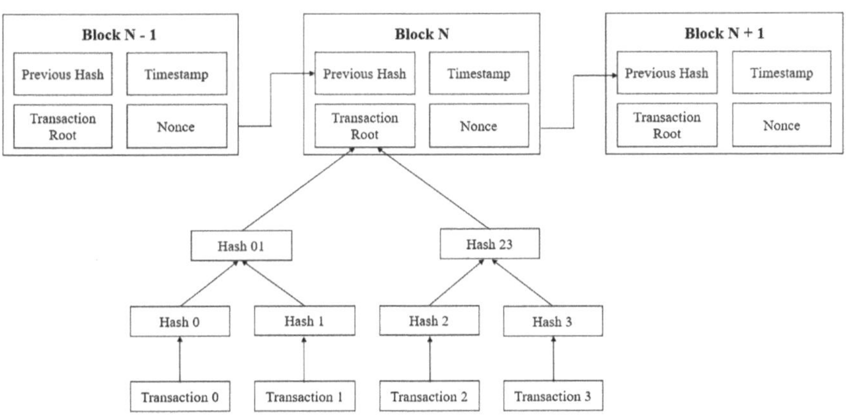

FIGURE 11.4 The structure of the blockchain.

hash functions. Some of the basic rules present in network are how ledger gets updated, when transaction can be considered as valid, which are verified nodes in the network, etc. The consensus algorithm in blockchain is the fault-tolerant mechanism to achieve necessary agreement on single state of network or single data value within the distributed system. There are three types of blockchain; they are, public, private, and consortium.

Public Blockchain Network: The public blockchain is an open network and everyone in the network can view and verify the transaction in the blockchain. Bitcoin and Ethereum are some examples of public blockchain network.

Private Blockchain Network: The private blockchain requires permission to participate and access the network, the access control will only be given to certain people and not everyone in the public can access the network. The private companies and organization use the private blockchain network, so that the access rights permission of network can be assigned only to employees of that particular organization. Hyperledger and Multichain are some examples of private blockchain network.

Consortium Blockchain Network: The consortium blockchain is a semi-private network, which is in the control of the various user group working across various organization, where some set of data may be set for public to view and some data are restricted only to the organizations. R3 Corda, TradeLens, and Energy Web Foundation (EWF) are some examples of consortium blockchain network.

3.2 Smart contracts

Smart contracts are digital transaction protocols that enable automatic execution of contractual obligations such as property transaction, contracts for financial repayment, loan agreement, etc., based on terms of the agreement or contract. The smart contract is used between two unknown people to exchange things of value in a secure and transparent way. The features of the smart contract include autonomy, back-up, safety, speed, and accuracy avoiding human error. The advantages of the smart contract are it eliminates the third party, middleman expenses, saves time, and increases the speed of the transaction. Smart contracts becomes more accessible and stronger when used along with the blockchain. The contract becomes tamper proof since verification can be done with hashed cryptographic keys that is stored on the blockchain ledger by anyone in the blockchain network. The terms and full details regarding the digital smart contract will only be known to the signed parties but anyone in the network can verify the authenticity. The blockchain makes the digital smart contract immutable and cryptographically secure.

3.3 Smart assets

In the past around the year 2000, the concept of smart asset was assumed as adding RFID tag to the assets makes them smart instead of barcode identifiers

(Lampe & Strassner, 2003). But recently, the idea has been evolved completely on including blockchain as a solution to track smart assets. In the supply chain management every product is tracked continuously along with its components or ingredients on real-time basis also updating every details regarding the product and its availability (Clauson et al., 2018). All these information is used to predict and transport where the product is needed in the near future and increase the availability of the product. All the products in smart city can be converted as smart assets, so that distribution and storage would improve efficiently. The objective of storage will be changed to just in time from just in case as the products can be managed and delivered to the respected shops at the right time and this saves more space that are maintained for stock just in case. Not only products even parking space in the smart city can be converted into smart assert. The blockchain can be used to monitor the free space and allocate parking space for the vehicle with or without drivers at the destination of the vehicle. This system can be automated along with the payment system saving time by directing drivers to find the parking space, easily distributing the traffic so that everyone finds parking space and punishing the parking offenders.

3.4 Digital identity

The unique attributes of the users such as fingerprints, retina, voice, facial recognition, hand signature, etc., can be cryptographically converted into digital identities with the help of blockchain protocol. The digital identities can be used to electronically sign a transaction or a smart contact in the blockchain. The digital identity can be combined with the smart assert to authorize the properties and things that a user owns and all these details can be registered in blockchain and the things themselves can initiate transaction. For example, if milk is running out, then the fridge itself can order milk, heating or cooling system in the smart home can pay their own respective utility bill, vehicle can schedule its own services and annual checkups, etc. Even without AI the IoT things can perform such activities on registering the things on blockchain which can be monitored by the user with the ownership of the things. If a product such as phone or even an umbrella registered on the blockchain is lost or forgotten somewhere, the current real-time location can be easily seen on the blockchain and thus the product can be found quickly and easily. This can also reduce the theft tremendously as the things can be tracked easily and brought back to the respective owners.

4. Characteristics of blockchain to develop smart city

The blockchain characteristics that are mainly used to secure and develop the smart city from privacy and security issues are immutability, security, robust, transparent, validation, and consensus mechanism.

Immutability and Cyber-Security: Once data are entered into the blockchain, they remain unchanged, thus reducing the fraud in the ledger across network. Immutable nature of the blockchain also ensures that no one can intrude or hack into the green smart city systems. Cryptographic algorithms provide more protection to the data avoiding data from suffering cyber-attacks and botnet attacks. The usage of blockchain in smart city reduces failure and avoids tampering at any point.

Robust: The blockchain technology is highly robust since all the data of the blocks are stored in multiple nodes distributed across the network. So, even if a node is attacked, the data can be recovered from other accounts, thus eliminating the problem of downtime and cope up with errors. Blockchain also solves the single point of failure in the green smart city that have the ability to disturb the whole system, since blockchain is not a centralized system.

Transparency and Validation: Any user of the public blockchain network can view all the data on the real time as soon as a block has been created by which data transparency is achieved. Transparency in the network also increases the interoperability, thus the data can be maximum utilized to provide more personalized smart city services to the citizens. Any changes made in the blockchain requires proper authentication from the nodes of the network, thus increase auditability. There are variety of fake news, deepfake videos, and forged photos are used to deceive the humans which can be eliminated by using digital identity and validation method.

Consensus Mechanism: It is the fault-tolerant mechanism that is used in distributed blockchain network for achieving agreement on a value or state of process by the nodes of the network based on the predefined procedures and rules called consensus protocol. There are different types of consensus protocols used in different blockchain platforms such as proof of work (PoW), Proof of Stake (PoS), Proof of Activity (PoA), Proof of Burn (PoB), Delegated Proof of Stake (DPoS), Byzantine Fault Tolerance (BFT), Proof of Elapsed Time (PoET), etc. The consensus protocol of blockchain is used in the smart city so that any data in the system cannot be changed so easily by anyone.

5. Smart city application integrated with blockchain

There are numerous amount of application and services that will be benefited on integrating the blockchain with the green smart city application, which are listed below.

5.1 Smart healthcare

Basically the hospitals are sponsored, managed, and owned by central authority and a single point of failure may lead to lots of problems. When considering all the data generated from the hospital, it is highly hard for a patient or a hospital to maintain the complete patient history of the patients.

Further, due to urbanization or some other reason if people migrate between cities and take treatment at various hospitals, then it is harder to keep track on the medical history of the patient. If the data are digitalized and are highly sensitive user data, it requires strong protection of the data. Since the data can be easily misused or unauthorized access or changes in the data can affect the patient health seriously and any change in the prescription may even lead to the death of the patient. Also there are variety of IoT sensors and health management equipment and devices to monitor the patient health on the real-time basis in the green smart city.

Thus, a strong system is needed to protect and manage all the patient medical history from various hospitals which can also be easily accessed by the patients whenever needed and for efficient diagnosis (Azaria, Ekblaw, Vieira, & Lippman, 2016). Blockchain can be the known best solution to manage all the growing records of data, to maintain efficient, sustainable, and intelligent smart healthcare in the green smart city also enhancing the data security. Smart healthcare does not only stop with hospitals but also includes smart wearable, smart ambulance, and emergency response system. The smart wearable collects all the user medical data and monitors the user health and gives intimation to the doctor or sends alert to the ambulance when there arises an emergency. Smart ambulance on connecting to the smart traffic control system, the system automatically changes light to green on the way of the ambulance so that it can reach the hospital soon. Also the patient condition and all other device's information are lively updated to the hospital, so that a doctor can prepare the treatment or the operation theater on time and start the treatment as soon as the ambulance arrives the hospital. The data sharing from the patient to the hospital is much important for a doctor to make a decision on real time depending on the condition of the patient at the remote location. Interoperability between hospitals, pharmaceutical, lab, and insurance can also be possible while using blockchain. The smart healthcare system integrated with blockchain technology is shown in Fig. 11.5.

5.2 Smart transportation

The Information and Communication Technology is developing, which leads to the development of smart vehicles in the recent past. The smart transportation objective is for providing more convenience to the drivers and passengers, improving travel efficiency, enhancing road safety, and managing traffic effectively. The blockchain improves sharing of the data and information, with which the vehicles can easily communicate with each other and overall system's robustness is also enhanced with the blockchain. When using blockchain technology in the smart transport system, documents such as driving license, vehicle registration certificate, insurance details of the vehicle, and other documents regarding vehicle can be maintained properly in the blockchain under the owner, where approval and customs clearance are made

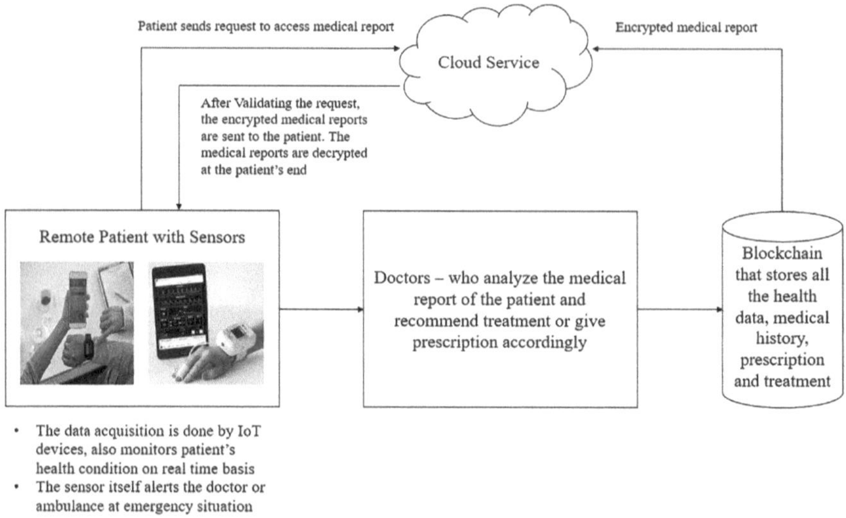

FIGURE 11.5 The smart healthcare system integrated with blockchain.

much faster and processing time are reduced. Hacking into the vehicles can get the hacker to access the location history of the user which can harm the vehicle system also the locations can be even used to assault the user, the blockchain solves the privacy and security issues in the smart transportation system.

Blockchain also provides network monitoring and data processing efficiently, improve the end user secure service delivery and cost mitigation. Smart transport not only includes the driverless vehicles and VANET but also the smart parking management, smart traffic light management, smart street lights, and smart roads. When reaching a destination, the smart parking system predicts and allocates a parking slot for the vehicle even before the vehicle reaches the destination. This is highly used to distribute the traffic flow and provide parking lot to all the vehicles. Since every vehicle has its connection with the Internet, the mobility and traffic flow can easily be monitored and managed by the smart traffic light management system. The map application in the smart phones is lined to the smart transport to provide the real-time updates regarding the traffic jams and accidents in the route of the vehicle where all the data are managed in the blockchain and distributed to the users at the right time when an alert arises in the system avoiding delay or traffic jams in the road. The purpose of smart roads is to give instant diversion details and warning in case of any sudden climate changes such as fog, land slide, etc., for avoiding accidents.

5.3 Supply chain management

Supply chain management is one of the most important applications of smart city when integrated with blockchain could solve numerous number of problems in the distribution and supply chain of the green smart city. The entities of the supply chain include individuals who are involved directly in service flow, products, information regarding the product and organizations. International companies and organizations throughout the world has complex and longer supply chains with numerous number of products in manufacturing and sale. Also there are lots of middleman involved in the supply chain such as transporter, distributor, supplier, retailer, etc. But the details on life cycle of the product from the manufacturing to the sale are very much limited. This information is important since it has the capability to enhance the consumer's trust on the product and its corresponding company. It also helps in predicting the trend in the market and to make any crucial decision on the business. Data sharing is the basic requirement for the supply chain which can be gathered by various IoT devices in the smart city. The efficient data management and sharing requires more advanced blockchain technology.

The blockchain can also be used for tracking all the product information details, location details, every single change is updated into the system. Thus, avoiding the forgeries and duplicate products in the market since all supply chain data are open to the citizens and also overall cost will be reduced, the stocking up of goods and selling them at a higher prize at the time of demand can be prevented completely. The smart contract can be used to track the availability of the product and the further production will only begin when the product availability reaches near the end, this will eliminate the problems in over estimating resource consumption and demand. The consumers can also directly connect with the manufactures without the middleman in between also improving the traceability and fairness in the business. For example, a coffee lover can scan the code on coffee bag to track their origin such as from which farm the coffee seeds are from, following what are all the process involved and which are all the manufacturing company doing the process, and finally from which shop the coffee bag has reached (Starbucks partners with microsoft azure blockchain for coffee traceability, Ledgeinsights, 2019). Likewise, farmers can also follow the product along the supply chain. Likewise, all the fruits, vegetables, rice, etc., can be tracked with the help of blockchain also assuring food safety in the agricultural sector. The supply chain can also be used in every sector where products are manufactured and reaches the customer like construction industry, vehicle manufacturing industry, watch manufacturing industry, etc.

5.4 Smart education

The education sector too consists of lots of information and pills of register where there are educational certificates, faculty records, students' records, etc. These records are shared among different stakeholders and needed to be ensured that the persons who have access to the records are trustworthy. Since these records are just managed in papers and can be tampered easily. It is also quite tough for the institute to maintain the records of all the years since the establishment of the institution and each and every record should be maintained accurately. The growing data management can be maintained easily and simply with the blockchain technology. The processes such as certificate verification, attestation, and in case a certificate are lost then duplicate certificate can also be applied and received easily with blockchain. In blockchain, all these educational asserts of the institutes are maintained in the single place and the students can access it from any part of the world.

5.5 Digital advertisement tracking

The advertisements (both in the television and the Internet) of the advertisers who offer more money is distributed to more audience for more number of times to influence the audience where in television the channel can be skipped if the advertisement is not more entertaining or influential. The viewers are mainly bombarded with frequent irrelevant advertisements. The introduction of the blockchain can change the whole system of the advertisements, whereas the AD can be tracked using the blockchain and AD with maximum audience reach are promoted better and thus increasing the quality of the AD and the product and also advertisers should make their AD more entertaining and worthy to be telecasted to maximum audience. The data collected from the audience help in promoting right advertisements to right set of audience. The blockchain in the digital advertisement could also eliminate frauds and increase transparency. For example, in June 2020, Zee Entertainment of India announced blockchain to be used for AD tracking (Emmanuel, 2020).

5.6 Smart voting

In this pandemic situation at 2020, people are advised only to stay at home not to gather massively on roads. Thus, election at this time seems impossible so it leads to the idea of digital voting at home hits a massive boost. The digital e-voting only consists of cryptographic mechanism to protect the vote where the private key is only known to the individuals. But these private keys can be sold for money and the private key can be used to change the votes. Thus, blockchain is necessary in the smart voting system, so that digital identity of the citizens is verified allowing their votes to be registered in the distributed ledger and thus preventing the identity thefts and frauds. The citizens can also track their voting at any time and the vote counting will also be verifiable. Any user can audit vote independently which shows no vote has been changed or removed from the network and all votes are legal. Already many countries are started considering blockchain for smart voting. For example, in June 2020, Russia used Blockchain-Based E-Vote System for their voting and hackers tried node attack on the E-Vote system. But as the votes are registered in the distributive blockchain, single node attack cannot cause harm to the system and the vote are safe in the blockchain thus the system withstood the attack (Stephen, 2020).

5.7 E-government

The government office is also filled with enormous amount of data files and most of the records are just papers and can be missed or misused, tampered, and the bribery took be few officials lead to lots of problems in the society. The IoT devices also collect huge data from the citizens for various smart city applications, and they have to be managed and protected. All these problems can be solved by the use of blockchain technology, which would also increase the citizen engagement in the planning process of the government to change a system. The decentralized blockchain can also make the government of the green smart city more trustable, reduce overall cost of the system, digital asserts and user identities (such as birth certificate, wedding certificates, passports, election ID, driving license, etc.) that are issued by the government can be managed easily especially in the country with larger population. Various governmental services can directly reach to the citizens of the country in a much transparent way and any citizen can directly contact the government regarding any support to the public needs. Any small, big, or global decision-making will involve citizen support and every citizen will be responsible for any decision of the country eliminating all the middleman. For example, in 2020, due to the pandemic situation, the people of the country are demanding for postponing JEE and NEET exams but the government is still in a debate thus increasing the anxiety and pressure among students which are even leading to death of some students. This situation can completely be sorted in the smart e-government where the citizens decision to postpone the exams can be implemented as per the decision of the citizens and not just by the decision of the political leaders. Even public petitions, juridical and legislative decision-making can be dependent of citizen's decision as democratic involvement improving the government—public relationship.

5.8 Publishing and copy rights management

The publishing industry has benefits from the blockchain technology. There are over millions of journals that are published every year across the world that cannot be maintained at a single place, which could be solved by the blockchain. The works of the authors can be credited under their name and the system can recognize the writers work so the authorship can be managed. The publishers can also print or publish book digitally and the copy rights agreement between the author and the publisher for publication can also be managed in the blockchain. A smart contract is an agreement where an author can sign it using electronic signature. Blockchain also supports self-publishing with micro payment and content monetization. The books of the author can be distributed to the readers directly in the blockchain avoiding the third party such as amazon. The readers can also pay to the authors directly for what they read with much transparency eliminating the third party. The publishing not only refer to the books and journals but also to the paintings, photographs. When a publisher accepts the content, then the authors can get the payment for their work. The frauds in the painting and artist circle can completely be reduced with the help of blockchain and the original creator of the art with get the credits correctly. The contracts between the author and the publishing house can also be traced easily, increasing the processing time of the payment and the payment accounting are made much easier.

5.9 Blockchain-as-a-service

The computing devices such as personal computer, smart phone, smart tablet, etc., work more efficiently and perform various countless tasks with the help of software and applications. Many developers create numerous number of software, whereas in certain projects, blockchain can be used to provide end-to-end solutions in developing the software, testing the software and deployment of the software. The features of the blockchain are added to the software-as-a-service platform to provide various utilities to the developer and the customers. Blockchain-as-a-service may even lead for a wider adaptation of the blockchain in the corporate. For example, Microsoft, R3, Amazon, etc., use blockchain-as-a-service for creating and managing the scalable blockchain network.

5.10 Smart energy

Energy industry also involves lots of middlemen. Thus, the smart contracts of the blockchain can be used by individual people for buying and selling the energy. The people can themselves generate solar energy with solar panels attached to the top of their house for their consumption at the time of power outage and if the energy is more, they can be sold which reduces the energy drain. The peer-to-peer trading of energy in smart grid with digital transactions are more secure and the customer's identity are also protected. Thus, smart grid is economical, sustainable, efficient, and secure. When blockchain is added along with the smart grid, the transaction and energy management becomes much easier, improving the stability, data security, and trust. Based on the availability and consumption profile, the smart energy rate varies dynamically, the consumption of the smart energy can be measured with smart meter which can be monitored with an application on the smart phone.

6. Conclusion and future directions

The green smart city consists of many serious problems which can easily be solved by using blockchain. The blockchain technology provides major cybersecurity boost to green smart city, improves the transaction speed, manages big data, simplifies verification and auditing requirements, enhances the performance of various industries, enhances immutability, the systems are highly tamper resisted to various attacks and the overall cost is minimized. The only problem in the blockchain adoption and smart city is that both are at its initial stage of the development and a government must approve the usage of the blockchain to improve the smart city purposes and more secure environment. Also the citizens of the green smart city must be willing to give their data where sensitive data can be managed privately and all other data can be used to study and analyze various smart services and applications. In this chapter, the present overview of green smart cities and its characteristics are studied to find the challenges in green smart city. Then blockchain and its components are explored along with various aspects such as smart contract, smart asserts, and digital identity and how various blockchain-based applications could solve the issues in a green smart city is discussed, creating a way to analyze the better usage of blockchain in the research and development of future green smart cities.

References

Alotaibi, S. S. (2019). Registration center based user authentication scheme for smart E-governance applications in smart cities. *IEEE Access, 7*, 5819–5833. https://doi.org/10.1109/ACCESS.2018.2884541.

Azaria, A., Ekblaw, A., Vieira, T., & Lippman, A. (2016). MedRec: Using blockchain for medical data access and permission management. In *Proceedings - 2016 2nd international conference on open and big data, OBD 2016* (pp. 25–30). Institute of Electrical and Electronics Engineers Inc. https://doi.org/10.1109/OBD.2016.11.

Barriga, J., Sulca, J., León, J., Ulloa, A., Portero, D., Andrade, R., & Yoo, S. G. (n.d.). Smart parking: A literature review from the technological perspective.

Brumfield, C. (2020). *Attempted cyberattack highlights vulnerability of global water infrastructure*. https://www.csoonline.com/article/3541837/attempted-cyberattack-highlights-vulnerability-of-global-water-infrastructure.html.

Cimpanu, C. (2020). *Emotet botnet returns after a five-month absence*. https://www.zdnet.com/article/emotet-botnet-returns-after-a-five-month-absence/.

Chaudhari, S., Rathod, P., Shaikh, A., Vora, D., & Ahir, J. (2018). Smart energy meter using Arduino and GSM. In *Proceedings - international conference on trends in electronics and informatics, ICEI 2017* (Vol. 2018, pp. 598–601). Institute of Electrical and Electronics Engineers Inc. https://doi.org/10.1109/ICOEI.2017.8300772.

Clauson, K. A., Breeden, E. A., Devidson., C., & Mackey, T. K. (2018). *Leveraging blockchain technology to enhance supply chain management in healthcare: Blockchain in healthcare today*. https://doi.org/10.30953/bhty.v1.20.

De Guimarães, J. C. F., Severo, E. A., Felix Júnior, L. A., Da Costa, W. P. L. B., & Salmoria, F. T. (2020). Governance and quality of life in smart cities: Towards sustainable development goals. *Journal of Cleaner Production, 253*. https://doi.org/10.1016/j.jclepro.2019.119926.

Emmanuel, O. O. (2020). *India: Zee entertainment taps blockchain for ad tracking*. https://btcmanager.com/india-zee-entertainment-blockchain-ad-tracking/?q=/india-zee-entertainment-blockchain-ad-tracking/&.

Fernández-Caramés, T. M., & Fraga-Lamas, P. (2018). A review on the use of blockchain for the internet of things. *IEEE Access, 6*, 32979–33001. https://doi.org/10.1109/ACCESS.2018.2842685.

Gharaibeh, A., Salahuddin, M. A., Hussini, S. J., Khreishah, A., Khalil, I., Guizani, M., et al. (2017). Smart cities: A survey on data management, security, and enabling technologies. *IEEE Communications Surveys and Tutorials, 19*(4), 2456–2501. https://doi.org/10.1109/COMST.2017.2736886.

Greenberg, A. (2017). *Crash override: The malware that took down a power grid*.

Gummalla, D. K., Raju, B., & Nandan, D. (n.d.). A review on the smart irrigation system. In: *2nd international conference on recent innovative trends in computer science and applications*.

Javaid, S., Sufian, A., Pervaiz, S., & Tanveer, M. (2018). Smart traffic management system using Internet of Things. In *International conference on advanced communication technology, ICACT* (Vol. 2018, pp. 393–398). Institute of Electrical and Electronics Engineers Inc. https://doi.org/10.23919/ICACT.2018.8323770.

Jignasa, S. (2018). *5 artificial intelligence-based attacks that shocked the world in 2018*. https://analyticsindiamag.com/5-artificial-intelligence-based-attacks-that-shocked-the-world-in-2018/.

Jill, B. (2018). *Autonomous vehicles and the threat of hacking*. https://www.cpomagazine.com/cyber-security/autonomous-vehicles-and-the-threat-of-hacking/.

Dr. Karamitsos, Y. (2017). *Blockchain & smart contracts building blocks for smart cities*. https://www.decentralized.com/wp-content/uploads/2017/12/John-Karamitsos-Blockchain-Smart-Contracts-Building-blocks-for-Smart-Cities.pdf.

Kenda, K., Rizou, S., Mellios, N., Kofinas, D., Ritsos, P., Laspidou, C., et al. (2018). In *Conference: Fragile earth at SIGKDD*.

Lampe, M., & Strassner, M. (2003). *The potential of RFID for moveable asset management*.

Leprince-Ringuet, D. (2020). *Self-driving cars: The hunt for security flaws steps up a gear*. https://www.zdnet.com/article/the-hunt-for-security-flaws-in-self-driving-cars-steps-up-a-gear/.

Mustafa, M. R., & Ku Azir, K. N. F. (2017). Smart bin: Internet-of-Things garbage monitoring system. In *MATEC Web of conferences* (Vol. 140). EDP Sciences. https://doi.org/10.1051/matecconf/201714001030.

Ocampos, J. (2020). *The future of IOT in 2020*. https://www.ingeniumweb.com/blog/post/the-future-of-iot-in-2020/4924/.

Pieroni, A., Scarpato, N., Di Nunzio, L., Fallucchi, F., & Raso, M. (2018). Smarter city: Smart energy grid based on blockchain technology. *International Journal of Advanced Science, Engineering and Information Technology, 8*(1), 298–306. https://doi.org/10.18517/ijaseit.8.1.4954.

Pramanik, P. K. D., Pareek, G., & Nayyar, A. (2019). Security and privacy in remote healthcare: Issues, solutions, and standards. In *Telemedicine technologies: Big data, deep learning, robotics, mobile and remote applications for global healthcare* (pp. 201–225). Elsevier. https://doi.org/10.1016/B978-0-12-816948-3.00014-3.

Sarkar, A.N. (2020). Smart cities: A futuristic vision. Retrieved August 30, 2020, from: https://www.thesmartcityjournal.com/en/articles/1333-smart-cities-futuristic-vision#:~:text=It %20is%20projected%20that%20over,Cities%20by%20the%20year%202020.&text=Common%20elements%20among%20these%20definitions,and%20to%20enhance%20urban%20systems.

Sharma, P. K., Moon, S. Y., & Park, J. H. (2017). Block-VN: A distributed blockchain based vehicular network architecture in smart city. *Journal of Information Processing Systems, 13*(1), 184–195. https://doi.org/10.3745/JIPS.03.0065.

Spec India. (2019). *Top internet of things (IoT) trends in 2020*. https://www.spec-india.com/blog/top-iot-trends-in-2020#:~:text=Conversion%20To%20Smart %20Cities%20Will%20Increase&text=Thanks%20to%20IoT%2C%20Smart%20Cities,leverage%20data%20access%20between%20entities.

Starbucks partners with microsoft azure blockchain for coffee traceability. (2019). Ledgeinsights.

Stephen, N. (2020). *Russia's blockchain-based e-vote system suffers node attack*. https://cointelegraph.com/news/russias-blockchain-based-e-vote-system-suffers-node-attack.

Sun, M., & Zhang, J. (2020). Research on the application of block chain big data platform in the construction of new smart city for low carbon emission and green environment. *Computer Communications, 149*, 332–342. https://doi.org/10.1016/j.comcom.2019.10.031.

Yusaku, F., Noriaki, Y., Akihir, T., & Naoya, O. (2013). *Smart street light system with energy saving function based on the sensor network*. https://doi.org/10.1145/2487166.2487202.

Zhang, K., Ni, J., Yang, K., Liang, X., Ren, J., & Shen, X. S. (2017). Security and privacy in smart city applications: Challenges and solutions. *IEEE Communications Magazine, 55*(1), 122–129. https://doi.org/10.1109/MCOM.2017.1600267CM.

Chapter 12

A novel Blockchain-based Access Control Manager to Electronic Health Records (EHRs)

Usharani Chelladurai, Seethalakshmi Pandian
University College of Engineering BIT Campus, Anna University Tiruchirappalli, Tiruchirappalli, Tamil Nadu, India

Chapter outline

Blockchain for Smart Cities. https://doi.org/10.1016/B978-0-12-824446-3.00003-X

1. Introduction

In the fast world, rapid improvements in the medical industry and advancement of biomedical research, healthcare perceives an inventive approach to disease diagnosis, prevention, and treatment that integrates an individual patient's inborn character, lifestyle, and environment. Advancement of technology has made vast databases of health information, healthcare applications, healthcare provided wireless tools to track health data and engaged individuals more in their own healthcare (Berner, 2004). Combining these advancements in the healthcare industry would foster transformative change in the field of health domain (Alexaki, George, Katos, & Nikolaos Petroulakis, 2018). However, the vast majority of these systems do not have the capacity to share their health data. Blockchain technology has the potential to address the interoperability challenges currently present in health systems and to be the technical standard that enables individuals, healthcare providers, healthcare entities, and medical researchers to securely share electronic health data (Ariani et al., 2017).

An effective access control mechanism and access control policies are needed to secure Electronic Health Records.

1.1 Fundamentals of blockchain technology

Blockchain is a peer-to-peer (P2P) distributed ledger technology for a new generation of transactional applications that establishes transparency and trust ("Bitcoin-NG: A scalable blockchain protocol," 2016). Blockchain is the underlying fabric for Bitcoin and is a design pattern consisting of three main components: distributed network, shared ledger, and digital transactions.

a. **Distributed Network:** Blockchain is a decentralized P2P architecture with nodes consisting of network participants. Each member in the network stores an identical copy of the blockchain and contributes to the collective process of validating and certifying digital transactions for the network (Barrow & Clayton, 1996).
b. **Shared Ledger:** Members in the distributed network record digital transactions into a shared ledger. To add transactions, members in the network run algorithms to evaluate and verify the proposed transaction. If a majority of the members in the network agree that the transaction is valid, the new transaction is added to the shared ledger (Hanley and Tewari, 2018). Changes to the shared ledger are reflected in all copies of the blockchain in minutes or, in some cases, seconds. After a transaction is added it is immutable and cannot be changed or removed. Since all members in the network have a complete copy of the blockchain no single member has the power to tamper or alter data (Jiang et al., 2018).
c. **Digital Transactions:** Any type of information or digital asset can be stored in a blockchain, and the network implementing the blockchain

defines the type of information contained in the transaction (Dubovitskaya, Xu., Ryu, Schumacher, & Wang, 2017). Information is encrypted and digitally signed to guarantee authenticity and accuracy. Transactions are structured into blocks and each block contains a cryptographic hash to the prior block in the blockchain. Blocks are added in a linear, chronological order.

1.2 Elements of blockchain technology

In this section, we will introduce some background knowledge related to the elements of blockchain.

1.2.1 Blocks in blockchain

Each block in the blockchain has a unique number which is in increased order. The first block of the blockchain is called the Genesis block. Fig. 12.1 illustrates a simplified blockchain, the number of blocks from beginning to end of the blockchain is called a block height. A single block can have one transaction or n number of transactions (Atoshi, 2008). In general, the blockchain constantly grows, and this continuous increase will require an enormous amount of computational power for the validation process and the blockchain data structure is presented in Fig. 12.1.

1.2.2 Blockchain transactions

A blockchain transaction is a movement of data value which is spread to the network and composed into blocks. The transactions are stored in a distributed, replicated public database (blockchain) with the consensus mechanism accomplished by the Proof of Work system called mining (Zyskind, Nathan, & Pentland, 2015). In blockchain a single transaction can contain one or more inputs and one or more outputs.

1.2.3 Hash function

A hash function H takes an input data D and returns a fixed-size string, which is called the hash value h. It is deterministic that this function provides that only the same input can result in the same hash value (Zhou et al., 2015). It is

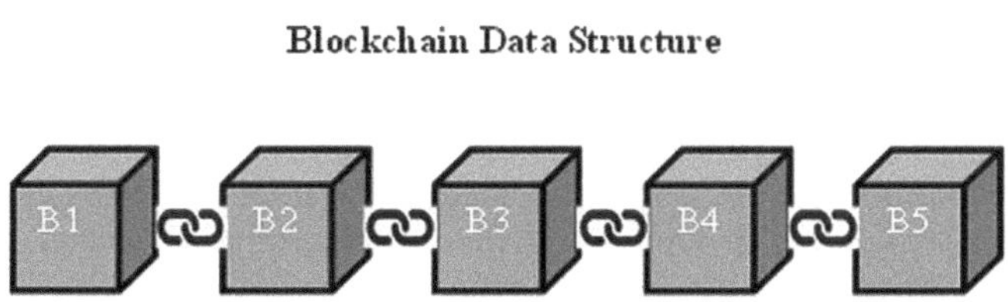

FIGURE 12.1 Blockchain Data Structure Chain of Blocks.

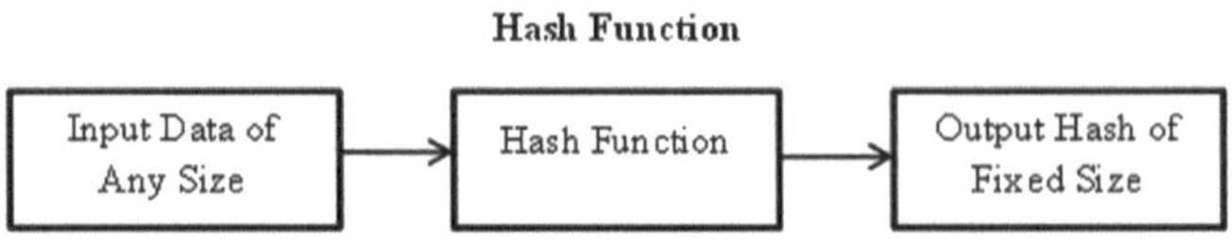

FIGURE 12.2 Structure of hash function.

pseudo-random that the input has a slight change, the hash value will be entirely different. It is a one-way function that is inferring the original data from the hash value is extremely difficult. The process of hash function is presented in Fig. 12.2.

1.2.4 Public-key cryptography

It is an asymmetric cryptography (Fig. 12.1). This cryptographic methodology utilizes key pairs, two different sets of keys are utilized for the encryption and decryption processes. The keys are public key and private key. A public key is publicly available to the world, and a private key is known only to the owner. In public-key cryptography, anyone can encrypt and send a message using the public key of the message receiver, and the message can be decrypted by only the receiver's private key (Fontaine & Galand, 2007).

1.2.5 Digital signature

A digital signature is a mechanism to check the origin/source of digital messages and documents using public-key cryptography. The digital signature provides authentication, nonrepudiation, integrity. Digital signatures are used for validating the source of data transaction and validating the addresses of the data owner in the blockchain network.

1.2.6 Peer-to-peer

A P2P network is a collection of various distributed resources that are connected by a network. From the application point of view, P2P is the opposite of Client/Server architecture. A distributed network architecture might be called a P2P network when the participants share a portion of their resources (processors, storage, network capacity, etc.). These resources should provide the service and content, which are accessible by other peers directly without going through intermediary entities, by the network. Bitcoin uses the P2P network that has contributed and has developed a complete distributed network to overcome the single point of failure.

1.2.7 Timestamp server

Bitcoin proposed that a timestamp server takes a hash of a block of transactions to be time stamped and publishes the hash broadly. The timestamp proves the connection sequence of transactions, and each timestamp is included in the previous timestamp in its hash.

1.2.8 Merkle trees

It is a binary hash tree. Each block of blockchain includes the summary of all transactions in the block using Merkle trees. A Merkle tree is a data structure to summarize and verify the integrity of large sets of data. A Merkle tree is similar to a binary tree, as each node will have a maximum of two child nodes. To create a Merkle tree, the hash function runs on a pair of nodes recursively until there is only one hash known as Merkle root. The cryptographic (Fig. 12.2) hash algorithm of Bitcoin's Merkle tree applies SHA256, and the Merkle tree is constructed bottom-to-top. Transactions will not be stored in the Merkle tree; rather, their data will be hashed, and the result of the hash will be in each child node. A Merkle tree requires an even number of child nodes because it is a binary tree. Fig. 12.3 illustrated a binary merkle tree with four transactions. Therefore, if there is an odd number of transactions, the last transaction will be duplicated to create an even number of child nodes.

1.2.9 Proof of work

A proof of work is a piece of data that is hard to (Fig. 12.3) generate, but it is easy to verify and check certain requirements. Generating the PoW is a random process with low chances, and the process goes through a lot of trial and error until it becomes a valid proof-of-work. The PoW normally refers to a mechanism to confirm that a person does certain work. A distributed ledger can store all transactions and ownerships. Bitcoin uses blockchain technology as a distributed ledger. However, distributed copies of blockchain data can create new chances for the users to cheat. Bitcoin uses the proof of work to prevent Sybil attacks. When the node validates transactions and broadcasts them, it needs to perform some proof of work to prove that it is a real identity. Therefore, the identification of the participant depends on the computational

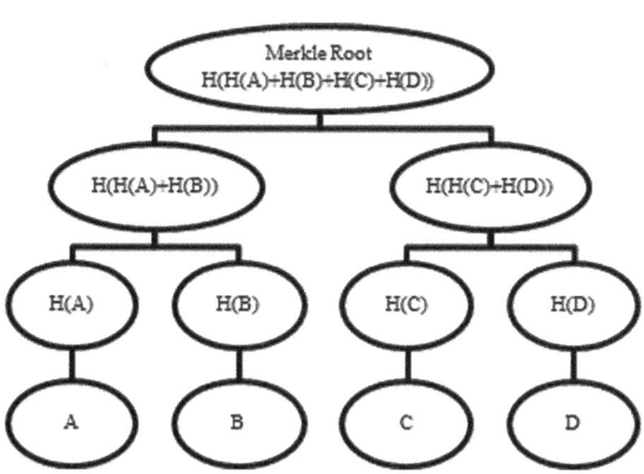

FIGURE 12.3 Merkle tree generation.

power, and not the number of identities, which can be fake. In Bitcoin, the difficulty of PoW's puzzle relies on the target value. If the target requires the more zeros, the puzzle becomes more difficult to solve.

1.2.10 Consensus mechanisms

Consensus mechanisms verify every single transaction in the public ledger. Once the transaction is stored in the chain and broadcasted to the network, it can never be altered or deleted (Mettler, 2016). Since the P2P networks do not depend on a central entity, blockchains have no single point of failure. If the network nodes are sufficiently distributed across different authorities and geographical locations, the blockchain becomes resilient to network outages by natural disasters, physical or cyber-attacks, and falsifications from authorities.

2. Related work

Samarati and de Vimercati (Swan, 2015) described Access Control as "the process of mediating every request to resources and data maintained by a system and determining whether the request should be granted or denied." The goal of blockchain access control manager is to check and verify the actions that authorized users can perform within a computer system.

Chapple et al. defined access control as "the formalization of those rules for allowing or denying access. Access controls define the allowable interactions between subjects and objects. It is based on the granting of rights, or privileges to a subject with respect to an object" (Koutsouris & Lazakidou, 2014). The following concepts are commonly used in the access control community.

Role-based access control (RBAC) was formally introduced by Sandhu et al. in 2000 and standardized by ANSI.

3. Blockchain as an Access Control Manager for EHRs

Access control manager includes what a user could perform directly as well as what programs can be executed on behalf of the users who are permitted to execute those programs. Access controls depend on cooperation with other existing security services.

3.1 Elements of Access Control Manager

In this section, the proposed Blockchain Access Control Manager for EHRs is presented and shown in Fig. 12.4. Fig. 12.4 illustrates the elements of the proposed system.

Object: Resource to which the subject requests access (file, database, network resource, cloud service, and device, etc.)

- **Subject:** in the form of user, device, process, or application requesting to access an object.
- **Operations:** A process initiated by a subject.

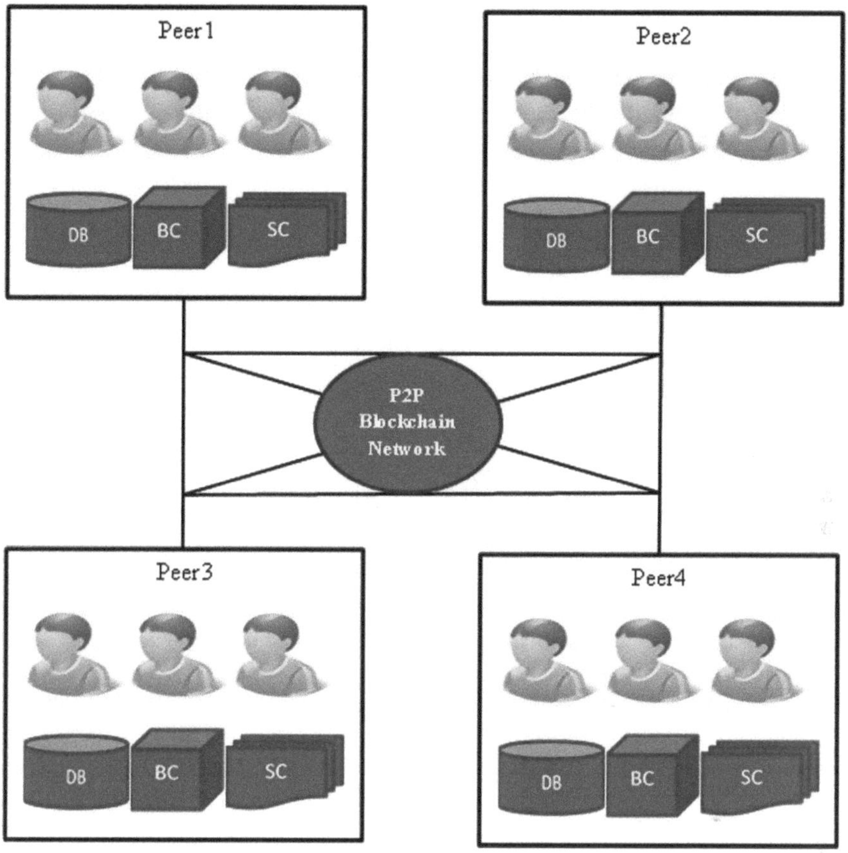

FIGURE 12.4 Proposed blockchain access control manage architecture.

- **Permission** (privilege)**:** The rules that govern who has access to which resources (For example, a manager may have permissions to see the dashboards, while an accountant may keep track of the transactions on the general ledger).
- **Access Control List (ACL):** Access control lists are a list of permissions that are associated with an object. It specifies who or which processes are granted access to the given object. For example, if Alice has read and write permissions, and Michael has to delete permission a given file, an access control list representation might be (Alice: read, write; Michael: delete).
- **Access control matrix:** The access control matrix is a two-dimensional table, which captures subjects as columns and objects as rows. Each cell shows the permission of the corresponding column (subject) and row(-object). The development of access control systems is typically a multi-stage based on the following concepts:
- **Security policy:** A set of high-level rules, which access control should be governed. (For example, managers can see to the dashboards.)

- **Security model:** A formal representation specified and enforced by security policies. It may establish a formal model of access rights, a model of computation, or model of distributed computing.
- **Security mechanism:** A low-level implementation based on the security policy and model. It can be hardware or software.

3.2 Access control models

In this section, the proposed system describes the Access Control Models (ACM). The models are namely MAC, DAC, RBAC, and ABAC. Three traditional access control models are mandatory to access control (MAC), discretionary access control (DAC), and role-based access control (RBAC). Additionally, another commonly used and more logical access control model is attribute-based access control (ABAC).

3.2.1 Mandatory access control

The mandatory access control policy grants access based on subjects and objects, which will be assigned security labels. These security labels are a combination of hierarchical classification levels and nonhierarchical compartments. The hierarchical order specifies the sensitivity of the data. For instance, an organization might have sensitivity levels: Top Secret, Secret, Confidential, and Unclassified. Every object is associated with a set of compartments (nuclear, weapon, aircraft weapon, etc.). As an example, if an object is related to the {aircraft weapon, nuclear}, it may be accessible by subjects, who need to know both about aircraft weapons and nuclear. In practice, a document might have the label (Top Secret, {aircraft weapon, nuclear weapon}) if it contains extremely sensitive information regarding aircraft weapons and nuclear weapons. Additionally, each paragraph of this document could be assigned a set of sensitivity levels and compartments, so the classification of this entire document would be the most restrictive classification given to each paragraph in the document. One of the well-known implementations of MAC is multilevel security that is commonly used on information technology systems deployed at highly sensitive government organizations. It is used in these scenarios as all security policies are centrally controlled by a security policy administrator, so users cannot override the security policies. In this case, MAC provides a high degree of security in MLS systems. Also, the underlying philosophy of MAC is that information belongs to an organization, so the organization should control the security policies of its information.

3.2.2 Discretionary access control

Discretionary access control (DAC) provides the possibility for data owners to control their data accessed by themselves rather than by a central authority.

DAC gives the possibility to users to grant and revoke access to objects, which are under their control. DAC is based on the notion that users are "owners" of objects; therefore, they will have complete control over their objects. Ownership usually refers to the creator of the object. DAC policies grant users access to information based on the user's identification and authorizations that state the user is allowed or denied access to information. Each user request will be assessed by these authorization rules. A mechanism of DAC policy implementation needs to answer the following question: does a subject (S) have access right (AR) for an object (O)? More practically, the presentation of this policy can be represented as an access control matrix. Each cell of the access control matrix includes a set of access rights.

3.2.3 Role-based access control

RBAC was initially proposed for enterprises where permissions will be granted to roles rather than individual users. A role is a combination of a set of users and set permissions. An organization can create roles based on various job functions and responsibilities, and users can be assigned these roles. The concept of roles logically maps to an organization's structure. The idea of grouping privileges was initially proposed by Baldwin. The core notion of RBAC is that users and permissions can be associated with roles. The idea of roles greatly simplifies the management of permissions because system administrators do not need to manage individual users and their permissions. Roles can be granted and revoked permissions to users of certain applications and systems as necessary. Specifically, when role and permissions' relationships are predefined, assigning users to the predefined roles will be easy for the system administrators. RBAC defines the following three main elements: (a) User—person, who interacts with a computer system; (b) Role—job function or responsibility within an organization; (c) Permission—an operation on an object.

3.2.4 Attributed-based access control

In the National Institute of Standard and Technology (NIST), attribute-based access control (ABAC) is defined by a logical model that evaluates access rights using policies and conditions against attributes of the subject, object, and environment relevant to a request. Attributes are characterized by entities (subjects and objects), environmental conditions, and requested actions. Traditional access controls primarily consider static authorization decisions based on the subject's privileges on the target resources (Hanley and Tewari, 2018).

3.2.5 Centralized and decentralized access controls

Access control administration can be centralized, decentralized, and hybrid, which uses both approaches. A centralized access control is a responsibility for granting or revoking an authorization to users, and it provides unified identity and authorization management across computer systems inside the organization. Centralized access controls simplify and help to manage the organization's security systems. One disadvantage of centralized access control is a single point of failure. Decentralized access control is when multiple places can grant access permissions for the users. This approach can be more stable than centralized access control, but it will require more administration and maintenance to protect several locations.

4. Access Control Policies

In this section, the process of access control policies of the proposed system is presented. Access Control Policies (ACPs) are the agreement between the participants in the blockchain network. ACP defines the privileges and rights of each nodes in the network. In the blockchain network, all data are transparent to authorized participants. Each provider in the proposed blockchain network has his own Access Control Polices. It decides whether access is certified or not. The proposed system is implemented with four access control policies namely Granting Access Rights (GAR), Granting Transfer Rights (GTR), Revoking Access Rights (RAR) and Revoking Transfer Rights (RTR). The proposed Access Control Mechanism is constructed based on the access control policies.

4.1 Granting Access and Transfer Rights

In this subsection, the process of granting and transferring Access Rights of the proposed system is presented. The access control policies consist of two granting policies namely Granting Access Rights (GAR) and Granting Transfer Rights (GTR). The steps involved in Granting Access Rights and Transfer Rights are as follows:

Step1. Resource owner of Peer1 grants Access Rights to Peer2.

Step2. Participant of Provider2 generates request for accessing data with public key.

Step3. Resource owner of Provider1 receives public key of requester1.

Step4. Resource Owner of provider1 grants Transfer Rights to requesting participant of provider2.

Step5. Requester1 of Provider2 can access the resource with read only rights.

4.2 Revoking Access Rights and Transfer Rights

In this subsection, the process of revoking rights of the proposed system is presented. The access control policies consist of two revoking policies namely Revoking Access Rights (GAR) and Revoking Transfer Rights (GTR). In the proposed system, the data owner has a special right to withdraw the granted access and transfer rights from the provider at any point of time. For the purpose of avoiding unauthorized access and restricting the users, the data owner enables revoking access rights and transfer rights with the access rights already granted is shown in Fig. 5.2. The data owner of P2 wishes to revoke the Access and transfer rights from P3 due to restrict the access. The steps involved in Revoking Access and Transfer Rights are as follows:

Step1. Resource owner of Provider2 Revokes Access Rights from Provider3.

Step2. Resource owner of Provider2 Revokes Transfer Rights from Provider3.

Step3. Participant of Provider3 restricted to generate request for accessing data.

5. Conclusion

A shared distributed infrastructure that provides a comprehensive view of an individual's health data across a lifetime is an equally essential component of interoperable health systems. Blockchain technology addresses interoperability challenges, is based on open standards, provides a shared distributed view of health data, and will achieve widespread acceptance and deployment throughout all industries. Utilization of the proposed health blockchain described in this chapter has the potential to engage millions of individuals, healthcare providers, healthcare entities, and medical researchers to share vast amounts of genetic, diet, lifestyle, environmental, and health data with guaranteed security and privacy protection. The acquisition, storage, and sharing of these data would lay a scientific foundation for the advancement of medical research and precision medicine, help identify and develop new ways to treat and prevent disease, and test whether or not mobile devices engage individuals more in their healthcare for improved health and disease prevention. Blockchain technology definitely has a place in the health ecosystem, using blockchain to promote the advancement of precision medicine.

References

Alexaki, S., George, A., Katos, V., & Nikolaos Petroulakis, E. (2018). Blockchain based electronic patient records for regulated circular healthcare jurisdictions. In *IEEE international workshop CAMAD* (pp. 1–6).

Ariani, A., Koesoema, A. P., & Soegijoko, S. (2017). Innovative healthcare applications of ICT for developing countries. In *Understanding complex systems* (pp. 15–70). Springer Verlag. https://doi.org/10.1007/978-3-319-55774-8_2. Issue 9783319557731.

Atoshi, N. (2008). *Bitcoin: A peer-to-peer electronic cash system satoshin@gmx.*

Berner, S. (2004). *Clinical decision support systems: Impacting the future of clinical decision making, health information management systems* (pp. 463–467). Springer.

Bitcoin-NG: A scalable blockchain protocol. (2016). In *Proc. 13th USENIX symp, networked systems (NSDI 16)* (pp. 45–59).

Barrow, R. C., & Clayton, P. D. (1996). Privacy, confidentiality, and electronic medical records. *Journal of the American Medical Informatics Association*, 139–148. https://doi.org/10.1136/jamia.1996.96236282

Dubovitskaya, A., Xu., Z., Ryu, S., Schumacher, M., & Wang, F. (2017). *Secure and trustable electronic medical records sharing using blockchain ArXiv170906528 Cs.*

Fontaine, C., & Galand, F. (2007). A Survey of homomorphic encryption for nonspecialists. *EURASIP Journal on Information Security*, 1–14.

Hanley, M., & Tewari, H. (2018). Managing lifetime healthcare data on the blockchain Smart-World/UIC/ATC/ScalCom/CBDCom/IoP/SCI 2018. In *Proceedings - 2018 IEEE SmartWorld, ubiquitous intelligence and computing, advanced and trusted computing, scalable computing and communications, cloud and big data computing, internet of people and Smart city innovations* (pp. 246–251). Institute of Electrical and Electronics Engineers Inc. https://doi.org/10.1109/SmartWorld.2018.00077.

Jiang, S., Cao, J., Wu, H., Yang, Y., Ma, M., & He, J. (2018). BlocHIE: A blockchain-based platform for healthcare information exchange. In *IEEE international conference on smart computing*.

Koutsouris, D., & Lazakidou, A. (2014). *A concepts and trends in healthcare information systems*. eBook- Springer International.

Mettler, M. (2016). Block chain technology in healthcare. In *IEEE 18th international conference on e-health networking, applications and services (healthcom)*.

Swan, M. (2015). *Blockchain: Blueprint for a new economy*. O'Reilly Media.

Zhou, J., Cao, Z., Dong, X., & Lin, X. (2015). TR-MABE: White-box traceable and revocable multi-authority attribute-based encryption and its applications to multi-level privacy-preserving e-healthcare cloud computing systems. In *Proceedings - IEEE INFOCOM, 26* pp. 2398–2406). Institute of Electrical and Electronics Engineers Inc. https://doi.org/10.1109/INFOCOM.2015.7218628

Zyskind, G., Nathan, O., & Pentland, A. S. (2015). Decentralizing privacy: Using blockchain to protect personal data. In *Proceedings - 2015 IEEE security and privacy workshops, SPW 2015* (pp. 180–184). Institute of Electrical and Electronics Engineers Inc. https://doi.org/10.1109/SPW.2015.27.

Chapter 13

Application of blockchain in automotive industry, waste management, and seed traceability

N.R. Rajalakshmi[1], V. Usha[1], Saravanan Krishnan[2]
[1]*Computer Science and Engineering, Vel Tech Rangarajan Dr. Sagunthala R & D Institute of Science and Technology, Chennai, Tamil Nadu, India;* [2]*Department of Computer Science and Engineering, Anna University Regional Campus, Tirunelveli, Tamil Nadu, India*

Chapter outline

1. Introduction

Blockchain is a well-known technology for its significant performance in various cryptocurrency systems. Satoshi Nakamoto in 2008 introduced blockchain technology and proposed a distributed ledger technology that permits money exchanges through the web without depending on trust or the requirement of a monetary establishment. The blockchain shares and records all transactions occurring in the network. Blockchain is created by linking valid nodes or blocks together, and the block preserves its shared state. A block contains its identifier, block hash, previous block hash, transactions hash, and the time stamp. A unique hash value is maintained in each block. Transactions are added to the timestamped block after validation by the network nodes that have a copy of ledger to verify the validity of transactions in the newly created block. After this process, the block is added to the

Blockchain for Smart Cities. https://doi.org/10.1016/B978-0-12-824446-3.00012-0

blockchain, which refers to the correct previous block. This makes blockchain resistant to change and traceable. The miner uses a consensus mechanism in the blockchain to mine the block details. The control is given to one organization to make a decision and validate the process in permissioned blockchain. But the permissionless blockchain allows any node to link to the network and any two nodes to execute transactions without being authenticated by a central authority based on a distributed open ledger technology. It uses a peer-to-peer model to ensure that the entire work load is split between similarly privileged peers. The probability of failure is reduced since there is no central server in between. The benefits of blockchain are enhancing transparency, tracking visibility, ensuring authenticity, establishing trust, and improving security. The data integrity and security in blockchain is done by using various cryptographic secure hash algorithm like SHA 256 and SHA 512, etc. Hence, the manufacturing and distribution of poor-quality products could be also monitored using blockchain.

Fig. 13.1 Smart contracts (SCs) are "orchestration and choreography protocols that run on the blockchain to facilitate, verify and enact with computing means a negotiated agreement between stakeholders" (Wood, 2014). A user can define data structures and functionality in SCs. Then, the SCs can be called through the transaction. An SC has a permanently stored state in the blockchain variables. The main characteristic of the SC is that it runs in an isolated environment. They are immutable and irreversible.

The stakeholders in the enactment of a smart contract are establishing the binding agreement and deploying the application using smart contracts on

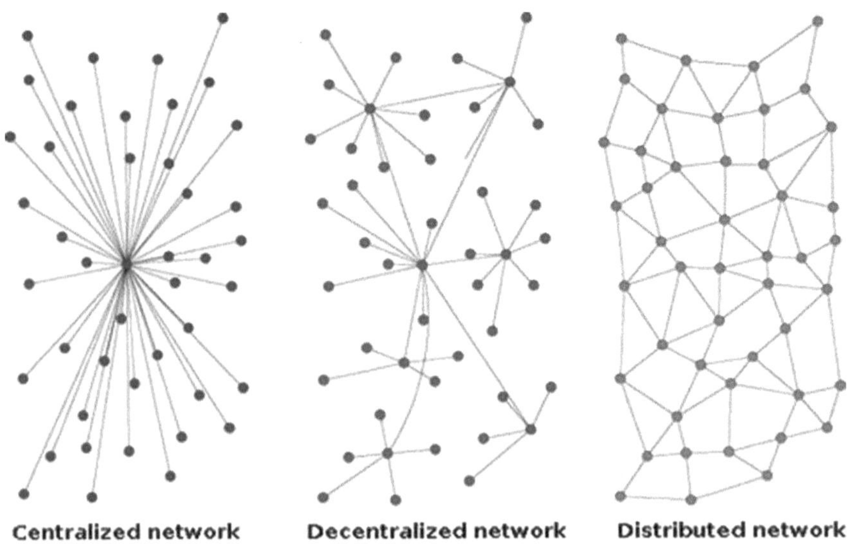

FIGURE 13.1 Blockchain network.

Ethereum Virtual Machine (EVM). The parties participating on SCs use the Ethereum address and application programming interface to communicate with each other.

Ethereum has been introduced to provide more functionality and flexibility in decentralized and permissionless blockchain-based platforms. Ether is used as a fuel for executing smart contracts on EVM.

The SC enhances the automation and synchronization in business process of property, value, and ownership transfer. SC plays an important role in ethereum because, it enforces legal contracts between the stackholders.

2. Literature review

Blockchain may be among the foremost important systems in terms of security, consistency, and speed. It consists of decentralized data storage that can hold a public ledger of transactions. The ledger can record many agreements: monetary transactions, property transfer, and even ballot storage. All transactions that occur on a typical blockchain are verified and signed with cryptography to ensure security and anonymity. The blockchain can support all kinds of agreements or SCs and financial transaction. It tracks the status of transactions, increases transparency, enhances security, and ensures trust and authenticity. Thus, all kinds of transactions, more specifically, trading of property rights, are benefited from such a trust-free, secure, and transparent system. The quality of manufacturing of automobile products could be guaranteed using blockchain. The main idea of blockchain is to form a decentralized ledger of things or records all connected into the form of a chain. Whenever a new block or person tries to get added it needs to be validated by using different methods like time stamp validation, proof of work, and so forth. For instance, a block can be created to record each payment transaction of the user. This type of visibility and transparency is important to improve security and to ensure the legitimacy and authenticity of payments. However, the vulnerabilities also started to appear to control consensus and voting (Rahouti, Xiong, & Ghani, 2018).

3. Blockchain in the automotive industry

Transportation systems, particularly in urban areas, are already overwhelmed by demand, causing strain on infrastructure and reducing traffic safety, according to the United Nations. Urban population will increase by 50% by 2050, making the effects of urbanization including pollution and traffic congestion incredibly difficult to manage in many countries. Vehicles tend to be the largest source of carbon emissions in addition to the environmental impacts; 1.25 million people have died in traffic accidents per year; across the globe, to mitigate the effects of vehicle congestion and overcrowding, cities need to develop new ways to manage road infrastructure and vehicle flow.

The mobility open blockchain initiative can leverage the convergence of rapidly maturing technologies, such as artificial intelligence (AI), Internet of Things (IoT), and blockchain to assign anything to a person, vehicle, or a piece of infrastructure. These digital identities can unlock the potential for a mobility network with intelligent entities that can perform transactions and communicate with each other. This in turn enables cities to build, operate, and finance their new infrastructure and new and radically better ways to enable and accelerate this transformation. t=These members develop the first-ever standard for a blockchain-based vehicle identity (VID). The VID can create a digital twin of a physical vehicle, unlocking the potential for vehicle to infrastructure (v2i) and vehicle to everything (v2x) payments from a car wallet, eliminating the need for a payment account associated with the person or business. With a car wallet, a vehicle can pay as it moves on a road. In addition to cities' road usage, the car wallet is allowing users to pay for infrastructure, congestion, and carbon emissions charges. The car owner can also settle transactions in an automated way if smart contracts are fulfilled. The driver or owner accepts regular or repeated settlement requests (e.g., road toll, parking zones, and fuel cost). Users can also receive credit in the car wallet for using green or clean fuel or taking an alternative route to receive charges and credits to their car wallet. The car owners can also charge the user fees to pay for repair maintenance and new construction. With a car wallet, a vehicle can pay the repair and maintenance cost to the garage (Fig. 13.2).

The vehicle wallet could also act as a digital tracker and communicator for the vehicle, having vehicle identities on a network. It allows vehicles to communicate to know the speed location, direction of travel, and braking to have an intelligent decision on coordinated mobility. It will permit vehicles to operate more safely and much faster. Car owners can sell services, data, and extra seats in their vehicles. The user can share with others for usage of seats in vehicles. The VID with car wallets are used more in real time to report traffic flows and infrastructure accidents. The complete history of all executed transactions are maintained in block, which are visible, verifiable, and

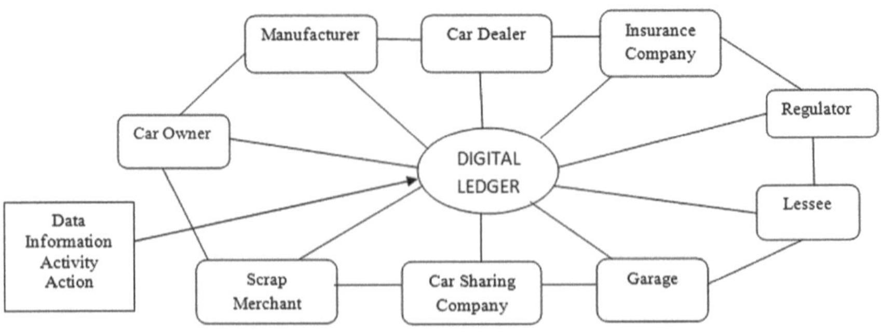

FIGURE 13.2 Blockchain in the automotive industry.

traceable. Blockchain records transactions between two parties in an efficient and verifiable manner. The recorded transaction could not be altered without agreement of the majority of network participants.

4. System model

The system is composed of several major components, that is, car owner, manufacturer, insurance company, garage, lessee, regulator, car dealer, and transaction. It is assumed in the model that a car owner may sell its used or new product to a car dealer, and a car dealer may have products from many car owners. Then there is a relationship between dealers and owners. A similar relationship is assumed between the car sharing company and the car owners. The system is designed after identifying data types and functionality of the system to change its state space. The possible constraints are identified over the data types. The relationships of system components are defined in terms of variants (Kumar, Choudhary, Raju, Chaudhary, & Sagar, 2019). Blockchain is a digital ledger containing a list of records linked together via cryptography. The transactions are created based on a set of rules that cannot be changed. This allows participants to verify and audit transactions permanently.

If the registered vehicle is verified in a database inventory then the regulator is notified the vehicle registered year, vehicle history sends the information to blockchain to update the inventory of vehicle and regulator to pay the road tax. The update inventory notification goes from blockchain to car owner and regulator. The same procedure is repeated for the flow of vehicle details from car owner to insurance company as explained and its sequence diagram is shown in Fig. 13.3.

The blocks in the Bitcoin blockchain store the details about a transaction like a sender, receiver, and balance as a number of coins. If the car owner or user initiated the transaction using car wallet, the transaction is flooded to all

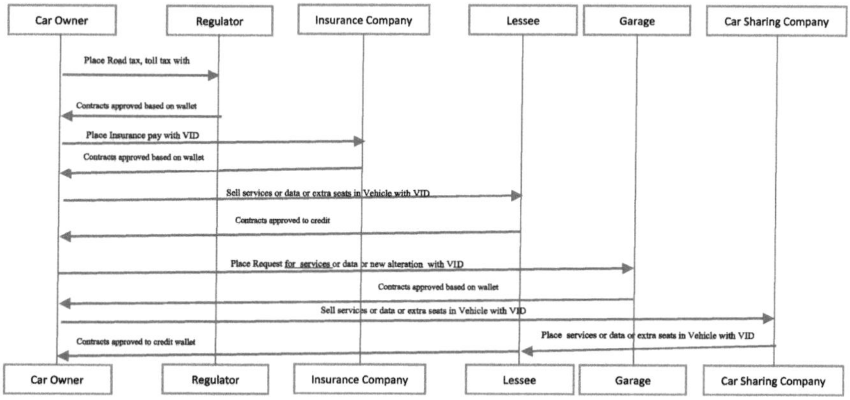

FIGURE 13.3 Sequence Diagram showing interactions of stakeholders in Automotive Industry.

members of the network by using a flooding protocol called Gossip protocol. Usually more than one member in the network validates the block, providing consensus. Then, the validated transaction is included in a block. At this point, the transaction is confirmed. The newly created block now becomes part of the ledger. The block containing current hash and hash of previous block and data is added in the blockchain. The next block links itself cryptographically back to this block. The link is called hash pointer, and at this point, after the transaction got a second confirmation, the block also gets a first confirmation. The hash of the block acts as a fingerprint as it gives a unique identification to the blocks and the contents it includes. A single copy of a shared ledger is updated and is available to all users. The transactions are then reconfirmed every time a new block is created. Usually, six confirmations is needed in the Bitcoin network to consider the transaction final.

The blockchain-based digital dozier were all data from a vehicle's life cycle that can be stored transparently and securely. The information can be accessed and used by different stack holders. This includes car manufacturer, importer, used car dealers, insurance company, regulator, registration authorities, and car owner. The blockchain technology covers all data that is being used and created during the life cycle of a car (He, Su, Gao, & Yue, 2020). Various gatherings locate a shared factor and speak with one another in a straightforward manner. The major functions to be considered in the vehicle life cycle of the blockchain-based model are (1) introducing a new vehicle product manufactured by a car manufacturer; (2) verifying logistics providers; (3) tracking all materials to make the car at the factory; (4) registering a new car dealer; (5) delivering products from the manufacturer to list of approved car dealer; (6) selling the product to the car dealer and private buyers; (7) registering vehicle using VID to track vehicle history; and (8) tracking vehicle by the regulator, user, and insurance agency.

In this work, the blockchain-based vehicle's life cycle system is proposed to track the vehicle's history. The main actors involved in the vehicle life cycle management system are the manufacturer, car dealer, car owner, regulator, and insurance company. The flow of information between actors are recorded in the blockchain. This type of visibility and transparency is important to improve the vehicle's traceability. Every vehicle is assumed to have a unique identifier to represent its identity. If a vehicle is falsely labeled, or more than one vehicles have the same identifier, then it can be identified using blockchain. The manufactured vehicle are supplied to the registered dealer. Then, the registered vehicle is delivered to the user. Through the blockchain, vehicle history is tracked by the regulator. The consortium blockchain-based model is proposed, the consensus process is handled by a preselected group of participants, and the read access can be public or is limited to its participants. In our model, only registered participants can perform manipulation in the blockchain, however, the read access is public because any consumer can verify the origin of a vehicle. The vehicle related notification goes from blockchain to regulator and insurance agency.

5. Blockchain in waste management

Each day, the world is suffocating under the expanding volume of waste, affecting human well-being, atmosphere, and climate. The developing wastage rates, the issues of finding new removal locales, exhaustion of landfill space, garbage removal, and the broad waste removal strategies are hampering sufficient waste administration. For productive outcomes, it is important to present an advanced waste administration framework to show a full pattern of waste administration, from how the waste was gathered to when and where they were discarded. Blockchain plays a major role on waste management (WM), that helps the specialists to accomplish feasible sterilization on squander. The WM methods are not equal when considering their environmental impact. The disposal of waste in landfills, for a long time the main and still widely spread WM method, has the biggest environmental impact. Disposal takes away land and may cause infiltration of leachate into soil and ground water, which creates unpleasant odors and attracts rodents. Therefore, disposal of waste is considered the least desirable option in the waste removal hierarchy. Waste recovery includes anaerobic processes such as digestion, incineration with energy recovery, gasification, and pyrolysis, as well as making fuel from waste, called refuse derived fuel (RDF). Waste recycling means turning waste into a new substance or product. For example, material from plastic bottles may be used for making clothing. Recycling also includes composting. Recycling is a more preferred method for WM than recovery and disposal. Reusing waste includes cleaning, repairing, and refurbishing whole items or spare parts. Finding somebody who would like to use already used item poses often the biggest challenge. All second hand and vintage shops serve well the goal of reusing waste.The best but hardest of all options is waste prevention. This means using less material already in the design and manufacture of products. Keeping products longer in use helps to prevent creating waste. Waste hierarchy is a good tool for roughly estimating the impact of a given WM system and planning changes to it (Fig. 13.4).

The waste comes from both commercial factory material, energies and house all are sorted and stored separately. The segregated waste after some process used for utilization, disposal, and landfill.

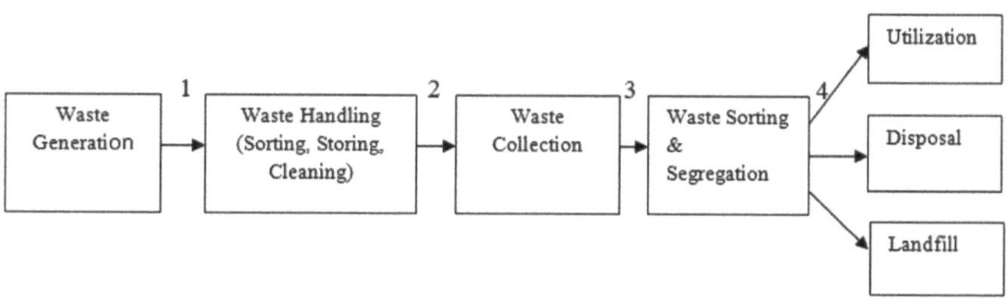

FIGURE 13.4 Architecture in waste management processes.

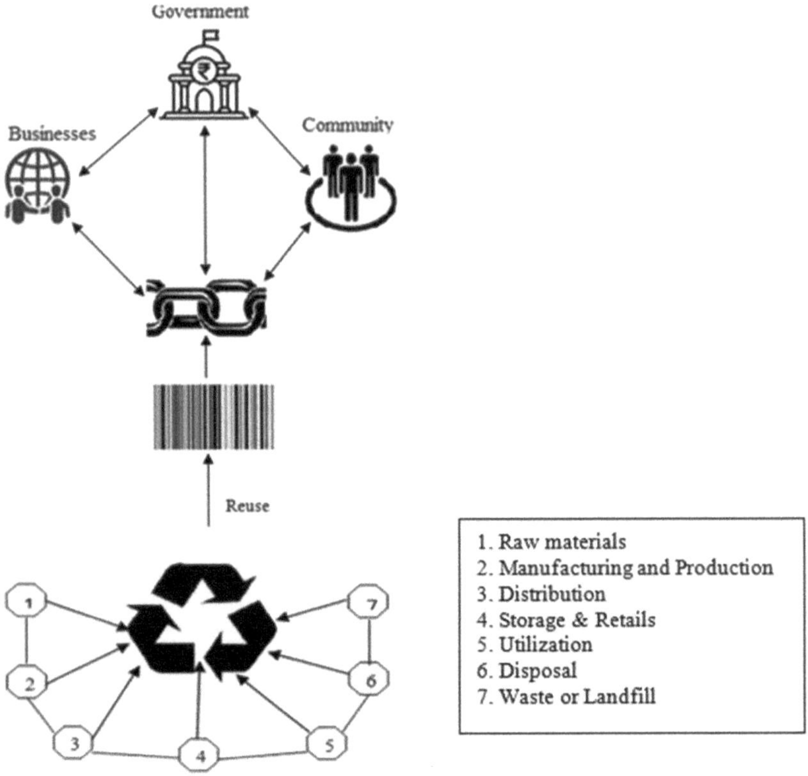

FIGURE 13.5 Blockchain in waste management processes.

The WM are collected by government service workers. The collected waste is stored in a particular place. The waste contractor and government workers sign the contract for collecting waste (Dinh et al., 2017).

Fig. 13.5 shows that, the WM is collected by government service workers. The collected waste is stored in a particular place. The waste contractor and government workers sign the contract for collecting waste.

Fig. 13.6 shows the interaction of various stakeholders of waste management. All the waste materials are reused. They assign separate digital IDs. that same ID is used by the government, business management, and the community to monitor all process about WM.

6. Seed traceability using smart contracts

Smart agreements can possibly change the security of farming and food items into an incorporated, keen framework that ensures the nature of the item conveyed to the end client. The implementation of the savvy contract capacities and system will be done by the large number of mining hubs that are

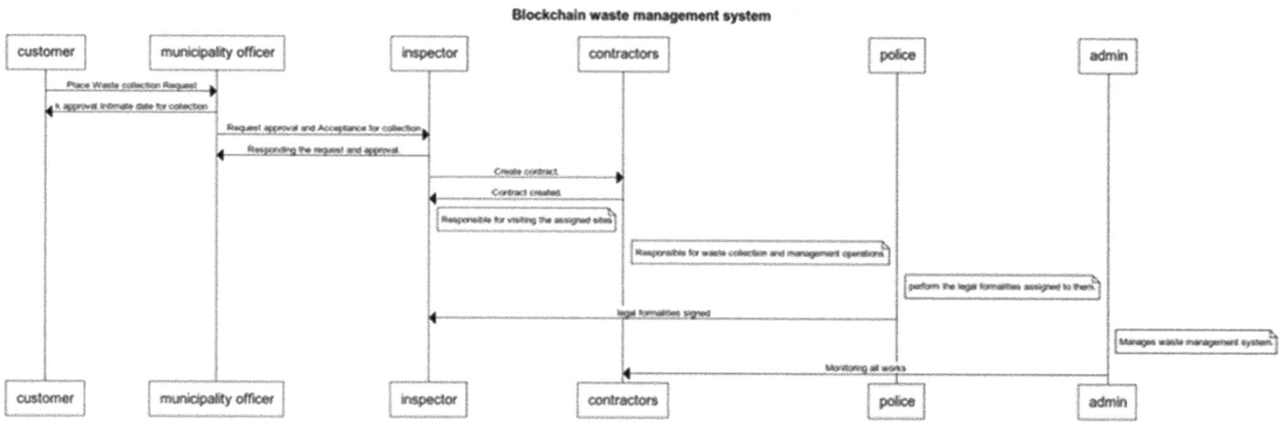

FIGURE 13.6 Sequence diagram showing interactions between the stakeholders of waste management.

circulated around the world, and the execution result is concurred by all of the mining hubs. It is important that mining hubs are what make up the blockchain network (Bogner, Chanson, & Meeuw, 2016). A mining hub can be any processing machine that gathers, approves, and executes exchanges. The hubs likewise store information and the results of these exchanges in a record that is recreated and synchronized by all mining hubs. In this manner a mining hub has the equivalent definite imitation of all other mining hubs. In the blockchain, savvy acquiescents will get exchanges in type of capacity calls, and furthermore, will trigger functions to potentiate the partaking of substances to sedulously screen, follow, and get plausible cautions when infringement transpires. Consequently, this avails in reestablishing the conditions to the ideal and reacting to the infringement that happens inside the seed esteem chain.

Fig. 13.7 shows the most important contributing entity containing the milt agency, cereal processor, agent, buyer, crofter, cereal conveyer, dealer, and smart contract. This agreement has both public and private key transaction times. Both are digitally signed, and the validation and verification performed.

Between the new agreement signed by the crofter and the milt agency, the crofter would buy milt from the milt agency. Milts are traded through a milt agency. The crofter is responsible for cultivating the milt and updating the crop information in the distributed system. The cereal conveyor calls the crofter to purchase a goods. The crofter trade the goods to conveyor (Li, 2013). Cereal processor purchases cereal from conveyor. The cereal conveyor trades the

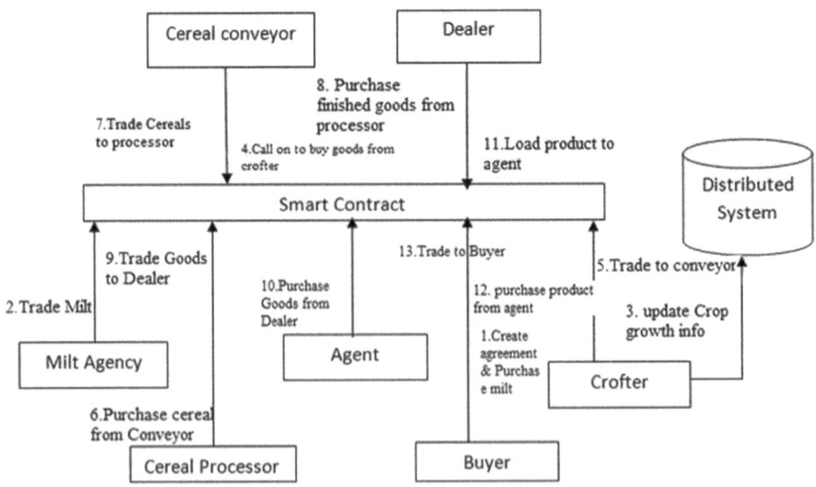

FIGURE 13.7 Overview of Seed Traceability.

cereal to processor. Dealer purchases a finished good from the processer. Milt agency trades goods to the dealer. Agent purchases goods from dealer. The dealer loads product to the agent. Buyer purchases the product from the agent. The Fig. 13.8 represents the UML digram representation to explain the relationships between stakeholders.

Fig. 13.9 represents the unified modeling language relationship chart that shows the savvy contract traits and capacities and the connection between the partaking substances and the keen agreement. Such metadata shows, furthermore, how relations are critical to execute savvy contracts. In the blockchain and distributed system, all pictures, information, and reports are carefully marked and ascribed to a specific entertainer. This implies that the entertainer for transferring pictures, for this situation the rancher transferring MPEG documents, is the undisputed proprietor of such activity and is answerable for incorrect or deceitful pictures or, on the other hand, MPEG records (Beck, Stenum Czepluch, Lollike, & Malone, 2016).

As per the Fig. 13.10, between the new agreement signed by the crofter and smart agreement using the method of CreateAgreement(). Once the agreement

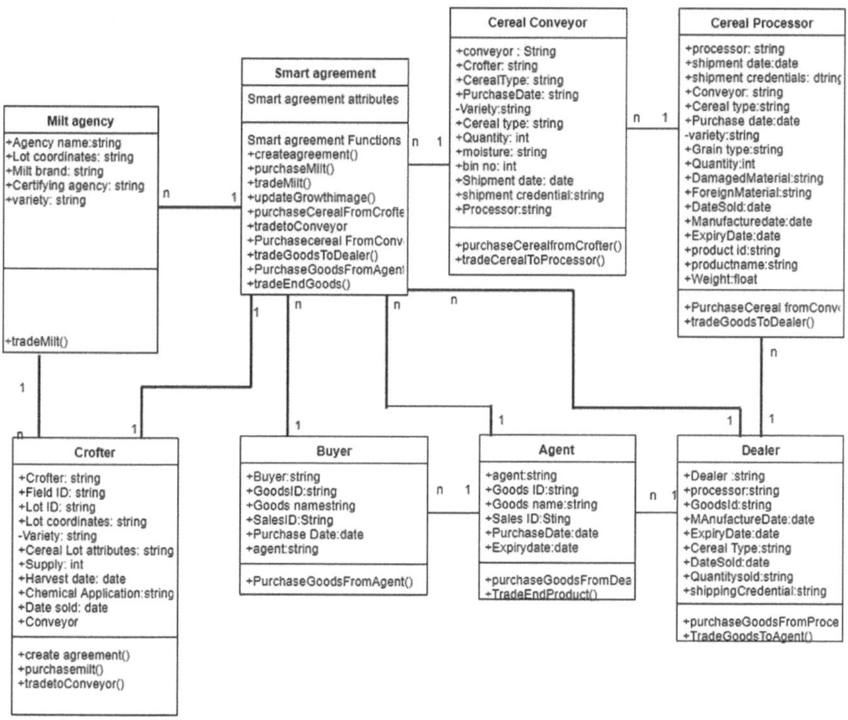

FIGURE 13.8 UML diagram representation.

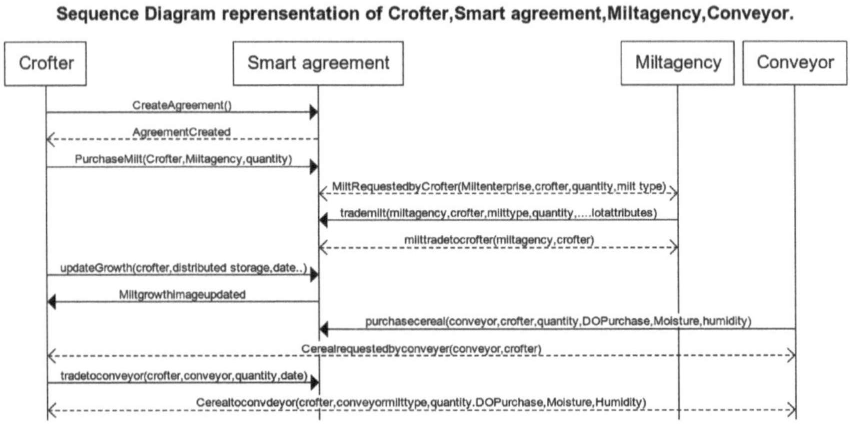

FIGURE 13.9 Sequence diagram representation of crofter, smart agreement, milt agency, and conveyor.

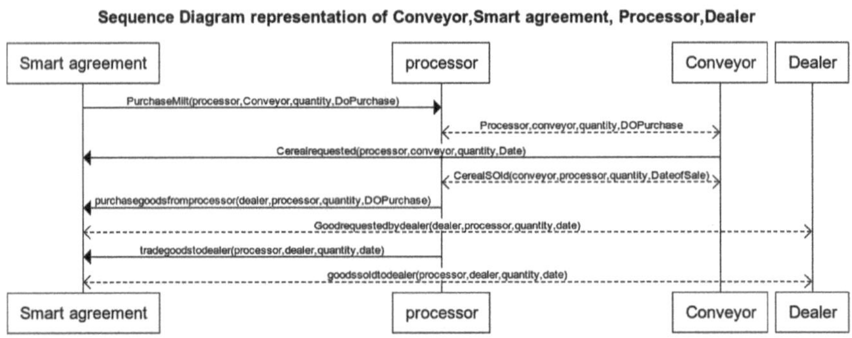

FIGURE 13.10 Sequence diagram representation of conveyor, smart agreement, processor, and dealer.

created, after that PurchaseMilt(crofter, miltagency, quality) method called to get the purchasing information of milt. Then, the method of MiltRequestedbyCrofter between smart agreement and miltagency. The updateGrowth() method called between crofter and smart agreement to know about the growth data. The method of PurchaseCereal() is called by cenveyor to purchase the cereal. The tradetoconveyor method is called by crofter to credit the details of cereal, conveyor, quality. Finally, smartcontracts are arised between all stakeholders.

Fig. 13.11 explains the relationship of dealer, Smart agreement, Agent and Buyer to make smart contracts. The method of Purchasegoodsfromagent() is sent by Dealer. The corresponding response forwarded from agent to dealer. TradeGoodsToAgent() which contain information about all buyers, agent,

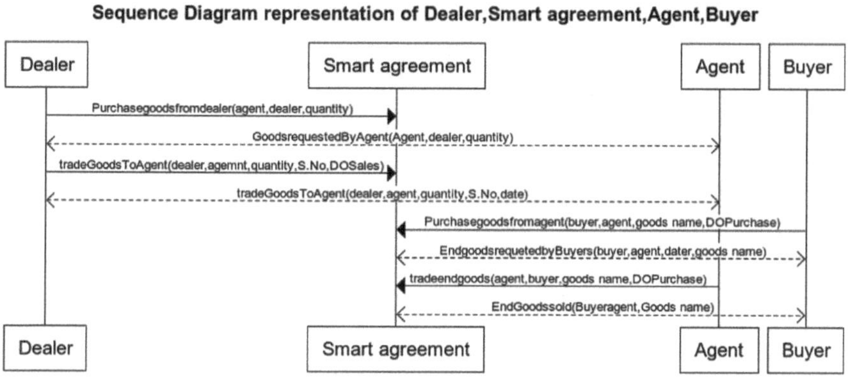

FIGURE 13.11 Sequence diagram representation of dealer, smart agreement, agent, and buyer.

quantity, date. Purchase requested and corresponding reply also performed. The EndGoods sold between buyer and Smart agreement. To conclude, the agent trade the goods to buyers by executing TradeEndGoods() function. The elegant agreement broadcasts the sale of the goods with the EndGoodsSold() event.

7. Electronic voting system

Blockchain may create one of the foremost prominent alternatives to traditional voting in terms of security, consistency, and speed. It includes decentralized data storage, which can hold a public ledger of transactions (Zhang, Kasahara, Shen, Jiang, & Wan, 2019). The ledger can record many agreements: monetary transactions, property transfer, and even ballot storage. All transactions that occur on a typical blockchain are verified and signed with cryptography to make sure security and anonymity. Fig. 13.12 The electronic voting system in which stealing votes or remaking votes are totally blocked. When a citizen casts a vote, the choice of voter is represented as a record of transaction in a block, and that transaction in a block is transferred to authorized agents such as election officers and chief electoral officers in the network to authorize the validity of a voter (Wang et al., 2019).

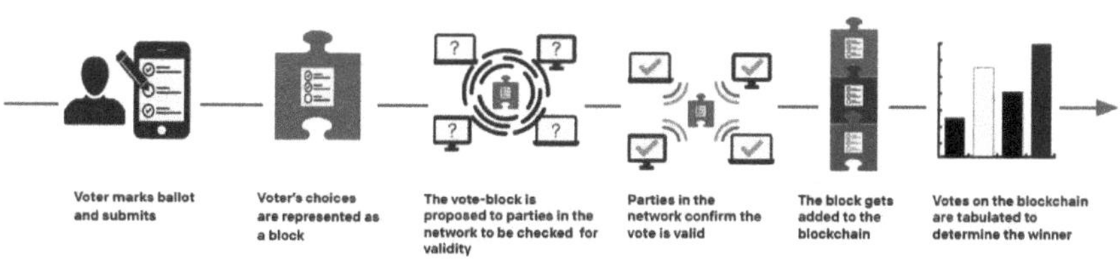

FIGURE 13.12 Workflow of the blockchain voting system.

8. Conclusion

Blockchain technology makes a stable and immutable record of every transaction. The blockchain and its derivatives are now engaged to protect any type of transaction. The adoption of blockchain is increased with the increasing global need for better security. This impassable digital ledger aids to preserve from fraud, hacking, data theft, and information loss. The technology could be used in all industries in the world, including manufacturing, real estate companies, retail, health care, and transportation. It is very much valuable to build smart cities, where trust is the main concern, to add an additional layer over the internet.

References

Beck, R., Stenum Czepluch, J., Lollike, N., & Malone, S. (2016). Blockchain – the gateway to trust-free cryptographic transactions. In *Twenty-fourth European conference on information systems (ECIS), Istanbul,Turkey* (pp. 1–14).

Bogner, A., Chanson, M., & Meeuw, A. (2016). A decentralised sharing app running a smart contract on the ethereum blockchain. In *ACM international conference proceeding series* (Vols. 07–09, pp. 177–178). Association for Computing Machinery. https://doi.org/10.1145/2991561.2998465.

Dinh, T. T. A., Wang, J., Chen, G., Liu, R., Ooi, B. C., & Tan, K. L. (2017). BLOCKBENCH: A framework for analyzing private blockchains. In *Proceedings of the ACM SIGMOD international conference on management of data* (Vol. 127746, pp. 1085–1100). Association for Computing Machinery. https://doi.org/10.1145/3035918.3064033.

He, G., Su, W., Gao, S., & Yue, J. (2020). TD-root: A trustworthy decentralized DNS root management architecture based on permissioned blockchain. *Future Generation Computer Systems, 102*, 912–924. https://doi.org/10.1016/j.future.2019.09.037

Kumar, A., Choudhary, D., Raju, M. S., Chaudhary, D. K., & Sagar, R. K. (2019). Combating counterfeit drugs: A quantitative analysis on cracking down the fake drug industry by using blockchain technology. In *Proceedings of the 9th international conference on cloud computing, data science and engineering, confluence 2019* (pp. 174–178). Institute of Electrical and Electronics Engineers Inc. https://doi.org/10.1109/CONFLUENCE.2019.8776891.

Li, L. (2013). Technology designed to combat fakes in the global supply chain. *Business Horizons, 56*(2), 167–177. https://doi.org/10.1016/j.bushor.2012.11.010

Rahouti, M., Xiong, K., & Ghani, N. (2018). Bitcoin concepts, threats, and machine-learning security solutions. *IEEE Access, 6*, 67189–67205.

Wang, Y., Cai, S., Lin, C., Chen, Z., Wang, T., Gao, Z., et al. (2019). Study of blockchains's consensus mechanism based on credit. *IEEE Access, 7*, 10224–10231. https://doi.org/10.1109/ACCESS.2019.2891065

Wood, G. (2014). Ethereum: a secure decentralised generalised transaction ledger. *Ethereum Project Yellow Paper, 151*, 1–32.

Zhang, Y., Kasahara, S., Shen, Y., Jiang, X., & Wan, J. (2019). Smart contract-based access control for the internet of things. *IEEE Internet of Things Journal, 6*(2), 1594–1605. https://doi.org/10.1109/JIOT.2018.2847705

Chapter 14

Blockchain-based health care monitoring for privacy preservation of COVID-19 medical records

J.V. Bibal Benifa[1], G. Venifa Mini[2], Saravanan Krishnan[3]

[1]*Department of Computer Science and Engineering, Indian Institute of Information Technology, Kottayam, Kerala, India;* [2]*Department of Computer Science and Engineering, Noorul Islam Centre for Higher Education, Nagercoil, Tamil Nadu, India;* [3]*Department of Computer Science and Engineering, Anna University Regional Campus, Tirunelveli, Tamil Nadu, India*

Chapter outline

Blockchain for Smart Cities. https://doi.org/10.1016/B978-0-12-824446-3.00005-3

1. Introduction

In December 2019, the pneumonia infection caused by coronavirus (COVID-19) occurred in Wuhan, China and spread rapidly in Wuhan areas and beyond (Wang, Hu, Hu et al., 2020). Potential stresses are now increasing on the global health care system. As the world fights the COVID-19 pandemic and its consequent economic and social effects, many governments and organizations around the world are joining the fight to minimize the spread of the virus while working fastidiously to develop a cure and a vaccine (Atri et al., 2020). Most of the countries are facing a dramatic spike in the number of medical patients, and it is very difficult to access primary doctors or caregivers. During COVID-19, the growth of IoT and wearable devices has increased the quality of patient care through remote health monitoring (RHM). In this pandemic situation, RHM is encouraged to provide self-distance and treatment outside hospitals. It also helps doctors (health care providers) to treat more patients. Using this ease of service, patients can remain linked to health care providers as needed. This also greatly reduces the medical costs and enhances the quality of health care.

The important components of RHM include monitoring devices, health data transmission to smart contracts, a smartphone with internet connectivity, and an application. This procedure is presented in Fig. 14.1, where the smartphone is a device that is used to fill the gap between individuals and organizations, and they could play a major role in resolving COVID-19 reactions. New technologies like blockchain have helped to diagnose many infected people, recognized in the areas in which COVID-19 spreads, and information can be tracked dynamically. In RHM, wearable devices and IoT play an important part to build secure medical records in smart cities. Wearable devices capture health data from patients and send it to hospitals to promote health monitoring, diagnosis of diseases, and medication.

In the health care systems, wearable devices are smart electronic devices with microcontrollers that can be implanted into clothing or attached as accessories on the body. They are seamless, user-friendly, and linked to advanced features such as wireless data transfer, real-time input, and system

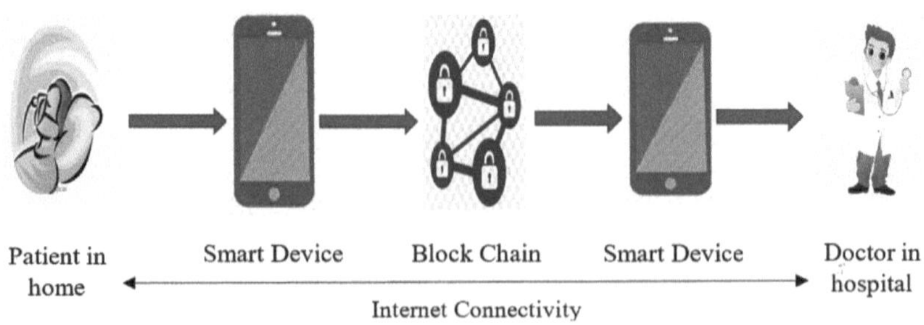

FIGURE 14.1 Remote health monitoring system.

built-in alerting mechanisms. These devices may provide health care providers with vital information such as blood pressure, blood glucose levels, and breathing habits. New wireless sensor wearable device be seated softly on the throat to track fever, cough, and respiratory activity shown in Fig. 14.2. The lightweight, flexible, and waterproof thin device about the size of a postage stamp sits just below the suprasternal notch and the noticeable dip at the base of the throat. The system tracks coughing intensity and patterns, movement of the chest wall, respiratory sounds, heart rate and body temperature, including fever, from this position. This is the first wearable device that keeps track of monitoring COVID-19 symptoms.

Health care devices can be divided into four different categories as illustrated in Fig. 14.3. Fixed medical devices can be fixed on a physical location. Medical implanted devices can be positioned inside the body. Wearable medical devices can be prescribed by doctors (health care providers). Consumer products like Fitbit and Fuelband are also used as wearable health monitoring devices.

IoT devices are embedded with software, electronics, sensors, actuators, and networking that enables the wearable user to link and share data (Haleem, Javaid, & Khan, 2020; Li et al., 2020). IoT devices are the combination of smart devices and sensors, and they have limited storage and computational resources. Smart cities trust deeply on sensors to observe parameters such as temperature, humidity, allergens, noise, traffic conditions, and power grid status. Such parameter values provide a context that helps to understand a citizen's state at any given time (Solanas et al., 2014). Further, responding to sensed data is strategically helping to make smarter health care services. By having access to this real-time COVID-19 information, the public services can react quickly to urgent health requirements and take decisions in critical situations. All countries, including India, are struggling with COVID-19 in the current pandemic situation and are still looking for practical and cost-effective

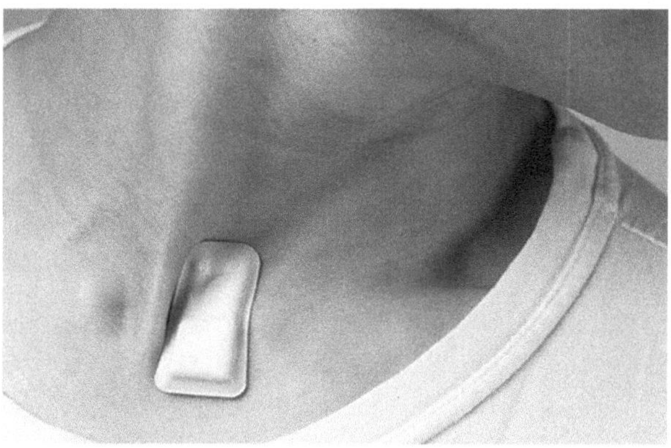

FIGURE 14.2 New wearable device.

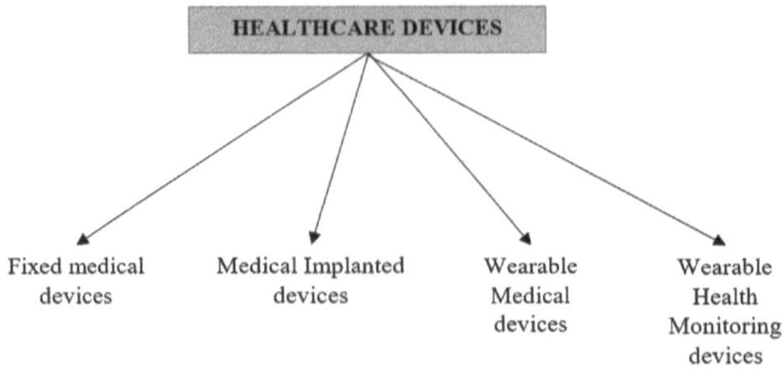

FIGURE 14.3 Health care devices.

solutions to the problems that arise in several ways. In the current typical case, almost all of the issues occur because of inadequate accessibility to the patients, which is the second most significant issue after the vaccine production question (Wang, Hu, Li et al., 2020). Using the IoT concept makes the patient's accessibility quite useful, which ultimately helps to give them substantial care so they can get out of this disease. One significant role of IoT is to acquire data through sensors and smart devices, which can be further processed and analyzed in cloud environments.

The proposed way of collecting and analyzing data from patients helps to improve the health of the civilization during this COVID 19 pandemic. Health care devices are utmost important to observe the symptoms of the patients in home by the wearable devices. The data collected by health care devices are protected and transmitted in a secure way. This will be achieved by cryptographic primitives. IoT devices encrypt the data and send it to cloud servers. The metadata of the original data is stored in blockchain. Thus, there has been a step toward combining IoT and the cloud to outsource data collection, processing and sharing capabilities.

Data Owner 1 (DO1) can store and share the data collected through the cloud with Data Owner 2 (DO2) to decrease costs as highlighted in Fig. 14.4. Since the cloud is semitrusted, there is a concern that user's data and privacy

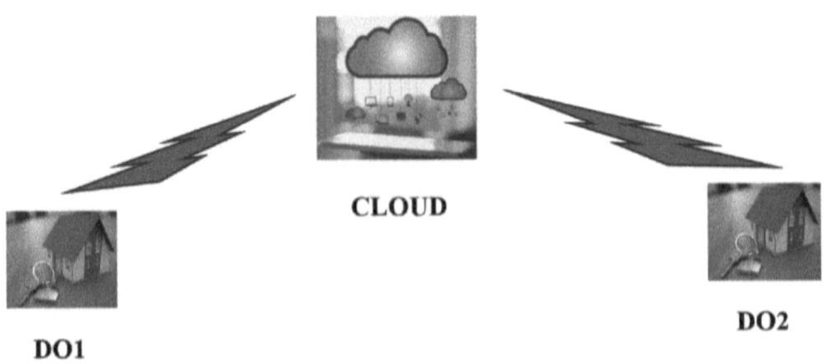

FIGURE 14.4 Sharing of data between DO1 and DO2.

can be leaked and breached in or outsiders. Though there are many techniques for privacy-preserving data processing using cryptographic access control tools in cloud storage, these access policies and data can be tampered with and breached by cloud service providers (CSP) or by data users (DU).

Data is an extremely important currency. Health care systems accumulate tons of sensitive information about individuals. Unfortunately, the data attracts hackers. The data is accumulated in one location and it is very easy for hackers to steal the data. When it comes to data, existing technologies use centralized storage. This can be changed in the future by storing data in a decentralized way. The decentralized way of storing data is achieved through the blockchain, becoming popular in the near future. Due to the decentralized existence of blockchain technology, hackers no longer have a single entry point, nor are they able to access entire databases in case of stealing data. Blockchain-based technology can also safeguard all data exchanges between IoT devices. It can be used to achieve safe data transfers near real-time and to ensure timely contact between devices located thousands of miles away (Jesus, Chicarino, Albuquerque, & de Rocha, 2018; Rifi, Agoulmine, Chendeb Taher, & Rachkidi, 2018).

The blockchain's irreversibility property becomes a resistance to a consent revocation feature, which allows users to deny permission for designated individuals from a certain action on the data. Blockchain's transparent property allows all network participants to view all data that can create a question of confidentiality. Blockchain's restricted storage becomes an issue toward the ease of use for the exponential growth of the different medical related data. Although private blockchain like Hyperledger Fabric (Nasir, Qasse, Abu Talib, & Nassif, 2018) have the potential to control the contribution in its blockchain network, the personal health record system gives access to a particular part of the system only to certain individuals. Due to that, a privacy disclosure is also an issue.

This chapter aims at resolving these blockchain disadvantages in smart cities and proposing an RHM model based on blockchain to solve issues in this pandemic situation. The proposed system is constructed using blockchain technology on IoT to reduce the probability of transmission, and increasing the possibility of better results. Proxy reencryption (PRE) and other cryptographic methods are used for privacy protection to protect the sensitive information of users. In other terms, users should determine individually who should access their data without violating the privacy of their data and identity.

The key contributions of this work are summarized as follows.

IoT device data were initially encrypted by advanced encryption standard (AES). Then, it is integrated with GateWay server (GW) to verify the validity of all authenticity and behavior inside the system.

Blockchain can be used to establish a privacy scheme that protects patient health information to promote the concept of nonrepudiation, transparency, and tamper resistance.

An access control framework is proposed by the use of PRE techniques to support features of fine-grained access control and consent revocation, while cloud storage is used to support the availability function.

The remainder of the chapter is structured according to this summary. Section 2 presents the background of blockchain and issues in the existing technologies, attribute-based encryption (ABE) and conventional health care systems on cloud environments. The proposed model is defined with a detailed architecture in Section 3. Section 4 investigates the safety measures and privacy of the proposed model, and Section 5 evaluates the efficiency and the discussion on the results is presented in Section 6. Finally, the chapter is concluded in Section 7.

2. Background and related works

Blockchain is an open, distributed ledger that can verifiably and permanently record transactions between two parties. It allows organizations to communicate without a central, trusted third party (Mayank, Danilo, & Katina, 2019).Furthermore, blockchain provides smart contracts without any key authority. At this date the blockchain Ethereum is the main facilitator of blockchain smart contracts (Buterin, 2014). A Blockchain is a diary which can be forged hardly. The main attribute of blockchain is decentralization, which means the information applied to blockchain is not regulated by any central authority. The blockchain consists of a number of nodes. A node is a virtual machine or physical machine. The entries sent to the blockchain are accepted in a peer-to-peer network using different consensus protocols. The protocols are designed by blockchain creators. One node can be identified by its unique IP address. Blockchain can maintain privacy in both the public key and private key. The user enters into the blockchain through public key cryptography and the messages are authenticated with a private key. Users interact without any clear reference to their identity with their private key and public key. Another main function of blockchain is that of persistence. Owing to the distributed ledger, maintained through several nodes (Zheng, Xie, Dai, Chen, & Wang, 2017), it is almost impossible to delete entries after being accepted into the blockchain. In addition, the prospect of anonymity is an attractive feature found in many blockchains.

Fig. 14.5 presents the basic structure of blockchain. A block is a basic element, which stores transactional data in the blockchain system. A block in the blockchain contains two parts, namely, header and body. The header of the block contains block number, hash code, previous block hash, nonce, metadata, timestamp, and Merkle root. The body of the block contains transactions and transaction counters. Each block is cryptographically connected to other blocks (Merkle, 1989). The blockchain retains an ever-growing collection of data entries, packed together into data blocks. Such data blocks are readable by all, viewed by everyone in the network, writable by everyone, and verifiable by

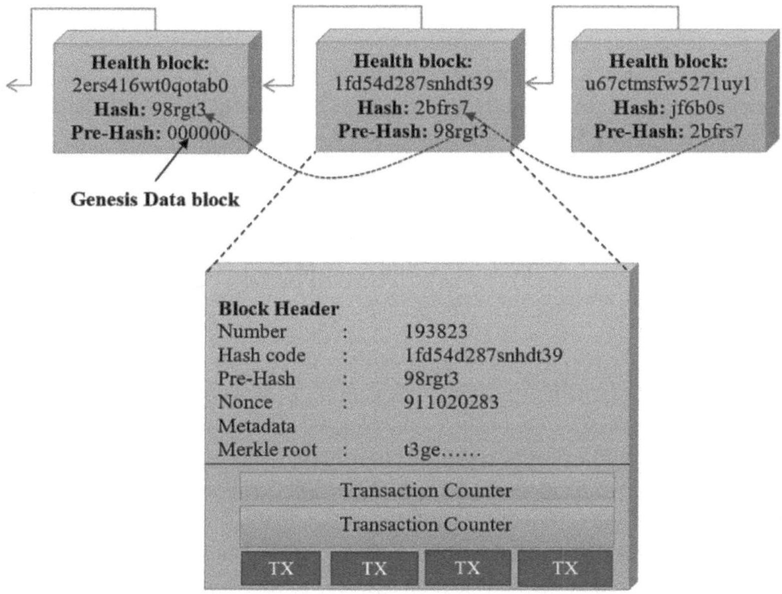

FIGURE 14.5 Basic structure of blockchain.

all. This allows decentralized data management and transactions. These properties of the blockchain receive a lot of publicity in various applications.

Blockchains allow traceability by linking a new block to the previous one using hash, and thereby creating a chain of blocks. The block transactions are generated in a Merkle tree (Merkle, 1989) as shown in Fig. 14.6, where each value of the leaf can be checked to the known root.

Au et al. (2010) offered a positive response to this issue by bringing forward a general structure for safe sharing of personal health records (PHRs). This program allows patients to securely store and share their data on the cloud server (such as their caregivers) and, in turn, practicing doctors can refer the

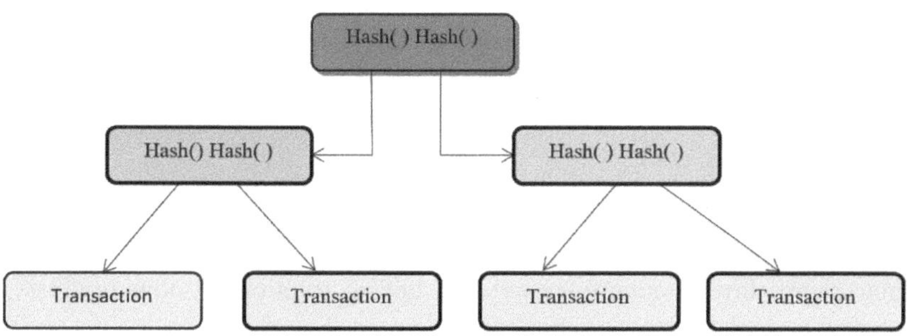

FIGURE 14.6 Merkle tree of transactions.

medical records of patients to specialists for testing purposes if appropriate, while ensuring that patient information remains confidential. Cloud technology has been seen as a popular candidate for the storage of confidential medical records in India, but the security protection given to date is still inadequate without affecting the practicality of the system.

Kaufman (2009) proposed an idea of data security in the world of cloud computing. Today, within the confines of the internet, a practice known as cloud computing technology enables use of scalable, distributed computer environments. In this modern computing environment, users are uniformly expected to acknowledge the principle of trust that underlies it. The virtual environment enables users in the cloud computing world to access computational power that exceeds that found within their own physical worlds. The data in a cloud vary from publicly accessible sources with minimal security issues to private data with highly confidential information. Like with other developments in technology, regulators are usually in a "catch-up" mode for defining regulation, governance and law. Cloud computing provides an extension of previously encountered issues with the internet. To ensure that these decisions are knowledgeable and suitable for the cloud computing environment, the industry itself should develop consistent and efficient policy and governance to define and enforce acceptable security methods.

Wangthammang and Vasupongayya (2016) suggested a centralized storage system for encrypted personal health record data for cloud storage purposes to handle encrypted PHR data. It offers an API for the upload/download of encrypted PHR data from a cloud storage network. The distributed storage design for encrypted personal health record (DSePHR) addresses Hadoop distributed file system (HDFS) namode memory problems by classifying the encrypted PHR data into small and large files while storing a lot of small files. HBase schema which is proposed in this work should manage the small files. The memory use and processing time of the proposed DSePHR was measured using actual data sets from various health care populations.

Liam, Buchanan, Jonathan, & Owen, 2018 addressed recent work in the fields of traceability, data sharing, clinical trials and methods for monitoring medications. There are many health care and wellbeing fields that could be improved using blockchain technology. Those include computer monitoring, clinical trials, prescription monitoring and insurance coverage. Hospitals may trace their properties through a blockchain network, and over a system's entire lifecycle, via patient monitoring. The collected information will then be used to enhance patient safety and include postmarket analyses to increase productivity savings.

Ivan (2016) addressed blockchain as an innovative approach to safe data storage for safety, implementation hurdles, and a strategy for increasing change from current technology to a blockchain solution. Various methods of capturing and exchanging medical data have a range of drawbacks that restrict patient access to their clinical information, minimize the availability of critical data to care providers and eventually pose an obstacle to turning US health care into a learning health system. These shortcomings can be remedied by storing patient health care data within a blockchain-based storage scheme.

Dubovitskaya, Xu, Ryu, Schumacher, and Wang (2017) proposed a system for the management and exchange of electronic medical records (EMR) data in patient care for cancer. Electronic medical records (EMRs) are important, highly confidential private health care information, and also need to be shared among peers. Blockchain offers a common, permanent, and transparent history of all transactions to create secure, accountable, and open applications. This presents a unique opportunity to build a stable and dependable EMR data management and sharing framework using blockchain. This proposed research will dramatically reduce the processing time for exchanging EMR, enhance patient care decision making, and minimize total costs. Kosba, Miller, Shi, Wen, and Papamanthou (2016) presented Hawk, a decentralized smart contract program that does not explicitly store financial transactions on the blockchain, thereby protecting the public's transactional privacy. A Hawk programmer can write an intuitive private smart contract without cryptography, and the compiler automatically generates an effective cryptographic protocol where contractual parties communicate with the blockchain using cryptographic primitives such as zero-knowledge proofs. To formally define and explain why these protocols are safe, the blockchain cryptography model was formalized. Nonetheless, existing systems lack transactional privacy. All transactions are revealed on the blockchain, including the flow of money between pseudonyms and the sum transacted.

Radanovi and Liki (2018) explored the possibilities of medicine for using blockchain technology. Cases of this technology have so far been used for cryptocurrency, digital contracts, financial and public records, and ownership of assets. Future uses are expected to spread to medicine, technology, education, intellectual property, and supply chain management. Blockchain technology is a decentralized database that stores a record of assets and transactions through a peer-to-peer computer network protected by cryptography and, over time, its history is locked into data blocks that are cryptographically connected and protected together. Medicine-related technologies may include electronic health records (HER), health care, biomedical testing, product distribution and procurement procedures, and medical education. The use of blockchain is not without its limitations and at present this technology is highly unstable and lacks public or even professional information, making it impossible to have a strong strategic view of its true potential for the future. There are currently issues with the scalability, smart contract protection and app acceptance.

Marek (2018) developed a ground breaking blockchain technology concept that addresses critical data protection, implementation and integration issues, providing a new strategy for the health care sector that has the ability to link providers while protecting sensitive data. In order to ensure ease of deployment in a hospital system, blockchain technology using a distributed architecture with microservices was built. This architecture enables user to encapsulate core functions with the system into discrete services that can be independently scaled depending on the requirements of a specific hospital system. As part of this architecture, core components for safe handling of cryptographic secrets, communicating with blockchain nodes, enabling large file sharing, allowing secondary-index dependent lookups, and incorporating external business logic that governs how users communicate with Smart Contracts (SCs). Through enabling data exchange between providers and EHR systems, blockchain technology has the ability to improve health care delivery. Major roadblocks however stand in the way of the widespread adoption of this technology in the health care industry. This blockchain-based data sharing solution solves two of the most important health data sharing problems associated with using blockchain: securing sensitive health records, and deploying and integrating blockchain applications through multiple hospital environments.

Clauson, Breeden, Cameron, and Mackey (2018) presented an overview of the opportunities and challenges associated with the implementation and deployment of blockchains for the health supply chain, focusing on pharmaceutical supply, medical equipment and services, the Internet of Healthy Things (IoHT), and the public health sector. Griggs et al. (2018) built an SC program that would enable patient tracking and medical procedures in real time by sending alerts to patients and medical practitioners, while maintaining a safe record of who initiated such activities. It will overcome several security issues associated with remote monitoring of patients and simplify the distribution of HIPAA-compliant alerts to all parties concerned. The proposed system is to use blockchain-based SC to facilitate safe analysis and management of medical sensors to handle the protected health information (PHI) created by those instruments. Using an Ethereum-based private blockchain, they built a network where the sensors interact with a mobile device that calls SCs and records all events on the blockchain.

Luis, Frank, & Ole, 2018 identified how blockchain technology can be used in public health surveillance by exchanging decentralized genomic data. A brief overview of why blockchain technologies are required in public health with a distinction between public and private blockchains is presented. Eventually, a plan for a network of blockchains is included, using the Cosmos architecture, along with decentralized storage frameworks such as IPFS and BigchainDB, to resolve the interoperability problems in the health sector. Luis, Frank, & Ole, 2018 aimed at exploring the role of blockchain in supporting data management in clinical trials and establishing proof-of-concept implementation of a patient-facing and research-facing framework. Blockchain-based SCs have been developed using Ethereum framework. The BlockTrial is a program that uses a web-based interface to allow users to run SC on an Ethereum network related to trials. Functions require patients to give researchers access to their data and allow researchers to query data that is stored off the chain. This program creates a permanent and clear log of these and other transactions as a form of distributed ledger. It could also encourage patients to become more involved and fully informed research partners.

Khan and Khan (2018) proposed a system for validating electrical transactions embedded in blockchain using multiple producers' signatures based on their assigned attributes. Such signatures are checked and supported by users who meet certain qualities without any details being released. The producers generate the public and private keys for these consumers, and the endorsement process using these keys ensures that these consumers are approved. No central authority is required in this approach. Producers are given a hidden pseudorandom seed function to avoid collision attacks. The comparative review shows the efficacy of the proposed solution over the current ones.

Azaria, Ekblaw, Vieira, and Lippman (2016) proposed a modern, decentralized record management system for managing EMRs using blockchain technology that provides patients with accurate, unchangeable logging and easy access to their medical details via providers and treatment sites. Taking advantage of special blockchain assets, MedRec handles authentication, confidentiality, transparency, and data sharing. A modular architecture combines existing, local data storage solutions from vendors, enabling interoperability and making our system simple and adaptable. MedRec thus allows data economics to emerge, offering big data to motivate researchers while at the same time involving patients and providers in the option of releasing metadata. The purpose of this short chapter was to show a working prototype by which was being examined and discussed this approach before field tests.

The IoT systems deliver various opportunities and provide successful cyber-security solutions. It has many obstacles such as providing a secure data sharing environment and maintaining privacy because of the vast amounts of data generated by IoT devices (either single devices or whole systems). Data owners continue to think about how their data is being used when the power is out of hand. The protection of their data in cloud computing is the most prominent problem, and this affects the efficiency of this paradigm (Mineraud, Mazhelis, Su, X., & Tarkoma, 2016; Tariq et al., 2019) Encrypting data until they are outsourced has proven to be a secure way to reduce security issues. It is impossible to share the data with users when the data is encrypted as the owner needs to share the decryption key with those users, thereby providing access to the data. Another concern that emerges from sharing the keys is the revocation of accounts, where certain accounts would be refused service. What data owners normally do is invalidate the current key by using a new key to reencrypt the entire collection of data and redistribute the key to (authorized) users (Agyekum et al., 2019; Xia et al., 2017). Such action often becomes tedious and requires tremendously when large volumes of data are outsourced, and the owner does not retain a copy of the outsourced data locally.

ABE, an encryption scheme initially suggested by Sahai and Waters, achieves both access control and data protection by granting users specific access rights based on their attributes. One of their functionalities is the removal of user's access privileges. Using attribute-based encryption is also intended to provide fine-grained access control, since it decides the user has access rights to which form of data (Guo, Zhuang, Jie, Ren, Wu, & Choo, 2016). ABE is an ideal method for enforcing complex access control policies: the data to be accessed is connected to a set of attributes, and the user's rights are defined by a logical expression over those attributes (Ostrovsky, Sahai, & Waters, 2007).

Chow, Weng, Yang, and Deng, (2010) proposed an efficient one-way PRE scheme (without resorting to pairings). Under the Diffie-Hellman theoretical theory, they achieved high efficiency and CCA-security using the token-controlled encryption method in the random oracle model and a relaxed but rational definition. PRE makes it possible for a semitrusted proxy to convert a ciphertext originally intended for Alice into one which encrypts the same plaintext for Bob. The proxy just needs Alice's reencryption key, and cannot know much about the encrypted plaintext. It provides versatility in different applications, such as sensitive email, digital rights management, and remote storage.

Rao (2017) proposed a proven, secure CP-ABSC system for cloud-based PHR sharing that provides fine-grained access control, confidentiality, authenticity, sign crypto privacy, and simultaneous public verification. This system takes advantage of concise monotonous boolean functions as predicates of signing and encryption, and understands protection in the standard model. At the positive side, this construction shows a short ciphertext size and involves fewer pairing computations compared to the current regional schemes.

Li, Li, Wen, Zhang, and Zhang, (2017) presented an improved ciphertext-policy attribute-based encryption (CP-ABE) scheme to create an encrypted data access control solution suitable for mobile users in a hybrid cloud environment. In hybrid cloud computing, encrypted data access control can provide organizations with a fine-grained method of access to policies which are similar to organizational policies. This scheme is secure, versatile and efficient for use in mobile hybrid cloud computing but not suitable for IoT devices. Gu, Jia, Wang, and Wen (2017) presented an attribute-based signature (ABS) system and outlined a comprehensive ABS protection model. Under this framework, in the standard model, an ABS scheme was presented for monotone predicates, where we select the Waters' signature scheme as the prototype of our ABS scheme. ABS is a novel cryptographic primitive that can render a message signed by the signing party with fine-grained control over information identification. Many ABS schemes had been proposed at the moment but most of them are not very successful.

Sangeetha and Vaidehi (2017) proposed a secure cloud-based PHR system for the sharing of PHRs between multiple users using ABE. Patients can encrypt their PHRs in this proposed system, and store them on semitrusted cloud servers. In addition, patients can retain control over access to their PHRs by allocating rights of fine-grained, attribute-based access to selected DUs. Ateniese, Fu, Green, and Hohenberger (2006) presented several efficient proxy reencoding schemes which offer security improvements over earlier approaches. The primary advantage of this schemes is that they are unidirectional (i.e., Alice can delegate to Bob without Bob needing to delegate to her) and do not allow delegates to show any of their hidden key to everyone. Proxy positions only a small amount of trust. For example, it is not able to decrypt the ciphertexts it reencrypts, and even though the proxy publishes all the reencryption information it knows, they have proved their schemes reliable. It allows for a range of applications that would not be possible if the proxy required full trust.

Private block chain (Hyperledger) is more efficient to reduce problem in mining process. The PHR data was protected using both the ABE scheme and the semitrusted servers. An access control mechanism is created by encrypting the sensitive PHR data. It increases the computational cost with the number of not revoked users if PHR contains huge volume of data. The issues related to blockchain are identified and privacy-preserving data sharing is projected as a potential solution.

3. Blockchain taxonomy

There are three major types of blockchains (Table 14.1): public, consortium, and private (Clauson et al., 2018; Zheng et al., 2017). They have different

TABLE 14.1 Type and properties of blockchain.

Property	Blockchain of public	Blockchain of consortium	Blockchain of private
Deciding consensus	All mineworkers	Selected group of nodes	By an organization
Read permit	Public	Public or limited	Private
Unchangeability	Almost impossible	Could be distorted	Could be distorted
Effectiveness	Low	High	High
Centralized role	No	Not fully	Fully
Consensus process	Unauthorized	Authorized	Fully authorized

features about who can access, write, and read the blockchain data. Everyone can access the data in a public chain, and everyone can join in and contribute to both consensus and modify the core program. The blockchain of public is used in cryptocurrencies. Cryptocurrencies such as Bitcoin (Nakamoto, 2019) and Ethereum (Buterin, 2014) are listed as public, permissionless blockchains. A blockchain of consortium may be considered as semicentralized, with only a small number of selected groups of organizations having access to viewing and participating in the consensus protocol. A private blockchain should be used as a central network, because it is fully controlled by one organization.

3.1 Popular blockchains

Existing blockchain technologies can be used to build decentralized applications. The most popular are Ethereum (decentralized platform) (Buterin, 2014) and Hyperledger (framework) (Cachin, 2016), both enabling developers to construct new blockchain applications on existing blockchains and to develop new test chains using their protocols.

3.1.1 Smart contract

SC is a digitized transaction protocol which enforces the contract terms (Macrinici, Cartofeanu, & Gao, 2018). It has long been proposed, and now it can be implemented in block chain technology. In blockchain, SC is a fragment of code that could be automatically executed by miners. SC have transformative potential in different sectors, such as financial services and IoT. The encrypted-only design of blockchain storage is perfect for storing data to facilitate a resistance function. Keeping data access permission can cause difficulties in providing a consent revocation feature. Transactions are publicly visible in a decentralized ledger (Kosba et al., 2016). It takes a lot of time for transactions and blocks to be propagated, because there are many nodes on public blockchain network. The transaction throughput is thus minimal, and the latency is high. A decentralized block generating phase requires a large amount of computing power, and energy. Consequently, the proposed model for RHM scheme should identify a method to deal with such limitations.

3.1.2 Proxy reencryption cryptosystem as a scenario

PRE is a type of asymmetric encryption (Public key cryptosystem) that enables a proxy entity to convert or reencrypt data from one public key to another, without accessing the underlying plaintext or private keys (Nunez, Agudo, & Lopez, 2017).The ciphertext which means plaintext is encrypted under the PRE scheme. The generated ciphertext can be retrieved in such a way that it can be decrypted by another user using his or her private key even though the cipher text is not initially encrypted with his or her public key. The scheme would therefore be a useful tool for providing a secure network for data sharing. The data owner must give the reencryption key to the proxy to transfer the data inside the proxy reencryption scheme. Moreover, the proxy will not be able to access any information from the reencryption key about the original data. The key to reencryption is generated by combining the secret key of the data owner and the public key of the intended-user. Therefore, the PRE scheme is flexible for the construction of an access control system in our proposed framework. In addition to the PRE proposal, four operations are required for the secure authentication of stored and transmitted data within the proposed framework: (1) encrypt and decrypt, (2) reencryption, (3) signing and verification (e.g., digital signatures), and (4) encryption and decryption verification.

Symmetric encryption cannot be used separately, because it means sharing the same key between suppliers, owners, and consumers. The conventional public key cryptographic system requires the data owner to be accessible online to reencrypt data when necessary. It is not always feasible, apart from being extremely inefficient. PRE can be used as a way to delegate decryption privileges, which then represents a natural candidate to create cryptographically enforced access control frameworks consistent with our motivating scenario. Moreover, the proxy cannot discover any information about the encrypted messages by any of the keys. The PRE cryptosystem is developed by Blaze, Bleumer, and Strauss, (1998). This cryptosystem contains some disadvantages of being bidirectional. Also, it is prone to collusion attacks. To resolve such a problem, the modified proxy reencryption scheme known as Ateniese, Fu, Green, and Hohenberger (AFGH) is used (Ateniese et al., 2016). Ateniese et al. work has been extremely influential as it offers the first formalizations of PRE syntax and notions of security as well as an initial description of PRE properties. The authors also offer multiple PRE applications, and in particular an access control server for a reliable file system. PRE enables the sender to encrypt the data just once and assign access to the data depends upon public keys for the recipients. This eliminates the data owner's requirement to be online, and also supports access revocation. The important properties of AFGH are (1) unidirectional ($X \rightarrow Y$ reencryption does not allow $Y \rightarrow X$ reencryption); (2) antiinteractive (trusted third parties' interactions does not require to generate a key for reencryption); (3) access (the reencrypted ciphertext will be decrypted by delegator); (4) optimal key (secret key size remains constant, whatever the number of delegations accepted); and (5) nontransitive (proxy alone cannot redelegate the right to decrypt). Such properties are suitable for the proposed model. The AFGH reencryption scheme will thus be used to establish the access control within the proposed model.

Fig. 14.7shows the correspondence between main actors in PRE. Alice (the data owner or sender) will assign decryption rights to Bob (the recipient or receiver) through a reencryption process performed by a group of reliable-minimized peers for any ciphertext designated for her. The delegator is

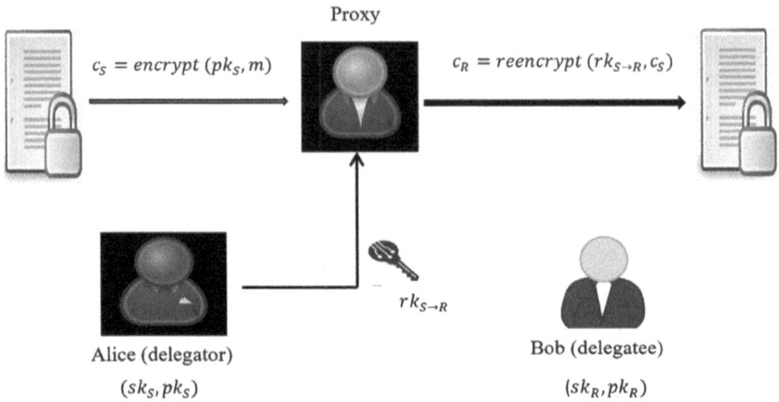

FIGURE 14.7 Correspondence between main actors in PRE.

Alice and delegatee is Bob. Delegator is the actor, who delegates his rights of decryption using proxy reencryption. The delegatee is allowed a reserved right to decrypt ciphertexts that were reencrypted for him with the permission of the original recipient, but not intended for him in the first place. Here, the encryption algorithm *encrypt* generates a ciphertext c_S when entering the public key pk_S and m message file. The reencryption algorithm *reencrypt* generates a second ciphertext c_S when a reencryption key rk_S and a ciphertext c_S is entered as input. By performing reencryption, Bob will combine these separate reencryptions and decrypt the original message using his private key sk_R.

3.2 Remote health monitoring on cloud

RHM is a technology that allows patient monitoring beyond traditional clinical environments, such as at home or in remote areas, which can improve access to treatment and decrease the costs of providing health care. In the medical sector cloud computing is rapidly becoming a requirement. It could be the solution to improve health care by exchanging patient details on urgent cases in real-time between medical providers. The deployment of the health record in a cloud environment provides many opportunities such as inevitable usability, versatile computing resources, high degree of intrusion detection and information sharing with other systems. As per HIPPA, the cloud providers are considered as noncovered organizations.

Like the other digital medical record system, health record data is handled and operated by the health record owner. The owners of health record can share their health data with others selectively while keeping other parts private. The cloud platform allows health record data to be accessed anywhere at every time. The health record program can also be assisted by the cloud to plan for medical appointments and create a more comprehensive image of personal health for communication, collaboration, and interaction.

3.3 Cryptographic primitives

The significant part of the proposed scheme depends upon pairing-based cryptography and cyclic group can be provided by bilinear mapping method. Let be a group which contains a collection of elements with binary operation, The properties of these primitives are presented as.

Closure property: Closure means an additional element in the set is the product of performing the operation on any two elements in the set. $\left(x, y \ \varepsilon \ G, \ \forall_{x,y} \ \varepsilon \ G\right)$.

Associative property: The order of the elements does not matter. $(x.y)z = x.(y.z), \ \forall_{x,y,z} \ \varepsilon \ G.$

Identity property: This property describes that $i \ \varepsilon \ G$ such that $x.i = x = i.x, \ \forall_x \ \varepsilon \ G.$

Inverse property: This property describes that $x^{-1} \ \varepsilon \ G$ such that $x.x^{-1} = i = x^{-1}.x, \ \forall_x \ \varepsilon \ G.$

The operations involved in group G are additive, multiplicative, or mapping function.

3.4 Bilinear mapping method

Let us define G_0 and G_1 as two cyclic groups on $F(Q)$, G_m is a multiplicative group with order of prime number q. The bilinear mapping method $e : G_0 \times G_1 \rightarrow G_m$ is a function, which should accept following properties.

Bilinearity: $\left(e\left(x^a, y^b\right) = e(x, y)^{ab}\right)$ such that $\forall \ x \ \varepsilon \ G_0, \ y \ \varepsilon \ G_1, \ a, \ b \ \varepsilon \ Zq.$

Nondegenerate: if $G_0 = < x >$, $G_{1=} < y >$, then $G_m = < e(x, \ y) >$

Efficient Computability: $\forall x \in G_1, \ \forall y \in G_2$ for any two elements x, y. This can be efficiently computed using $e(x, y)$.

4. Proposed work

Fig. 14.8 illustrates the architecture of an IoT-based remote health monitoring system that can be used in smart cities or the home. In these systems, information related to patients are recorded by body-worn or implanted sensors, in which the patient is prepared to monitor multiple metrics in a personal way. IoT is an advanced technology that ensures quarantine for all infected persons suffering from this virus. This is beneficial for a secure monitoring system during quarantine. Using the internet-based network, all high-risk patients are easily monitored. This system is used for measuring biometrics such as blood pressure, heart rate, and levels of glucose. Data obtained from IoT devices will help doctors to diagnose patients with the appropriate treatment plan and achieve the desired outcomes. Because of the sensitivity of the health record, the data should be protected at rest and on transmission using cryptographic primitives.

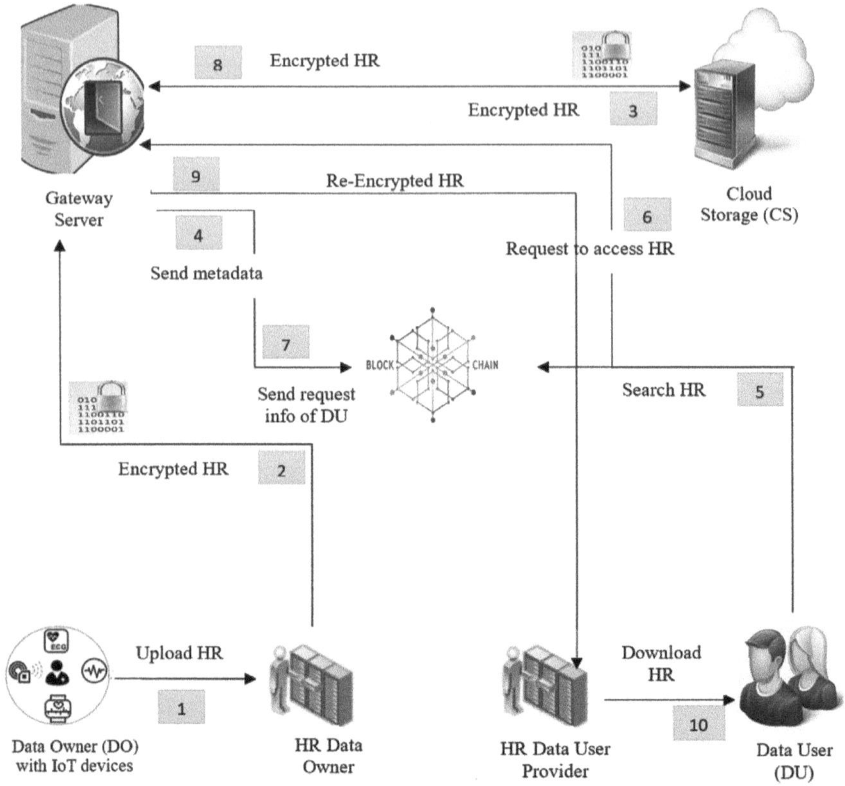

FIGURE 14.8 Architecture of an IoT-based remote health monitoring system.

The health record is exchanged through a proxy reencryption mechanism. The reencryption keys and other information required for the authentication process will then be stored on the gateway server proxy. Health record metadata will be stored on a private blockchain to enable search and resistance characteristics. The health record can be used by other health record owners such as health care providers, e.g., physicians and nurses. The following six major entities will be part of our model.

4.1 IoT devices

The IoT-enabled devices have made remote monitoring possible in the health care sector, the ability to keep patients safe and secure and inspiring doctors to provide untouchable treatment. The interaction between patient and doctor has become efficient and easier. IoT has changed the lives of people, especially corona patients, by allowing for continuous monitoring of health conditions. It has a huge influence on the people and their families quarantine in this pandemic situation. Monitoring system sends alerts to family members and health care providers concerned about any disruptions or changes in a patient's daily activities. The spreading of infections is of serious concern to hospital patients, who will be in active infection. Hygiene monitoring IoT devices helps to prevent patients from being contaminated. Patients are the owners of their medical data and are responsible for authorizing, refusing or revoking access to data to any other entity, such as health insurance or health care providers.

4.2 Data owner (DO)

The data owner (DO) is a person possessing the health record (HR) data and required to access or store their health data. DO has full control over its HR results. DO must specify an access control policy for its HR data, and DO can allow or disallow such access permissions on its HR data to others. DO must then create the reencryption keys and the metadata for its HR. DO may also allow some users to add more data on their behalf and DO may re-check access to the newly added HR data.

4.3 Data user (DU)

The Data User (DU) is an entity that demands access to the HR data with the corresponding DO authorization. DU may include health care professionals, health insurance companies, and caregivers from many categories. DU can scan and get the metadata through the blockchain, and can later request the gateway server about HR information. DU can create/upload new HR data into program with a delegated authority.

4.4 Gateway server (GWS)

The duty of the gateway server (GWS) is to verify the validity of all authenticity and behavior inside the network. These activities include reencrypting the HR files, storing the metadata, and accessing the stream log. The private blockchain is also operated by GWS. All communications between GWS and other entities are made through a secure SSL/TLS channel. GWS is considered to be a semitrusted server in this work; the server obeys the procedure specified in the work but is curious to know the data.

4.5 Blockchain network

The blockchain network will make the transition from medical interoperability to patient-centered interoperability. It is defined as a private blockchain. The trusted third parties such as hospitals, institutions, researchers, and government agencies have the power to manipulate the blockchain within a private network. Blockchain technology enables patients to delegate access rules for their medical data. For instance, allowing different researchers to access portions of their data for limited span of time. Patients can connect to other hospitals with blockchain technology, and collect their medical data automatically.

4.6 Cloud storage (CS)

Cloud storage (CS) is responsible for storing real PHR data that is encrypted. CS is also called a server with a semitrusted feature. The data owner and users (data subjects) must enroll with the IoT gateway server for registration to make an initial agreement. The private and public key of the owner and users are generated by bilinear mapping function. The generation of key pair is based on the cyclic groups G_0 and G_1 with order of prime number q for the mapping function $e : G_0 \times G_1 \rightarrow G_m$. Let $x \in G_0$ and $(x, y) \in G_1$ be the parameters of system. A randomly selected number $a \in Z_q$ is taken as private key sk. The public key is generated with computation $pk = x^a$. The private key is preserved locally and the personal identifiable information and public key is transferred to the gateway server for registration. The enrollment details for registration is stored by gateway server and provide access control privileges for the data subjects are stored in Hyperledger (private blockchain). As a patient-centric system, RHM provides patients with a detailed view of their medical record, restoring authority through interaction. It facilitates information collection, alteration, generation and review, maintains data integrity, authenticates sense of identity, encourages unambiguous exchange, and executes access rights given to users through SC. RHM is not designed to replace PHRs, but to serve as a gateway between patient and trusted parties.

4.7 Uploading HR to cloud

The procedure for storing a HR or medical data in the proposed model is shown in Fig. 14.9.Patient created his/her own medical data (m) by acquiring data through IoT devices and responsible for allowing or denying access to data. Once it is created, patients connected to the HR data owner provider and submit the medical data with updated information. The consistency, reliability and accuracy of data should be secured over its entire life cycle. To maintain the consistency of data, the message digest $mdig = msgdig(m)$ is computed using SHA-2 hashing algorithm. Message digests are used to preserve the integrity of a piece of data to detect modifications and alterations to any part of a document.

Public key cryptography is used to convert the medical data to encrypted form. According to HIPAA act, the health information of the patient should be encrypted before transmission. As per the guidelines of National Institute of Standards and Technology 800-11, the AES, or any public key cryptography is applicable for data encryption. Traditional AES and public key cryptography

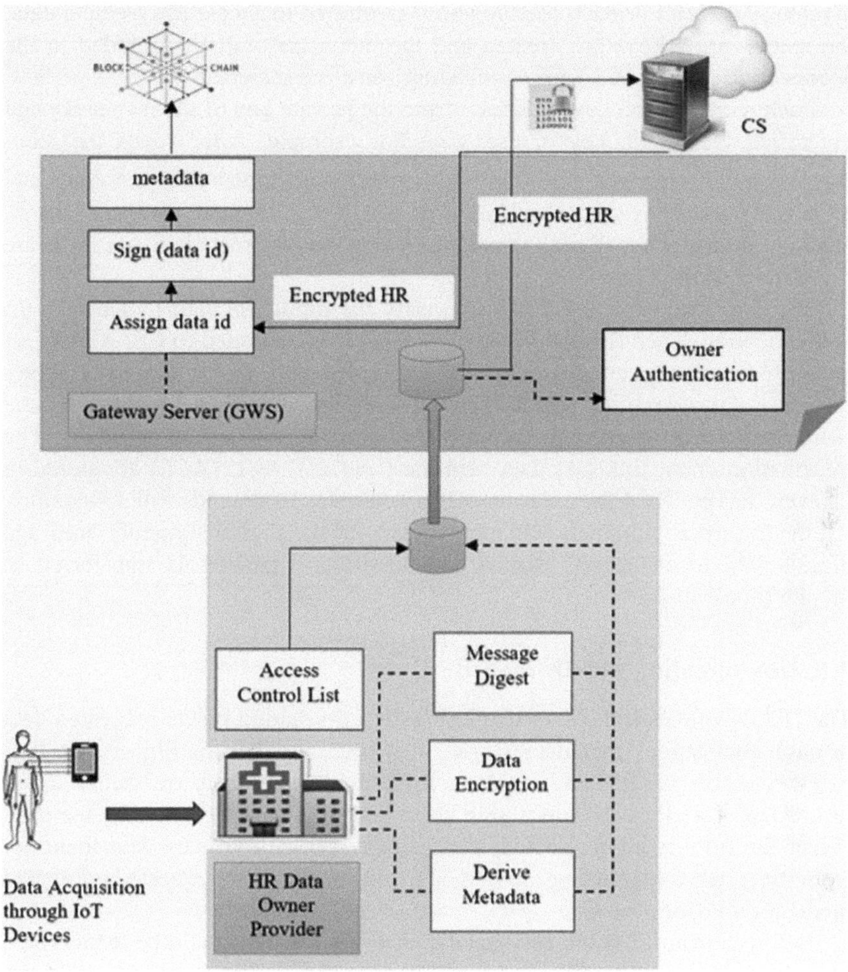

FIGURE 14.9 Uploading HR to cloud.

are infeasible for IoT devices. In the proposed model, PRE is acceptable as a public key cryptography. The medical data of the patient are encrypted by the public key pk_S and exchanged without revealing sensitive information. Encryption provides the data privacy and data confidentiality. The ciphertext is calculated using $c_s = encrypt(pk_s, mz^n)$, where '$n$' is a random element and $pk_s = x^{an}$.

A record of medical data can be searched by means of metadata. Patient metadata consists of message digest, owner identification, digital sign, type and format, date, description and data identifier. To ensure the authenticity of the message, digital signature is generated using data owner's private key

$digsign_{sk_s}(mdig)$. For each person who is permitted to access the medical data, the reencryption keys are created and the individual will be included in the access control list (ACL) for exchanging sensitive information.

Each reencryption key is derived from the private key of the owner sk_S and the respective public key of user pk_R. For example, Alice sends the reencryption key $rk_{S \to R} = x^{b/a} \in G_0$ to Bob, where a is the secret key of Alice and x^b is the public key of Bob. Then a package of encrypted medical data c_s, message digest *mdig*, digital signature *digsign*, patient metadata and ACL are transferred to the GWS.

The important job of GWS is to verify the digital signature of owner for authentication. Then the encrypted medical data is uploaded to CS. A link (lk) is created and it is connected to the encrypted data block. Gateway server creates a data identifier (md_id) and it is assigned to data block. After the assignment, it is mapped to the link (lk) of data, which resides in CS. The information about link (lk) data identifier (md_id), ACL (ACL) are stored in gateway server. Then the gateway server generates its data identifier signature. All the information namely patient metadata, owner digital signature, message digest, data identifier and server data identifier signature is transferred to private block chain.

4.8 Downloading health record

The DU downloaded the data by getting the information of the requested data through the patient metadata stored in private blockchain Fig. 14.10. The gateway server verifies and validates the user with the help of his/her signature. Also, it verifies the timestamp appended with data identifier by the user. When the user is authenticated, the gateway server retrieves data identifier from the local storage using the (md_id), and retrieves the requested encrypted medical data from the CS.

Before transmitting the encrypted data to the user, it should be reencrypted by gateway server. The gateway server derives the reencryption key to change the encrypted data of owner to ciphertext that is to be decrypted by the user. For instance, Alice original ciphertext can be reencrypted to ciphertext that is to be decrypted by Bob, the reencryption key is obtained from ACL maintained by gateway server $rk_{S \to R} = x^{b/a}$. Alice's original cipher text $c_S = encrypt(pk_s, mz^n)$ is converted to $c_R = \left(mZ^n, Z^{bn}\right)$, where $Z^{bn} = e\left(x^{an},\ x^{b/a}\right)$. The reencrypted ciphertext is transferred to Bob. Bob uses his private key sk_R to decrypt the ciphertext.

4.9 Data user revocation

The revocation of a user is included in the RHM system. CS is available for the storing of real encrypted HR files. The authenticated HR is queried by all users

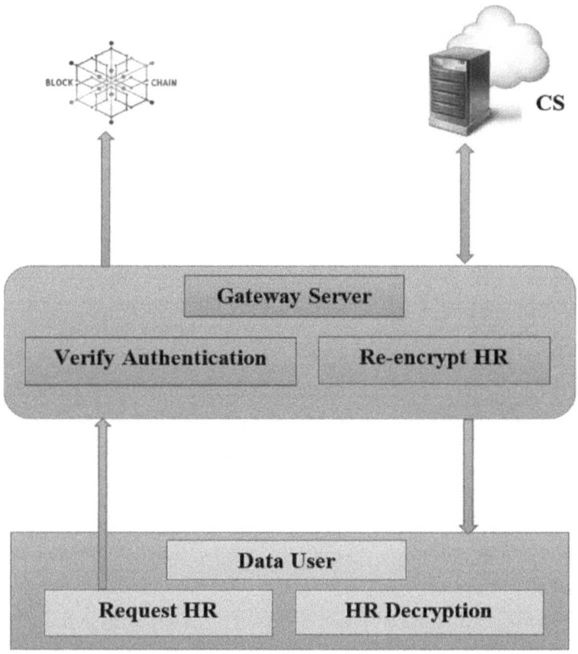

FIGURE 14.10 Downloading health record.

through the gateway service. The gateway server will validate each request under the authority of the requester according to the predefined access list provided by the owner of the HR. Once the request has been successfully verified, the gateway server will store transaction log information on the blockchain and reencrypt the HR with the corresponding reencryption key for the authorized users. As a result, by updating the ACL and retaining control of HR, the HR owner may revoke any access to his/her HR data.

If the gateway server violates the presumption that the gateway server follows the protocol (semitrusted) described in this work, it may be audited on blockchain. When a user is discovered to have an illegal activity on blockchain, the HR owner must update his/her details and encrypt the data again. This will include the encryption method for completely untrusted gateway servers.

4.10 Protocol on access control

Under the proposed model, proxy reencryption of the access control protocol reduces the gateway server requirement. As such, in the proposed model, the gateway server may be viewed as a semitrusted object. The real HR data is securely encrypted with the public key of the HR owner, and a group of approved users can access the ciphertext according to the ACL. Updating the ACL can easily revoke the access to actual HR data. The delegated user may also add new HR data on behalf of the HR owner and create the corresponding metadata. The list of access controls is stored in the local gateway server database. The corresponding secret keys are however protected as the secret key belongs to the owner of the HR. Only the HR owner can generate the reencryption keys that are used by the gateway server. In addition, the reencryption keys only allow the gateway server to reencrypt the original ciphertext for the authorized user. Consequently, the gateway server cannot access the actual HR data, because the actual PHR data can never be decrypted on the gateway server.

4.11 Security analysis

In every model, there are five key security criteria need to be addressed by model architects, data confidentiality, data integrity, data authentication, user authentication and data availability. Data confidentiality guarantees that data are detained under the policies of access control and it should be accessed by authorized users.

Integrity of data is highly achieved using robust cryptographic protection. Integrity is responsible for nonchangeable messages sent to the destination. Data authentication is a process of verification services that provides the permission to access the protected data and data availability enables that the different kinds of information should be available and accessed at all time when they are necessary. The features of cloud data storage are low cost, consistency, high performance, availability and durability. This model can be evaluated under various threats. The adversary in this model can be a home device or any other node in the server network or providers become an attractive target for data breach. These adversaries may remove transactions, sniff messages, generate false transactions, or alter or remove information from the storage.

The security of the model is evaluated by different attacks which could be possible with this model Table 14.2. For that, this model considers some assumptions that the cyclic group generator x is not computed from (x, x^a) with undeniable probability. Next, the gateway server and CS are semitrusted, they have the curiosity to understand the data. A successful cryptosystem can withstand known attacks of all kinds which can be addressed in the following.

Theorem 1. Adversary hack the channels of public contact and get confidential information, such as medical data. The proposed RHM system is secure against the tampering of medical data by adversary.

Proof:

The medical data is encrypted and stored in the CS. The link for the storage block is only known to the gateway server. The adversary has no chance to modify data, because the message digest in block chain can track this kind of

TABLE 14.2 Security evaluation mode.

Evaluation mode	Solution
Data confidentiality	Public key
Data integrity	Hashing
Data authentication	Public key, digital signature
User authentication	Public key
Data availability	Acceptable data transactions

actions. If an adversary enters to change the record, he has to change the hash of all previous blocks. But it is very difficult to forge the data, because block chain ensures the characteristics of immutability and tamper resistant.

Theorem 2. Assume that attack occurs when two or more parties agree to reveal some secret information, illegally. For example, a cloud server and a gateway server can combine to reveal some critical information about the data owner's data. This can affect the entire privacy and security of all of the data stored in the cloud, thus severely affecting the trust of the data owner with regards to using the cloud for data sharing and collaboration purposes.

Proof:

The ACL is stored locally in the gateway server. The ACL does not contain any information about secret keys. The gateway server therefore cannot gain access to the encrypted PHR data. Since the secret key is in the control of the HR data owner, the HR data owner can only produce the reencryption keys that are used by the gateway server. The gate way server has no idea to create the reencryption keys for an adversary.

Theorem 3. The adversary can trace the transaction details of a legal user from block chain network or penetrating the channel and retrieve the information on transit, will lead to replay the information for data theft.

Proof:

The time stamp is appended with the packed data block, and it should be verifiable by gate way server. Also, it verifies the digital signature. Initially, it verifies the time stamp, if it is incorrect, the gateway server will not send any response to the adversary (DU). Consider, if the time stamp is correct, then also the adversary cannot retrieve the encrypted data. Thus, replay attack is not possible.

Theorem 4. The adversary intercepts and update or retrieve the medical data in an unauthorized access.

Proof:

The ACL plays a vital role to protect data against unauthorized access. Based on the policies, the DU seeks the data identifier from private block chain. If the adversary does not have the valid data identifier, then he will not be able to access data. After the verification, the gateway server sends the reencrypted data with a decryption key.

Theorem 5. The adversary forges the credentials of data owner and tries to communicate with gate way server.

Proof:

Once the system can securely communicate with the gateway server or data owner, their signatures must validate one another's identity. If the attacker impersonates a valid device and sends the credentials to the gateway server, this cannot be authenticated and the adversary can revoke access to the device that is malicious or compromised. Thus, impersonation attack is impossible in this model.

5. Efficiency of proposed model

The performance of the proposed system is analyzed based on the RHM by IoT devices, cryptographic operations, the storage of the COVID-19 data set in the cloud, the metadata storage, and the access control in the blockchain. PRE is deployed for the cryptographic operations. The estimation of store and retrieve operation on 4G/5G network is done in the blockchain model for the access control signature storage.

The dataset used is the freely available COVID-19 Open Research Dataset (CORD-19) which gives the complete information regarding of the COVID patients. The CORD-19 dataset is included with the COVID19_line_list_data.csv (358.85 KB) contains the attributes such as reporting date, summary, location, country, gender, age, symptom_onset, If_onset_approximated, hosp_visit_date, international_traveler, domestic_traveler, exposure_start, exposure_end, traveler, death, recovered, symptom, source, and link. This dataset also includes covid_19_data.csv (5.14 MB) which updates the daily level information on the number of COVID-19 affected cases across the globe. time_series_covid_19_confirmed (197.03 KB) which gives the time series data on the total confirmed cases, time_series_covid_19_death (140.7 KB) displays the time series of the total data of death and time_series_covid_19_recovery data (172.32 KB) shows the time series of the total recovered cases.

5.1 Experimental setup

The experiment of the proposed model is done by the VMware workstation with Ubuntu OS. The host system is equipped with Intel(R) Core (TM) i7-4510U CPU, 2.60 GHz, 8 GB RAM, running Windows 10. Using Eclipse IDE (oxygen), Java 1.8, Java security library, Oracle Commons Lang 3.6 (OCL), Java Pairing-Based Cryptography (JPBC) the PRES encryption algorithm is implemented. For the blockchain service testing, the Hyperledger blockchain network is created with node.js. The blockchain network contains peer nodes. For this experiment, CORD-19 is used.

CORD-19 consists of data of varying sizes. These data with the various sizes are used for the experiment which contains 64, 128, 256, 512, 1024, 1536, 2048, 2560, 3072, and 4096 KB size. The RHM scheme must respond with variety of users, the performance analysis of the proposed system is also examined with varying number of users from a minimum of 1 to a maximum of 16 users.

The performance analysis is performed using the proxy reencryption scheme, gives sufficient data storage for the IoT devices that are used for data acquisition from the patients than the attribute-based encryption scheme. PRE is implemented using the AFGH algorithm and AES (Nechvatal, Barker, Bassham, Burr, Dworkin, Foti, and Roback, 2001). Under PRE scheme the encryption algorithm used is symmetric cipher AES. The effective access control scheme for the peer nodes in the block chain is achieved by using the symmetric key for encryption. This provides a ciphertext with the combination of encrypted data and the encrypted symmetric key. Both the encryption and decryption process use the same symmetric key.

5.2 Experimental results

The RHM experiments are done by varying the data sizes, varying number of users with the real scenario for evaluation. The encryption and decryption time for the IoT device data is performed by the PRE method 50 times by documenting the average processing time. First the performance of the encryption and decryption operations are done by sharing the varying data sizes with one user. Then the analysis is done by sharing the data sizes with 2, 4, 8, 16, and 32 users, respectively. Finally, the store and retrieve time is analyzed for the entire system.

5.2.1 Encryption and decryption for varying data sizes

The performance analysis is done by sharing the CORD-19 data sizes to one user for the cryptographic operations such as encryption and decryption. The PRE method used consists of the various phases such as key pair generation, reencryption key generation, encryption, reencryption, and decryption processes. The time taken by the key pair generation phase is 2.94 ms, the time consumed by the reencryption key generation is 17.32 ms and the time for the proxy reencryption is 17.92 ms. The ABES method consists of different phases such as owner key generation, user key, encryption and decryption. The time taken to generate owner key is 360.12 ms. The time taken to generate user key is 360.80 ms. The encryption and decryption time for varying data sizes for the proposed model is calculated. Fig. 14.11 portrays the encryption time and decryption time of PRE with varying data sizes. Fig. 14.12 shows that as the

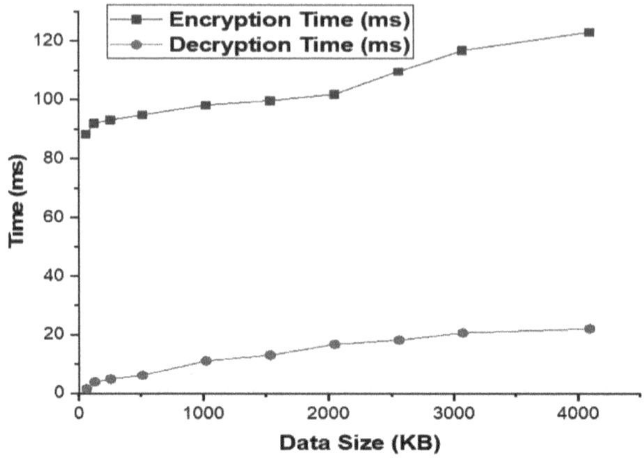

FIGURE 14.11 PRE-encryption and decryption time versus data size.

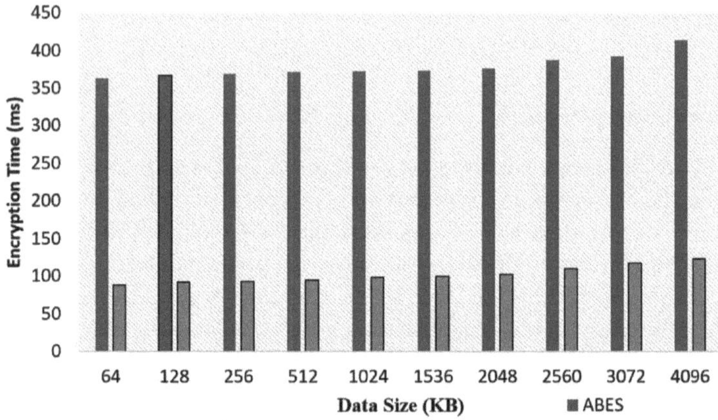

FIGURE 14.12 Encryption time for proposed and existing method.

data size increases, the encryption time gradually increased in the proposed model. Similarly, Fig. 14.13 shows the decryption time for the proposed and existing method. It is an appreciable improvement in the computation obtained from the proposed model. Based on the results presented in Fig. 14.14, the storing time and retrieving time of medical data is more efficient for better transactions. Many ABE schemes, require great computational resources for storage and retrieval of data. Given the limited capabilities of the sensors, it may not be a good idea to apply existing ABE schemes to sensors. Computational time can be calculated by summing the time of encryption and decryption as portrayed in Fig. 14.15.

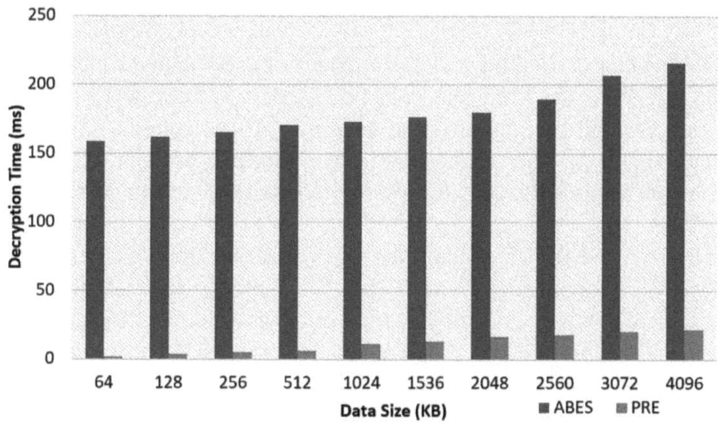

FIGURE 14.13 Decryption time for proposed and existing method.

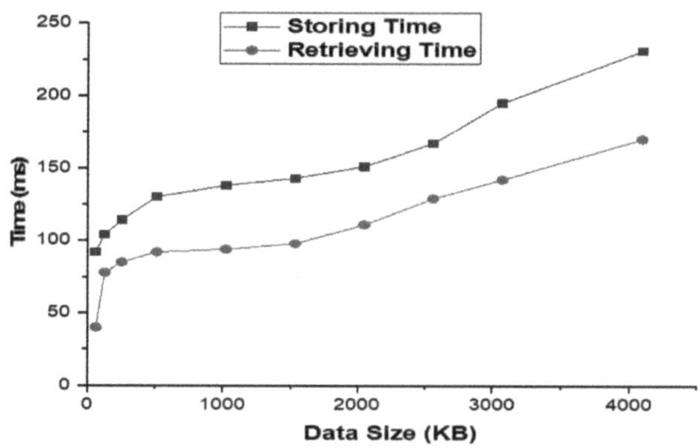

FIGURE 14.14 Storing and retrieving time for proposed method.

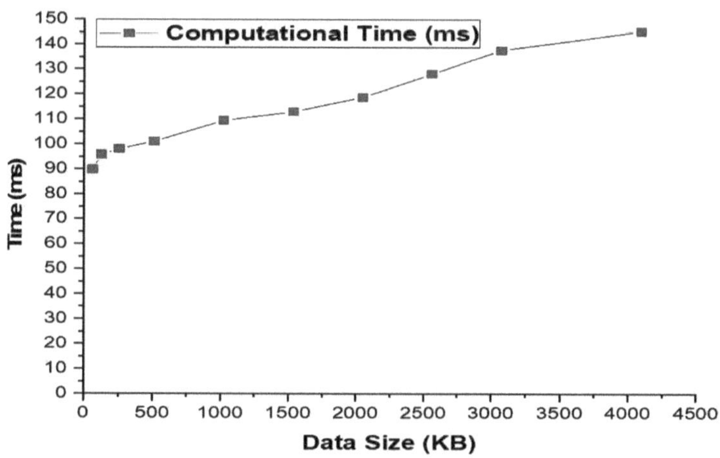

FIGURE 14.15 Computational time for proposed method.

5.3 Varying the number of users

The RHM system is used by various numbers of users and variety of data sets. The data stored in the cloud need to be shared to multiple users during the storing process and assigns different access control mechanisms in the block chain and shares as per requested in the retrieve process. These various operations are analyzed using different number of users with 4 MB CORD-19 data.

In the proposed RHM system, the proxy reencryption key is generated by the data owner for storing the HR data. As per the request, the users can retrieve the data using the proxy reencryption and the decryption process. If there is an additional authorization given by the data owner, then additional reencryption process is mandatory. Fig. 14.16 portraits that, as the number of users increase, the time taken to generate the reencryption key also increases. This is due to the time taken for the change in attribute set and policy as there is an increase in the number of authorized users. This change leads to the increase in storing operation but the retrieving operation remains with little change and is given in Fig. 14.17.

As per the results estimated, the implementation is conducted for increasing number of users such as 1, 2, 4, 8, 16, 32, and 64. Under PRE model, the storing time is increased with the varying number of users. The time taken for storing operation for 1 user is 234 ms and it is increased to 257 ms for two users. Also, it increases to 1049 ms for 64 users. The retrieval time is 71, 83, 89, 90, 93, 95, and 96 for 1, 2, 4, 8, 16, 32, and 64 users. The time taken for storing operation gets increased with the increase in number of users is due to the time taken for performing the master key generation and the encryption process. The retrieving process performs the user key generation

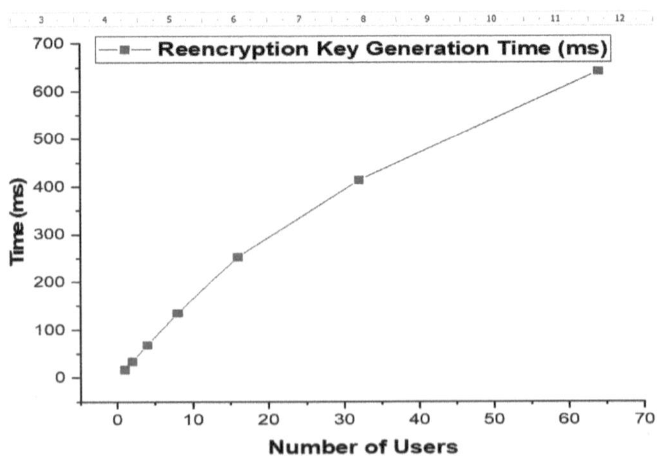

FIGURE 14.16 Reencryption key generation time.

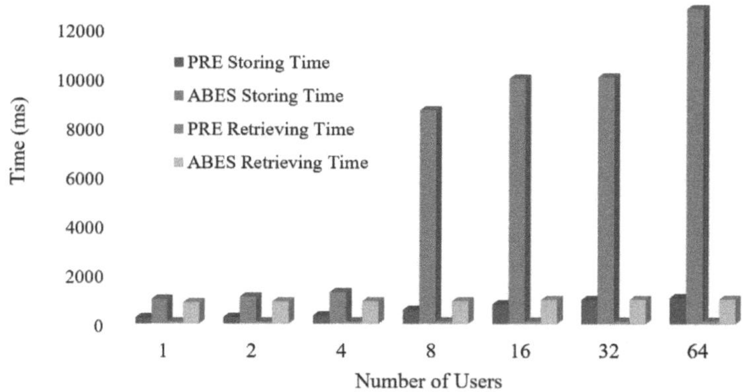

FIGURE 14.17 Storing and retrieval time versus varying number of users.

and decryption process which requires less time. The comparative results show that, there is a great difference between PRE storing time and ABES storing time. The storing time of PRE for two users are 257 ms and for ABES it takes 1073 ms. 816 ms difference obtained for both schemes shown in Fig. 14.17. The number of computational operations conducted in ABE reflects the overhead of storage and retrieval of medical data.

5.4 Throughput

Throughput can be calculated by using the number of users from 50 users to 500 users (with a time of 20–50), who use the device and perform its different functions. From the experiments, it is observed that as the number of users and requests increases, the system throughput considerably increased in a linear manner. It is increased from 118 KBps for 50 users to 865 KBps for 500 users. The linear increment of throughput indicates the efficiency of the proposed PRE model. The following Fig. 14.18 offers a summary of the proposed framework's throughput.

5.5 Latency

Latency as described earlier is the delay or time gap when one system component sends out a request and some other system component generates a response. The difference is defined as latency between those two actions. The latency is evaluated for number of users varying from 50 to 500 given in Fig. 14.19. The delay occurred for 50 users as 3.5 ms, 100 users as 5.1 ms, 150 users as 6.8 ms, and so on. The greatest delay occurred as 10.1 ms for 500 users. Fig. 14.19 offers a summary of the system's average latency.

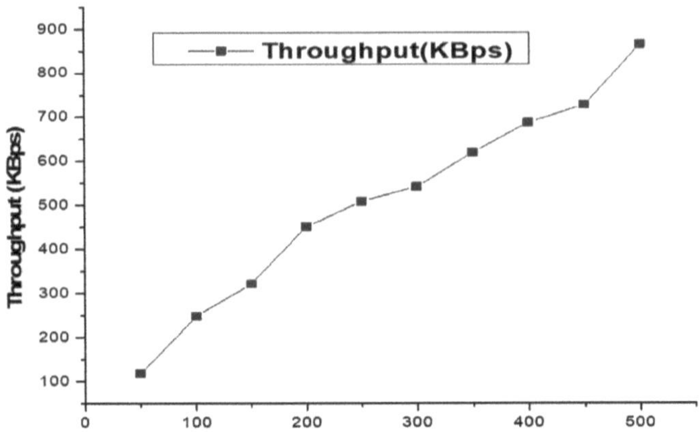

FIGURE 14.18 Throughput versus number of users.

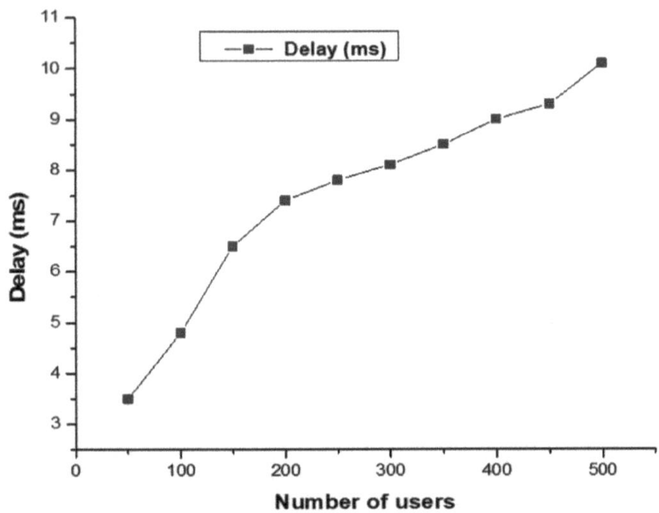

FIGURE 14.19 Number of users versus latency.

5.6 Discussion

In this experimental study, we suggested a blockchain-based access control paradigm to safeguard HR privacy. Our suggested architecture uses a proprietary database, CS and other cryptographic methods, including proxy reencryption, hashing, and digital signature to provide HR device resistance and anonymity. The proposed model can strengthen the highlights: (1) the HR owner can safely store and offer his/her HR information; (2) the HR owner can deny an entrance directly on any HR information effectively; and (3) all clients including the HR owner can advantageously check the trustworthiness of the information. The blockchain technology is used to attain the tamper-resistance property of our platform. Consequently, problems relating to the use of blockchain in the HR system such as the usability, withdrawal of consent, and confidentiality of stored data are discussed in this research work.

The proposed model allows blockchain hold only the limited metadata to tackle the availability problem of blockchain. To maintain security, the encrypted actual HR is stored in the CS. Users can search the HR using blockchain metadata, and can order encrypted HR from the gateway server. As a result, all access to HR data should be stored on blockchain to enable an unchanging audit trail. To address the problem of blockchain secrecy, the real HR is encrypted with the HR owner's public key and only the metadata is exposed on the blockchain itself. To deal with the problem of consent revocation, the ACL, which includes the reencoding keys of the permitted users, can limit the users of a certain service. By updating the access list (revoking the reencryption key) and retaining data ownership, the HR owner can revoke access to his/her HR data.

The proposed model is evaluated from a security and performance viewpoint to ensure that the priorities are achieved. When dealing with a single individual, the theoretical model is tested on various data sizes. The model suggested outperforms when compared to the existing method for increasing data size; however, the average running time is identical for different data sizes. The existing method operational time increases explosively, while the operational time of proposed model increases linearly. The suggested solution is also not only more effective but much better suited to the HR method.

6. Conclusions

In this chapter, the blockchain-based RHM system is proposed to facilitate a privacy model that retains access control of the medical data in the COVID-19 pandemic situation. The approach suggested addresses the requirements of the HR system and the challenges of using blockchain technology in the implementation of HR systems. The protection of privacy in the HR system needs a qualitative necessity, namely the storage of tamper resistance and functional requirements, namely revocable access control. In addition, there are other concerns such as restricted storage and protection of nonchain data for the use of blockchain in HR development. Prior to IoT, patients' associations with specialists were restricted to visits, and tele and text correspondences. There were no chance to specialists or medical clinics could screen patient's well-being consistently and make suggestions in like manner. IoT empowered gadgets have made far checking in the medicinal services area conceivable, releasing the possibility to keep patients protected and solid and enabling doctors to convey standout care. Also, the proposed model expanded patient commitment and fulfillment as collaborations with specialists have become simpler and more productive. The ACL is built to be applied using existing cryptographic techniques and private blockchain network in such a way that it can manage blockchain concerns for HR system development and determines the privacy and access control prioritization. Then the privacy and security of the proposed model is analyzed with threat models, namely a manipulative attack, a replay attack, a collusion attack, a malicious attack and an impersonation attack, to ensure that the original objectives of proposed system are met. This model is suggested not only to address HR privacy issues, but also to look forward to ongoing research into the applications of blockchain for data security and privacy protection in health care. IoT majorly reduces human services costs altogether and improves treatment results.

References

Agyekum, Xia, Q., Sifah, E. B., Gao, J., Xia, H., & Du, X. (2019). A secured proxy-based data sharing module in IoT environments using blockchain. *Sensors, 19*(5).

Ateniese, G., Fu, K., Green, M., & Hohenberger, S. (2006). Improved proxy re-encryption schemes with applications to secure distributed storage. *ACM Transactions on Information and System Security, 9*(1), 1–30. https://doi.org/10.1145/1127345.1127346

Atri, D., Hasan, K., Siddiqi, J., Lang, V., Nauffal, D. A., Morrow, E. A., et al. (2020). COVID-19 for the cardiologist: A current review of the virology, clinical epidemiology, cardiac and other clinical manifestations and potential therapeutic strategies. *Journal of the American College of Cardiology: Basic to Translational Science, 5*(5), 518–536.

Au, M. H., Yuen, T. H., Liu, J. K., Susilo, W., Huang, X., Xiang, Y., et al. (2017). A general framework for secure sharing of personal health records in cloud system. *Journal of Computer and System Sciences, 90*, 46–62. https://doi.org/10.1016/j.jcss.2017.03.002

Azaria, A., Ekblaw, A., Vieira, T., & Lippman, A. (2016). MedRec: Using blockchain for medical data access and permission management. In *Proceedings - 2016 2nd international conference on open and big data, OBD 2016* (pp. 25–30). Institute of Electrical and Electronics Engineers Inc. https://doi.org/10.1109/OBD.2016.11

Blaze, M., Bleumer, G., & Strauss, M. (1998). Divertible protocols and atomic proxy cryptography. In: Advances in Cryptology—EUROCRYPT '98 (Espoo). *Vol. 1403 of Lecture Notes in Computer Science* (pp. 127–144). Berlin, Germany: Springer.

Buterin, V. (2014). *A next-generation smart contract and decentralized application platform* (Vol. 3).

Cachin, C. (2016). Architecture of the hyperledger blockchain fabric. In *Workshop on distributed cryptocurrencies and consensus ledgers* (Vol. 310).

Chow, S. S. M., Weng, J., Yang, Y., & Deng, R. H. (2010). Efficient unidirectional proxy re-encryption. In *Lecture notes in computer science (including subseries lecture notes in artificial intelligence and lecture notes in bioinformatics)* (Vol. 6055, pp. 316–332). https://doi.org/10.1007/978-3-642-12678-9_19

Clauson, K. A., Breeden, E. A., Cameron, D., & Mackey, T. K. (2018). *Leveraging blockchain technology to enhance supply chain management in health care: Blockchain in health care today*. https://doi.org/10.30953/bhty.v1.20

Dubovitskaya, A., Xu, Z., Ryu, S., Schumacher, M., & Wang, F. (2017). Secure and trustable electronic medical records sharing using blockchain. In *AMIA Annual symposium proceedings. AMIA symposium, 2017* (pp. 650–659).

Sangeetha, D., & Vaidehi, V. (2017). A secure cloud based Personal Health Record framework for a multi owner environment. *Annals of Telecommunications*, 95–104. https://doi.org/10.1007/s12243-016-0529-4

Griggs, K. N., Ossipova, O., Kohlios, C. P., Baccarini, A. N., Howson, E. A., & Hayajneh, T. (2018). Health care blockchain system using smart contracts for secure automated remote patient monitoring. *Journal of Medical Systems, 42*(7). https://doi.org/10.1007/s10916-018-0982-x

Gu, K., Jia, W., Wang, G., & Wen, S. (2017). Efficient and secure attribute-based signature for monotone predicates. *Acta Informatica, 54*(5), 521–541. https://doi.org/10.1007/s00236-016-0270-5

Guo, C., Zhuang, R., Jie, Y., Ren, Y., Wu, T., & Choo, K. K. R. (2016). Fine-grained database field search using attribute-based encryption for e-healthcare clouds. *Journal of Medical Systems, 40*(11), 235.

Haleem, A., Javaid, M., & Khan, I. H. (2020). Internet of things (IoT) applications in orthopaedics. *Journal of Clinical Orthopaedics and Trauma, 11*, S105–S106. https://doi.org/10.1016/j.jcot.2019.07.003

Ivan, D. (2016). *In moving toward a blockchain-based method for the secure storage of patient records.*

Jesus, E. F., Chicarino, V. R. L., Albuquerque, C. V. N. de, & de Rocha, A. A. A. (2018). *A survey of how to use blockchain to secure internet of things and the stalker attack.*

Kaufman, L. M. (2009). Data security in the world of cloud computing. *IEEE Security and Privacy, 7*(4), 61–64. https://doi.org/10.1109/MSP.2009.87

Khan, S., & Khan, R. (2018). Multiple authorities attribute-based verification mechanism for blockchain mircogrid transactions. *Energies, 11*(5). https://doi.org/10.3390/en11051154

Kosba, A., Miller, A., Shi, E., Wen, Z., & Papamanthou, C. (2016). Hawk: The blockchain model of cryptography and privacy-preserving smart contracts. In *Proceedings - 2016 IEEE symposium on security and privacy, SP 2016* (pp. 839–858). Institute of Electrical and Electronics Engineers Inc. https://doi.org/10.1109/SP.2016.55

Li, B., Dawei, Y., Xun, W., Lin, T., Xiaodan, Z., Nanshan, Z., et al. (2020). Chinese experts' consensus on the Internet of Things-aided diagnosis and treatment of coronavirus disease 2019 (COVID-19). *Clinical eHealth*, 7–15. https://doi.org/10.1016/j.ceh.2020.03.001

Li, W. M., Li, X. L., Wen, Q. Y., Zhang, S., & Zhang, H. (2017). Flexible CP-ABE based access control on encrypted data for mobile users in hybrid cloud system. *Journal of Computer Science and Technology, 32*(5), 974–990. https://doi.org/10.1007/s11390-017-1776-1

Liam, B., Buchanan, W. J., Jonathan, C., & Owen, L. (2018). Applications of blockchain within health care. *Blockchain in Health Care Today, 1*. https://doi.org/10.30953/bhty.v1.8

Luis, B. C. J., Frank, M. A., & Ole, L. (2018). Public health surveillance using decentralized technologies. *Blockchain in Health Care Today, 1*, 1–14. https://doi.org/10.30953/bhty.v1.17

Macrinici, D., Cartofeanu, C., & Gao, S. (2018). Smart contract applications within blockchain technology: a systematic mapping study. *Telematics and Informatics, 35*(8), 2337–2354.

Marek, A. C. (2018). Blockchain as a foundation for sharing health care data. *Blockchain in Health Care Today, 1*, 1–6. https://doi.org/10.30953/bhty.v1.13

Mayank, R., Danilo, G., & Katina, K. (2019). SoK of used cryptography in blockchain. *IEEE Access*, 148550–148575. https://doi.org/10.1109/access.2019.2946983

Merkle, R. C. (1989). A certified digital signature. In *Conference on the theory and application of cryptology* (pp. 218–238).

Mineraud, J., Mazhelis, O., Su, X., & Tarkoma, S. (2016). A gap analysis of Internet-of-Things platforms. *Computer Communications, 89–90*, 5–16. https://doi.org/10.1016/j.comcom.2016.03.015

Nakamoto, S. (2019). *Bitcoin: A peer-to-peer electronic cash system.*

Nasir, Q., Qasse, I. A., Abu Talib, M., & Nassif, A. B. (2018). *Performance analysis of hyperledger fabric platforms.* Security and Communication Networks. https://doi.org/10.1155/2018/3976093

Nechvatal, J., Barker, E., Bassham, L., Burr, W., Dworkin, M., Foti, J., & Roback, E. (2001). Report on the development of the Advanced Encryption Standard (AES). *Journal of Research of the National Institute of Standards and Technology, 106*(3), 511.

Nunez, D., Agudo, I., & Lopez, J. (2017). Proxy re-encryption: analysis of constructions and its application to secure access delegation. *Journal of Network and Computer Applications, 87*, 193–209.

Ostrovsky, R., Sahai, A., & Waters, B. (2007). Attribute-based encryption with non-monotonic access structures. In *Proceedings of the ACM conference on computer and communications security* (pp. 195–203). https://doi.org/10.1145/1315245.1315270

Radanovi, & Liki, R. (2018). Opportunities for use of blockchain technology in medicine. *Applied Health Economics and Health Policy, 16*, 583–590. https://doi.org/10.1007/s40258-018-0412-8 [Medline: 30022440].

Rao, Y. S. (2017). A secure and efficient ciphertext-policy attribute-based signcryption for personal health records sharing in cloud computing. *Future Generation Computer Systems, 67*, 133–151. https://doi.org/10.1016/j.future.2016.07.019

Rifi, N., Agoulmine, N., Chendeb Taher, N., & Rachkidi, E. (2018). Blockchain technology: Is it a good candidate for securing IoT sensitive medical data? *Wireless Communications and Mobile Computing, 2018.* https://doi.org/10.1155/2018/9763937

Solanas, A., Patsakis, C., Conti, M., Vlachos, I., Ramos, V., Falcone, F., et al. (2014). Smart health: A context-aware health paradigm within smart cities. *IEEE Communications Magazine, 52*(8), 74–81. https://doi.org/10.1109/MCOM.2014.6871673

Tariq, N., Asim, M., Al-Obeidat, F., Farooqi, M. Z., Baker, T., & Hammoudeh, M. (2019). The security of big data in fog-enabled IoT applications including blockchain: A survey. *Sensors, 19*(8).

Wang, D., Hu, B., Hu, C., Zhu, F., Liu, X., Zhang, J., et al. (2020). Clinical characteristics of 138 hospitalized patients with 2019 novel coronavirus-infected pneumonia in Wuhan, China. *JAMA - Journal of the American Medical Association, 323*(11), 1061–1069. https://doi.org/10.1001/jama.2020.1585

Wang, Y., Hu, M., Li, Q., Zhang, X. P., Zhai, G., & Yao, N. (2020). *Abnormal respiratory patterns classifier may contribute to large-scale screening of people infected with COVID-19 in an accurate and unobtrusive manner.*

Wangthammang, M., & Vasupongayya, S. (2016). Distributed storage design for encrypted personal health record data. In *2016 8th international conference on knowledge and smart technology, KST 2016* (pp. 184–189). Institute of Electrical and Electronics Engineers Inc. https://doi.org/10.1109/KST.2016.7440505

Xia, Q., Sifah, E. B., Asamoah, K. O., Gao, J., Du, X., & Guizani, M. (2017). MeDShare: Trust-Less medical data sharing among cloud service providers via blockchain. *IEEE Access, 5*, 14757–14767. https://doi.org/10.1109/ACCESS.2017.2730843

Zheng, Z., Xie, S., Dai, H., Chen, X., & Wang, H. (2017). An overview of blockchain technology: Architecture, consensus, and future trends. In *Proceedings - 2017 IEEE 6th international congress on big data, BigData congress 2017* (pp. 557–564). Institute of Electrical and Electronics Engineers Inc. https://doi.org/10.1109/BigDataCongress.2017.85

Chapter 15

A reliable blockchain and edge–cloud architecture for facilitating fault-tolerant IoT applications

A. Amuthan
Department of CSE, Pondicherry Engineering College, Puducherry, India

Chapter outline

1. Introduction

The advent of edge computing has increased the utilization of ubiquitous technologies specifically in the domain of Internet of Things (IoT) (Massonet et al., 2017a,b). In particular, ubiquitous technologies have been utilized in the industrial environments, thereby increasing the velocity, variety, and data volume handled in the application (Cao, Tang, Guo, Li, & Zhang, 2020). As a result, Cyber-Physical Systems (CPS) need to extend their storage and

Blockchain for Smart Cities. https://doi.org/10.1016/B978-0-12-824446-3.00013-2

computing capabilities based on their reliability over the cloud (Yousefpour et al., 2019). But, the exchange of data enabled between the cloud and CPS/IoT has the possibility of suffering from the limitations of reliability, security, bandwidth cost, and latency (Ghanta et al., 2018). The majority of the data are processed as the edge which is very close to the data sources for the purpose of preventing the limitations of data exchange facilitated during the interaction between the cloud and the CPS/IoT (Massonet et al., 2017a,b). Moreover, edge computing being a new computing paradigm aids in utilizing intelligent computing infrastructure and makes physical objects more smart through the connection established between them and virtual world through the use of the IoT environment (He, Liu, Jin, & Dai, 2017a,b). These data processing edge—cloud architecture is generally comprised of five important phases such as data acquisition, data management (data transmission and processing), data utilization and analytics, and data storage (Qiong, Kaiwen, & Xu, 2020). In the data acquisition phase, the real-time data are collected from sensors, actuators, IoT devices, and other cooperative information systems (Cui, Tso, & Jia, 2020). The data management phase consists of data transmission and data processing phases for the purpose of facilitating data processing locally on the equipments before been forwarding them to the cloud for transmitting them to the cloud infrastructure for subsequent computation (Comma-Di et al., 2018). Then, the information systems are the end users that in common utilize them in the phase of data utilization (Yuen, 2020). The utilization of data can be the structure of graphical user interface and diversified responses that are associated with the user-typed commands for achieving the required operations such as video surveillance or controlling traffic signals (Jiang & Ding, 2018). In the phase of data storage, the application data if needed need to be permanently stored in the cloud infrastructure for future exploration (Gupta & Quamara, 2018a,b). This five phases are employed in the architecture of the edge—cloud as it forms the basis behind the minimization of network latency that conserves a huge amount of network bandwidth (Peinl, Holzschuher, & Pfitzer, 2016). In some specific scenario, the edge computing capabilities may be insufficient and thereby requiring the processing of data to be transferred to the back-end of the cloud which is closer to the edge. Moreover, this category of data processing and placement need to be chosen at execution time based on the availability of computing resources existing on cloud and edge side. In addition, the challenges that need to be handled are listed as follows: (i) unique data semantics that need to be included for preventing the issue of data corruptions and duplication, (ii) fault-tolerant systems integrated with data replications need to be designed for enhancing the degree of availability (Costa & Beko, 2019; Nambi, Sarka, Prasad, & Rahim, 2014).

In this chapter, a Reliable Blockchain Edge—Cloud Architecture named CEFIoT architecture is contributed to support the fault-tolerant IoT applications. This CEFIoT architecture is proposed for high preservation of data that prevents the duplication or corruption of data. It was proposed with the characteristics of Multi-Transaction Mode Consortium Blockchain (MTMCB) for the purpose of gaining the network control for achieving maximized access, maximized storage, and minimum computation uniformly distributed over the edges. It aids in guaranteeing increased large scale of data storage, validity computation, and network servers at the near proximity of the users in a more secure way. Moreover, it prevented malicious harming of edge equipments or other hostile environments in which the IoT applications are deployed for the objective of monitoring.

2. Related work

An architectural model named SAT-IoT was proposed by Lopez and Munoz (Lopez Pena & Munoz Fernandez, 2019) for attaining maximized fault tolerance in the fog–cloud environment. It was proposed with three major ideas such as (i) the topology management of IoT supporting global scope from the global point of communication infrastructures and hardware, (ii) integration and automation of IoT visualization systems for the purpose of visualizing the data in the real time, (iii) support for current data flow paths and the current IoT topology, (iv) the transparency in the process of computing data over the edge–cloud paradigm, (v) incurs less change in the behavior of computing nodes that changes dynamically without the intervention of the administrator. It was proposed with the concept and merits that covers the demands of it, in order to safeguard the services of the architecture through the inclusion of blockchain. The throughput, frequency, memory utilization, and CPU utilization of this architecture was determined to be greater than the baselines schemes. However, the mean number of computations utilized in the fault-tolerance process is comparatively higher on par with the state of the schemes. Then, an Integrated Federated Cloud-based IoT (IFCIoT) architecture was proposed by Munir, Kansakar, and Khan et al. (2017) for facilitating maximized federated cloud services between the youth devices based on the intermediary fog computing environment. This IFCIoT architecture provided transparent and fault-tolerant services from the federated cloud that inherits the external and internal cloud servers for satisfying the needs of application and business. It also ensured that each individual fog node of the environment is autonomous for guaranteeing undistrupted services for the services or facility, it tries to offer to the end users. The throughput, frequency, memory utilization, and CPU utilization was determined to be greater than the baselines schemes. However, the degree of performance facilitated by this IFCIoT architecture is still considered to possess a room for improvement on par with the existing approaches.

In addition, an IoTEF architecture was proposed by Javed, Robert, Heljanko, and Främling et al. (2020) for supporting fault tolerance in IoT applications that aids in adopting edge and fog environment together. It included the platform of Apache Kafka publish/subscribe for addressing the issue of fault tolerance. It included the benefits of Kubernetes for managing fault tolerance with the deployed federated approach that offers a unique management interface by facilitating the process of automatic configuration for pipelining the process of data processing. It was formulated with the merits of functional and nonfunctional requirements for comparing different IoT architectures. The experimental results of the IoTEF architecture confirmed its performance in terms of conserving network bandwidth with minimized delay. It was also developed for handling the failure of network and hardware connectivity.

3. Reliable blockchain and edge–cloud architecture for facilitating fault-tolerant IoT applications

In this section, the proposed CFIoT architecture is first presented with complete details of its fault-tolerant edge computing platform. Then, the role of Multi-Transaction Mode Consortium Blockchain (MTMCB) in facilitating security to the Edge–Cloud Architecture is presented.

3.1 Proposed CFIoT architecture

The proposed CFIoT architecture is proposed through the integrated benefits of the microservices that aids in achieving node and network fault tolerance. It is developed as a unified management systems that support both the edge and cloud servers in the IoT systems. This proposed CFIoT architecture integrated the modular properties of microservices that reduce the complexity involved in the process of node and network fault tolerance. This architecture is proposed with a single data semantic concept for preventing the failure of nodes or networks in a multicluster large-scalability environment. This architecture utilized limited amount of available resources at the edge by including lightweight containers instead of using conventional virtual machines. It is designed with separate clusters for both the cloud and edge to permit the edge devices to independently function when the connection is terminated from the cloud back-end. It is also propounded as a fault-tolerant approach that functions even when the connection of the cloud is considered to be lost. Moreover, the functional and nonfunctional requirements of the proposed CFIoT architecture are listed as follows:

It is proposed for continuing the processing of data through the active nodes without influencing the complete processing pipeline even during the failure of the nodes.

It will result in data corruption or loss in the event of system crash or network disconnection.

It is contributed for processing data exactly once by replicating the data n-times in the k-nodes clusters without any kind of data duplication.

It facilitates efficient data computation with the potentiality of scaling the number of processing stages.

It reduced additional delays and minimized latency peaks in the event of moving the stages of processing the data within and across the clusters.

It is capable of monitoring, managing, and deploying data management with the placement of computation that is achieved based on federated management system without the dependence of the underlying hardware.

It also supports the persistence of local data at the edge whenever the event of fault tolerance or malfunctioning of node occurs in the node clusters.

In addition, the proposed CFIoT architecture is constructed logically through the deployment of inherent cloud technologies that can be developed for edge computing with the abstractive four layers such as (i) Layer of unified placement and management layer, (ii) Distributed OS, (iii) Data transportation, and (iv) Isolation of application.

3.1.1 Application isolation layer

This layer puts processes into distinct containers and organizes them to function as a sole application. Wrapping is attained by using Docker that eases suitable software positioning and permits the dispensation of data streams regardless of the hardware. For machines with unlike architectures, the containers implement similar code in the application level (Java, Python, or Spark) without any alterations in the source code. Docker is selected for application segregation as it is widely used in development. It supports software portability and version control with reduced CPU overhead and improved performance. It also segregates numerous IoT applications deployed in the unchanged edge and cloud nodes.

3.1.2 Data transport layer

This layer offers a framework for pub/sub messaging wherein data streams are saved and reproduced in the cluster. It permits the architecture to support logical data flow as containerized procedures utilizing pub/sub topics for transport medium. Data handling is dispersed proficiently, thus offering flexibility of positioning computations on the edge or on the cloud. This form of pipeline used in data handling preserves fault tolerance in the network by permitting local data buffering at the edge, as Internet connectivity is reorganized. An Apache Kafka platform is chosen to support the abilities of the layer by dealing with providing fault tolerance of data stream by guaranteeing both online and offline message consumption. It offers a united better data replication scheme for real-time handling with reduced latency and better data rates. The data are available in the cluster nevertheless some processing stages are reconfigured. Kafka transactional API is used to guarantee data semantics among producers and consumers supporting data handling exactly once in a cluster. The existing Kafka applications do not provision transactions among many clusters. The edge- and cloud-side clusters include three Kafka topics in addition to four PCs.

3.1.3 Distributed OS layer

This layer is accountable for allocating cluster processing phases to the nodes. IT eases the issue of failures in nodes by rescheduling unsuccessful processing phases on added accessible nodes, thus offering node as well as network fault forbearance at the hardware. Further, it is proficient in evenly distributing the workload among nodes. The Kubernetes framework to arrange the assignment of processing phases, as it allows fault-tolerant organization and offers better accessibility solution. The key design concepts are to have no particular point of failure. All configuration details are presented as a replica permitting Kubernetes to persist any single point of failure in a cluster with three nodes.

3.1.4 Unified federated management layer

This layer offers a scheme for dealing with numerous cloud and edge clusters from only management interface. Data handling pipeline are observed, guaranteeing that application placement occurs across numerous clusters. This layer has a distinct control plane for generating and organizing containerized applications. Further, the layer is efficient in taking decisions and including more number of clusters to the federation system. Kubernetes Federation scheme is used to offer the desirable functionality. This united organization is capable of overcoming hardware or network connectivity–based issues by permitting automatic reconfiguration of the handling pipeline. The associated control plane achieves clusters on the edge and cloud. Primarily, this layer monitors the Distributed OS layer to organize data communiqué stacks using Kafka. The handling phases of an IoT application are bundled into containers and organized in multiple clusters.

3.1.5 Construction of multi-transaction mode consortium blockchain

The Multi-Transaction Mode Consortium Blockchain (MTMCB) is constructed based on the different characteristics derived from the current technologies of blockchain that corresponds to secure encryption, point-to-point encryption, distributed storage, and relative decentralization. These features of MTMCB help in generalizing the transaction that could be achieved through blockchain technology into processing and results of and reconstructs "the mechanism of verifying the transaction" through "block distribution and storage."

Fig. 15.1 presents the view of the transaction and regulatory node system used in the design of MTMCB used in the proposed CEFIoT architecture. The

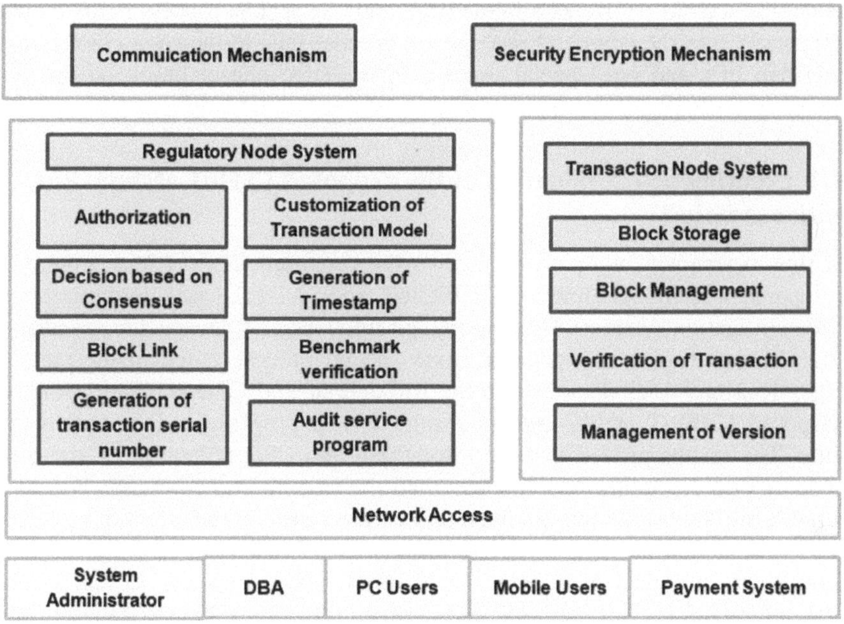

FIGURE 15.1 Constructed MTMCB used in the proposed CEFIoT architecture.

regulatory node system is included in the server side of the service providers of the CEFIoT architecture for achieving the process of initialization, auditing, and transaction processing. However, the operation of initialization is performed only once. On the other hand, transaction nodes can represent the mobile devices, PCs, and payment machine that could be used for checking the availability of CEFIoT architecture, performing online reservation and completing payment to the service providers.

4. Simulation results and discussion

The performance of the proposed CEFIoT architecture with the security blockchain is deployed in the use case of the smart buildings. It is considered that the three types of sensors considered for implementing the proposed CEFIoT architecture is capable of estimating the values of fault tolerance, throughput, and latency (Cominardi et al., 2018). The complete data determined in the environment are processed, replicated, and transmitted toward the cloud environment for the purpose of fault tolerance, futuristic analysis, and permanent storage (Ganchev et al., 2014). The complete investigation of the proposed CEFIoT architecture is attained only based on the image data, since the camera sensor can generate huge amounts of data that refer to the worst case in most of the scenarios (Liu et al., 2014; Wu et al., 2020). The simulation experiments of the proposed CEFIoT architecture are evaluated based on three scenarios such as (i) five PCs in the edge and one special process in the cloud, (ii) five PCs in the edge and one special process fails in the edge cluster, and (iii) five PCs and one special process in the edge cluster based on the parameters of throughput, CPU Utilization, Memory Utilization, and frequency.

4.1 Experiment 1: evaluation of the proposed CEFIoT architecture with scenario-I

In this experiment, the performance of the proposed CEFIoT architecture is compared with the baseline IoTEF, IFCIoT, and SAT-IoT based on throughput, CPU Utilization, Memory Utilization, and frequency with respect to scenario-I (five PCs present in the edge and one special process failed in the cloud). Figs. 15.2 and 15.3 demonstrate the throughput and CPU utilization of the proposed CEFIoT architecture determined with progressive increase in the time. The results proved that the throughput and CPU utilization of the proposed CEFIoT architecture is better than the baseline schemes, since used a reliable and fault-tolerant approach that concentrates on the target of end user satisfaction. The throughput of the proposed CEFIoT architecture is considered to be improved by 5.21%, 7.84%, and 8.31%, superior to the benchmarked IoTEF, IFCIoT, and SAT-IoT architectures. The CPU utilization of the proposed CEFIoT architecture is considered to be improved by 4.62%, 5.91%, and 6.74%, superior to the benchmarked IoTEF, IFCIoT, and SAT-IoT architectures.

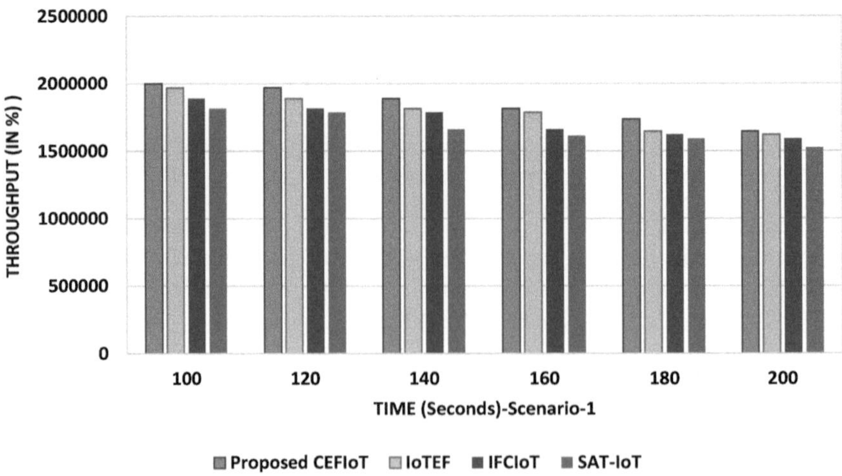

FIGURE 15.2 Proposed CEFIoT: throughput with increasing amount of time (Scenario-1).

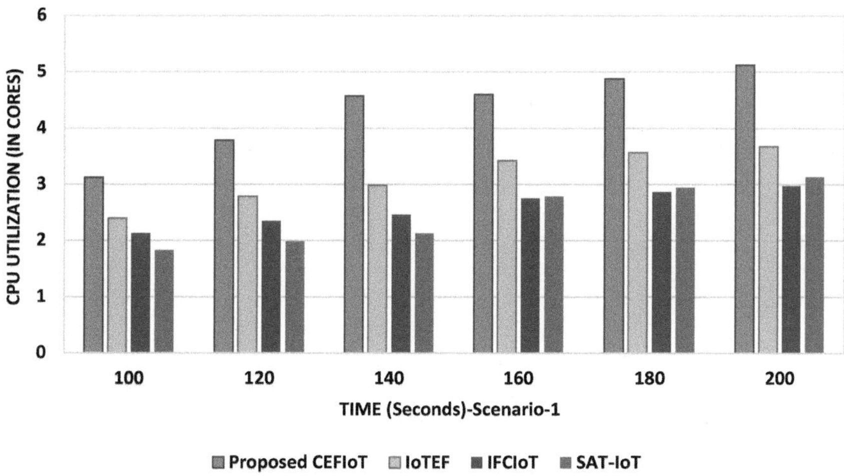

FIGURE 15.3 Proposed CEFIoT: CPU utilization with increasing amount of time (Scenario-1).

Fig. 15.4 portrays the memory utilization of the proposed CEFIoT architecture determined with progressive increase in the time. The memory utilization of the proposed CEFIoT architecture was considered to be improved due to the inclusion of adaptive fault-tolerance approach that determines a predominant role in preventing node and network failure. The memory utilization of the proposed CEFIoT architecture is considered to be improved by 4.94%, 6.92%, and 7.72%, superior to the benchmarked IoTEF, IFCIoT, and SAT-IoT architectures. Like the frequency in the process of preventing failure with increasing latency is also determined to be improved due to the flexible fault-tolerance policy included into the implementation process. The CPU utilization of the proposed CEFIoT architecture is considered to be improved

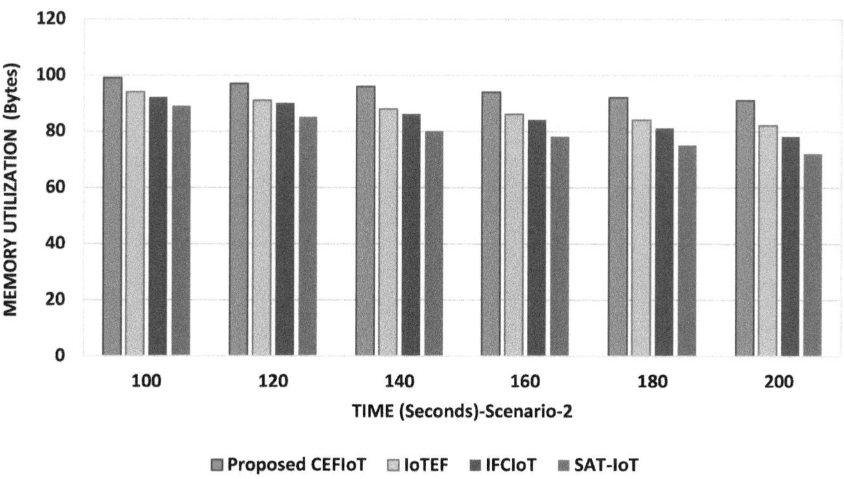

FIGURE 15.4 Proposed CEFIoT: memory utilization with increasing amount of time (Scenario-1).

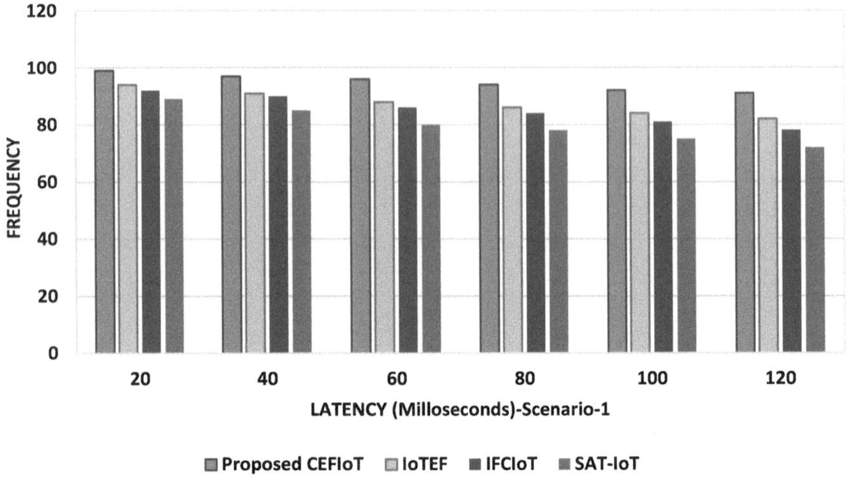

FIGURE 15.5 Proposed CEFIoT: frequency in preventing failure with increasing latency (Scenario-1).

by 5.94%, 6.82%, and 8.26%, superior to the benchmarked IoTEF, IFCIoT, and SAT-IoT architectures (Fig. 15.5).

4.2 Experiment 2: evaluation of the proposed CEFIoT architecture with scenario-2

In this experiment, the performance of the proposed CEFIoT architecture is compared with the baseline IoTEF, IFCIoT, and SAT-IoT based on throughput, CPU Utilization, Memory Utilization, and frequency with respect to scenario-2 (five PCs in the edge and one special process fails in the edge cluster). Figs. 15.6 and 15.7 exemplar the throughput and CPU utilization of the

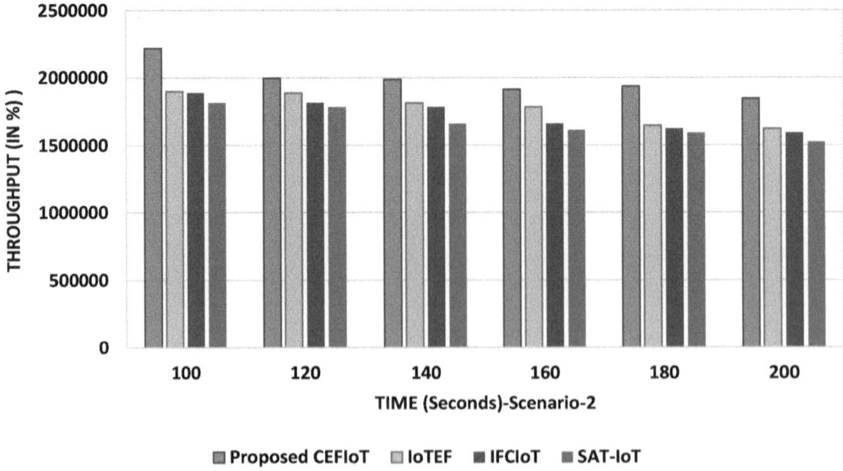

FIGURE 15.6 Proposed CEFIoT: throughput with increasing amount of time (Scenario-2).

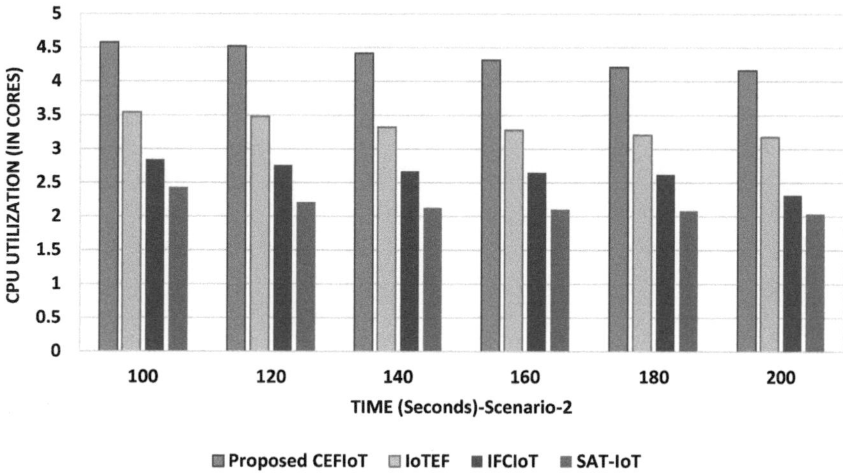

FIGURE 15.7 Proposed CEFIoT: CPU utilization with increasing amount of time (Scenario-2).

proposed CEFIoT architecture determined with progressive increase in the time. The results proved that the throughput under scenario-2 of the proposed CEFIoT architecture is better than the baseline schemes, since the inclusion of blockchain not only provided security but also facilitated a reduced degree of transactions to compensate with the failures. The throughput of the proposed CEFIoT architecture is considered to be improved by 6.74%, 8.92%, and 9.41%, superior to the benchmarked IoTEF, IFCIoT, and SAT-IoT architectures. The CPU utilization of the proposed CEFIoT architecture is considered to be improved by 5.21%, 6.72%, and 7.546%, superior to the benchmarked IoTEF, IFCIoT, and SAT-IoT architectures.

Fig. 15.8 presents the memory utilization of the proposed CEFIoT architecture determined with progressive increase in the time. The memory utilization of the proposed CEFIoT architecture under scenario-2 was estimated to be enhanced by the flexible resource utilization policy that played an anchor role in the prevention of node and network failure. The memory utilization of the proposed CEFIoT architecture is considered to be improved by 5.92%, 6.74%, and 7.42%, superior to the benchmarked IoTEF, IFCIoT, and SAT-IoT architectures. Fig. 15.9 highlights the frequency in preventing failure with increasing latency with scenario-2. The frequency in preventing failure with increasing latency of the proposed CEFIoT architecture is considered to be improved by 4.82%, 5.69%, and 6.21%, superior to the benchmarked IoTEF, IFCIoT, and SAT-IoT architectures.

4.3 Experiment 3: evaluation of the proposed CEFIoT architecture with scenario-3

In this experiment, the performance of the proposed CEFIoT architecture is compared with the baseline IoTEF, IFCIoT, and SAT-IoT based on throughput,

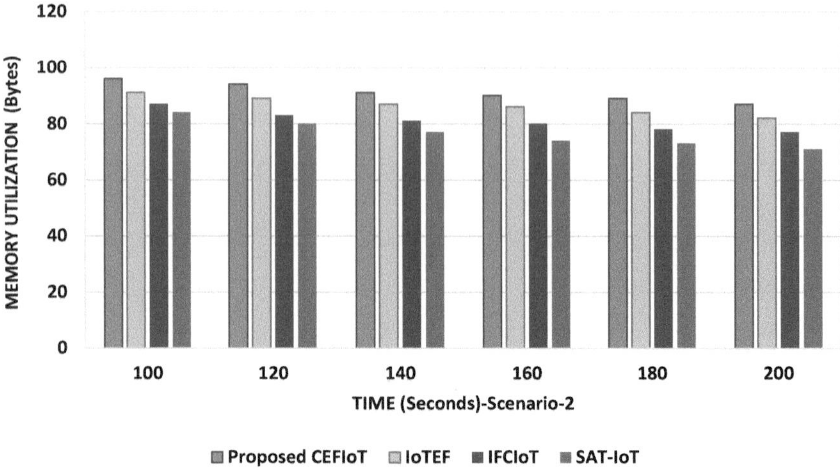

FIGURE 15.8 Proposed CEFIoT: memory utilization with increasing amount of time (Scenario-2).

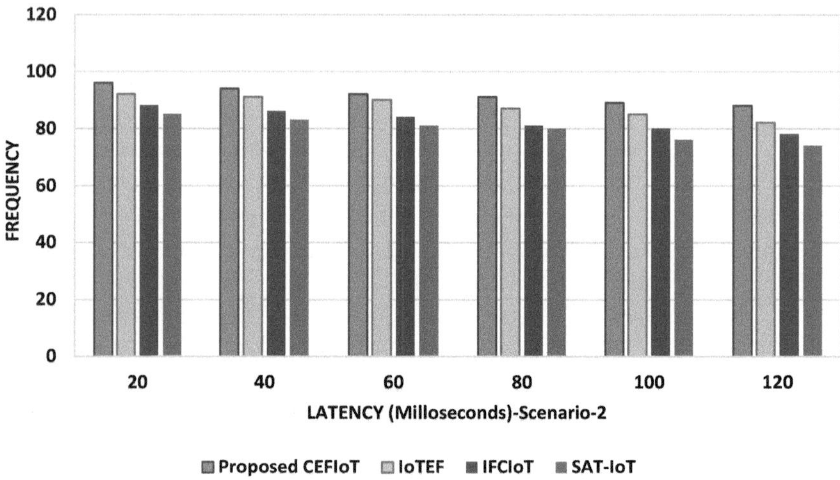

FIGURE 15.9 Proposed CEFIoT: frequency in preventing failure with increasing latency (Scenario-2).

CPU Utilization, Memory Utilization, and frequency with respect to scenario-3 (five PCs and one special process in the edge cluster). Figs. 15.10 and 15.11 depict the throughput and CPU utilization of the proposed CEFIoT architecture under scenario-3 determined with progressive increase in the time. The results proved that the throughput of the proposed CEFIoT architecture under scenario-3 is also superior as it inherited some functional and nonfunctional requirements for handling fault tolerance in order to attain a maximized degree

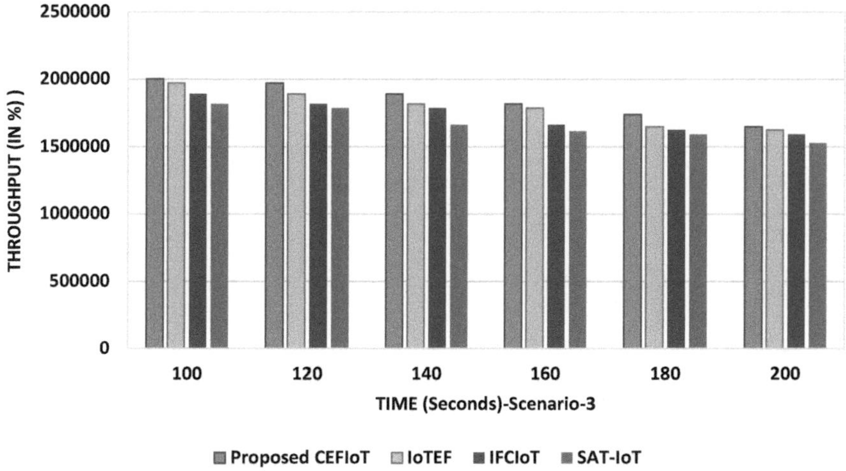

FIGURE 15.10 Proposed CEFIoT: throughput with increasing amount of time (Scenario-3).

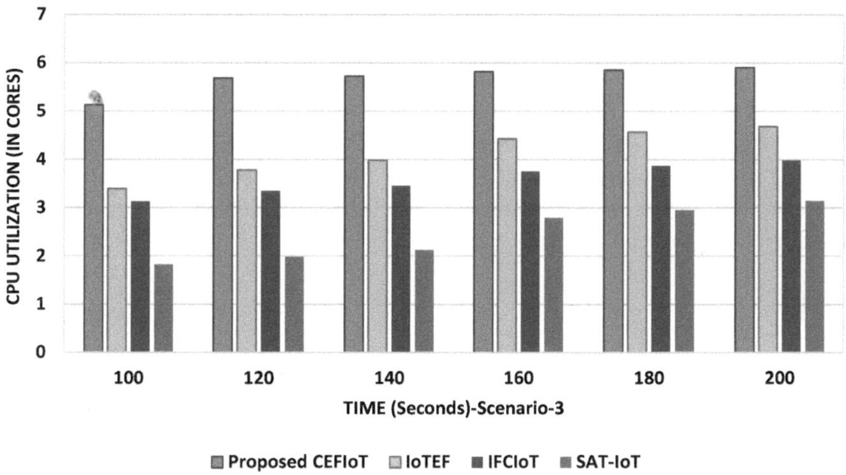

FIGURE 15.11 Proposed CEFIoT: CPU utilization with increasing amount of time (Scenario-3).

of end user satisfaction. The throughput of the proposed CEFIoT architecture under scenario-3 is considered to be improved by 5.82%, 6.91%, and 7.12%, superior to the benchmarked IoTEF, IFCIoT, and SAT-IoT architectures. The CPU utilization of the proposed CEFIoT architecture is considered to be improved by 5.68%, 6.26%, and 7.04%, superior to the benchmarked IoTEF, IFCIoT, and SAT-IoT architectures.

Fig. 15.12 presents the memory utilization of the proposed CEFIoT architecture determined under scenario-3 with progressive increase in the time. The memory utilization of the proposed CEFIoT architecture was considered

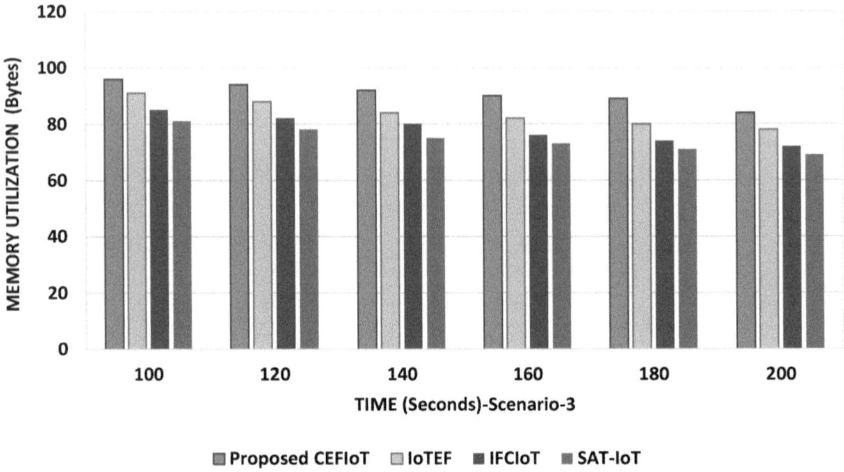

FIGURE 15.12 Proposed CEFIoT: memory utilization with increasing amount of time (Scenario-3).

to be improved due to the inclusion of adaptive-fault tolerance approach that determines a predominant role in preventing node and network failure. The memory utilization of the proposed CEFIoT architecture is considered to be improved by 5.63%, 6.87%, and 8.18%, superior to the benchmarked IoTEF, IFCIoT, and SAT-IoT architectures. Fig. 15.13 depicts the frequency in preventing failure with increasing latency with scenario-3. The frequency in preventing failure with increasing latency of the proposed CEFIoT architecture is considered to be improved by 5.68%, 6.42%, and 7.84%, superior to the

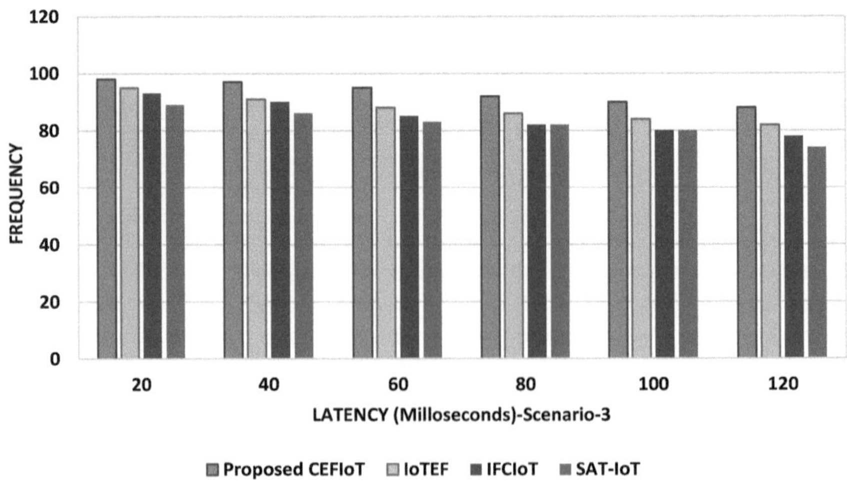

FIGURE 15.13 Proposed CEFIoT: frequency in preventing failure with increasing latency (Scenario-3).

benchmarked IoTEF, IFCIoT, and SAT-IoT architectures. The aforementioned predominant performance is mainly visualized due to the reliable blockchain characteristics included into the proposed architecture.

5. Conclusions

In this chapter, a Reliable Blockchain Edge–Cloud Architecture was contributed proposed for facilitating fault-tolerant IoT applications in which the data are highly preserved from being duplicated or corrupted during the scenario of intermittent connectivity over the longer distance data transmission. The integration of edge computing with blockchain is utilized for facilitating control over the network and for attaining reliable access, computation, and storage uniformly distributed at the edges. This integration of the blockchain also facilitates the option of enabling large scale of data storage, validity computation, and network servers at the near proximity of the users in a more secure way. The throughput of the proposed CEFIoT architecture is considered to be improved by 5.21%, 7.84%, and 8.31%, superior to the benchmarked IoTEF, IFCIoT, and SAT-IoT architectures. The CPU utilization of the proposed CEFIoT architecture is considered to be improved by 4.62%, 5.91%, and 6.74%, superior to the benchmarked IoTEF, IFCIoT, and SAT-IoT architectures.

References

Cao, X., Tang, G., Guo, D., Li, Y., & Zhang, W. (2020). Edge federation: Towards an integrated service provisioning model. *IEEE/ACM Transactions on Networking, 28*(3), 1116–1129. https://doi.org/10.1109/TNET.2020.2979361.

Cominardi, L., Abdullaziz, O. I., Antevski, K., Chundrigar, S. B., Gdowski, R., Kuo, P., et al. (2018). Opportunities and challenges of joint edge and fog orchestration. In *2018 IEEE wireless communications and networking conference workshops (WCNCW)* (Vol. 2, pp. 67–78), 1.

Comma-Di, L., Abdullaziz, O. I., Antevski, K., Chundrigar, S. B., Gdowski, R., Kuo, P. H., et al. (2018). Opportunities and challenges of joint edge and Fog orchestration. In *2018 IEEE wireless communications and networking conference workshops, WCNCW 2018* (pp. 344–349). Institute of Electrical and Electronics Engineers Inc. https://doi.org/10.1109/WCNCW.2018.8369006..

Costa, P. A., & Beko, M. (2019). Dependable and secure voting mechanism in edge computing. *Future Internet, 11*(12), 262. https://doi.org/10.3390/FI11120262.

Cui, L., Tso, F. P., & Jia, W. (2020). Federated service chaining: Architecture and challenges. *IEEE Communications Magazine, 58*(3), 47–53. https://doi.org/10.1109/MCOM.001.1900627.

Ganchev, I., Ji, Z., & O'Droma, M. (2014). A generic IoT architecture for smart cities. In *25th IET Irish signals & systems conference 2014 and 2014 China-Ireland international conference on information and communities technologies (ISSC 2014/CIICT 2014)* (Vol. 2, pp. 45–59), 2.

Ghanta, S., Subramanian, S., Sundararaman, S., Khermosh, L., Sridhar, V., Arteaga, D., et al. (2018). Interpretability and reproducability in production machine learning applications. In *2018 17th IEEE international conference on machine learning and applications (ICMLA)* (Vol. 2, pp. 23–34), 1.

Gupta, B. B., & Quamara, M. (2018). Multi-layered cloud and fog based secure integrated transmission and storage framework for IoT based applications. In *2018 5th international conference on signal processing and integrated networks (SPIN)* (Vol. 3, pp. 21–35), 1.

Gupta, B. B., & Quamara, M. (2018). Multi-layered cloud and fog based secure integrated transmission and storage framework for IoT based applications. In *2018 5th international conference on signal processing and integrated networks, SPIN 2018* (pp. 462–467). Institute of Electrical and Electronics Engineers Inc. https://doi.org/10.1109/SPIN.2018.8474278.

He, X., Liu, J., Jin, R., & Dai, H. (2017). Privacy-aware offloading in mobile-edge computing. In *GLOBECOM 2017 - 2017 IEEE global communications conference* (Vol. 2, pp. 89–99), 1.

He, X., Liu, J., Jin, R., & Dai, H. (2017). Privacy-Aware offloading in mobile-edge computing. In *2017 IEEE global communications conference, GLOBECOM 2017 - proceedings* (Vol. 2018, pp. 1–6). Institute of Electrical and Electronics Engineers Inc. https://doi.org/10.1109/GLOCOM.2017.8253985.

Javed, A., Robert, J., Heljanko, K., & Främling, K. (2020). IoTEF: A federated edge-cloud architecture for fault-tolerant IoT applications. *Journal of Grid Computing, 18*(1), 57–80. https://doi.org/10.1007/s10723-019-09498-8.

Jiang, Y., & Ding, S. (2018). A high performance consensus algorithm for consortium blockchain. In *2018 IEEE 4th international conference on computer and communications (ICCC), chengdu, China* (pp. 2379–2386). Institute of Electrical and Electronics Engineers Inc. https://doi.org/10.1109/CompComm.2018.8781067.

Liu, L., Zhang, M., Lin, Y., & Qin, L. (2014). A survey on workflow management and scheduling in cloud computing. In *2014 14th IEEE/ACM international symposium on cluster, cloud and Grid computing* (Vol. 2, pp. 45–59), 1.

Lopez Pena, M. A., & Munoz Fernandez, I. (2019). SAT-IoT: An architectural model for a high-performance fog/edge/cloud IoT platform. In *2019 IEEE 5th world Forum on internet of things (WF-IoT)* (Vol. 2, pp. 67–78), 2.

Massonet, P., Deru, L., Achour, A., Dupont, S., Croisez, L., Levin, A., et al. (2017). Security in lightweight network function virtualisation for federated cloud and IoT. In *2017 IEEE 5th international conference on future internet of things and cloud (FiCloud)* (Vol. 3, pp. 45–58), 1.

Massonet, P., Deru, L., Achour, A., Dupont, S., Levin, A., & Villari, M. (2017). End-to-end security architecture for federated cloud and IoT networks. In *2017 IEEE international conference on smart computing (SMARTCOMP)* (Vol. 2, pp. 34–46). Institute of Electrical and Electronics Engineers Inc. https://doi.org/10.1109/SMARTCOMP.2017.7947005, 1.

Munir, A., Kansakar, P., & Khan, S. U. (2017). IFCIoT: Integrated fog cloud IoT: A novel architectural paradigm for the future internet of things. *IEEE Consumer Electronics Magazine, 6*(3), 74–82. https://doi.org/10.1109/MCE.2017.2684981.

Nambi, S. N., Sarkar, C., Prasad, R. V., & Rahim, A. (2014). A unified semantic knowledge base for IoT. In *2014 IEEE world Forum on internet of things (WF-IoT)* (Vol. 3, pp. 12–25), 1.

Peinl, R., Holzschuher, F., & Pfitzer, F. (2016). Docker cluster management for the cloud - survey results and own solution. *Journal of Grid Computing, 14*(2), 265–282. https://doi.org/10.1007/s10723-016-9366-y.

Qiong, W., Kaiwen, H., & Xu, C. (2020). Personalized federated learning for intelligent IoT applications: A cloud-edge based framework. *IEEE Open Journal of the Computer Society*, 35–44. https://doi.org/10.1109/ojcs.2020.2993259.

Yousefpour, A., Fung, C., Nguyen, T., Kadiyala, K., Jalali, F., Niakanlahiji, A., et al. (2019). All one needs to know about fog computing and related edge computing paradigms: A complete survey. *Journal of Systems Architecture, 98*, 289–330. https://doi.org/10.1016/j.sysarc.2019.02.009.

Yuen, T. H. (2020). PAChain: Private, authenticated & auditable consortium blockchain and its implementation. *Future Generation Computer Systems, 112*, 913–929. https://doi.org/10.1016/j.future.2020.05.011.

Wu, Q., He, K., & Chen, X. (2020). Personalized federated learning for intelligent IoT applications: A cloud-edge based framework. *IEEE Open Journal of the Computer Society, 1*(3), 35–44.

Chapter 16

Applications of blockchain technology for smart cities in a nutshell

V.V. Satyanarayana Tallapragada
Tallapragada Department of ECE, Sree Vidyanikethan Engineering College, Tirupati, Andhra Pradesh, India

Chapter outline

1. Introduction

Blockchain provides services such as decentralization, security, preservation of privacy, immutability, and authenticity. In terms of IoT, this entirely relates to and is targeted toward the development of Smart Cities. A Smart City (Application of blockchain technology in smart city infrastructure, n.d.) practices information technology to assimilate and accomplish body, social, and business setups. It can offer its occupants better services while preserving the well-organized and optimal exploitation of existing possessions. In this juncture, blockchain technology assists in identifying the blockages for smart city implementation making ease in delivering a reliable, transparent, robust, and decentralized environment for real-life implementation. It also provides exceptional support that smart cities can leverage to expand the value of life, administrative processes, and sustainability of the environment. Blockchain technology is nearly a decade old with its introduction by Satoshi Nakamoto (2008) in 2008 which was proposed primarily for the use of cryptocurrency. It is a known fact that Bitcoin is one such type of cryptocurrency which is

Blockchain for Smart Cities. https://doi.org/10.1016/B978-0-12-824446-3.00008-9

widely used. This is provoked to have a replacement for the monetary system, which is thought of as the main reason for the economic crisis in 2008. The potential of blockchain technology has made foundations in smart cities giving decentralized solutions for handling data that have the property of immutability. In the recent past, the extensive emphasis is made by researchers to develop a variety of applications for supporting IoT gadgets for the upgrowing digitized community, Smart City (Ahmed, Shah, & Wakil, 2020). After the Internet, it is observed that blockchain technology is the second throttle that humans are about to perceive. Internet is developed for the transfer of information wherein blockchain is for transactions. Due to the lack of understanding and the stigma that has developed in adapting to the new technology has made blockchain to take the last place till now. But as the technology is adaptive and the users who use this are adaptive will be using this. The use of blockchain for smart cities (Bhushan et al., 2020) has covered a broad range of scenarios viz., education, mobility, home, electricity, healthcare, business, alongside security and privacy (Theodorou, Sklavos, Rawat, & Ghafoor, 2019).

Now, let us define what is this blockchain. To have the applications to get defined for this particular topic, it is quite necessary to understand what is blockchain. It is a list of entries that are encrypted using cryptography, called blocks. These blocks can be documents or any messages that are encrypted using cryptographic techniques (Licheng, Xiaoying, Jing, Jun, & Yixian, 2019) that preserve timestamp of the data when it was encrypted, data used for transactions, etc. This maintains all the entries or documents as a shared document. This shared document or entry is called the shared ledger which is also called the Distributed Ledger Technology (DLT) (Roeck, Sternberg, & Hofmann, 2020) that creates replicates of the shared documents or entries that are synchronized over the network. This means that the data are spread across several systems that are heterogeneous in nature. Hence, a question arises for the availability of a central level administrator who controls all these tasks. There is no such administrator available to control all these functions. Fig. 16.1 shows the blocks that are interconnected which is the basis for blockchain technology.

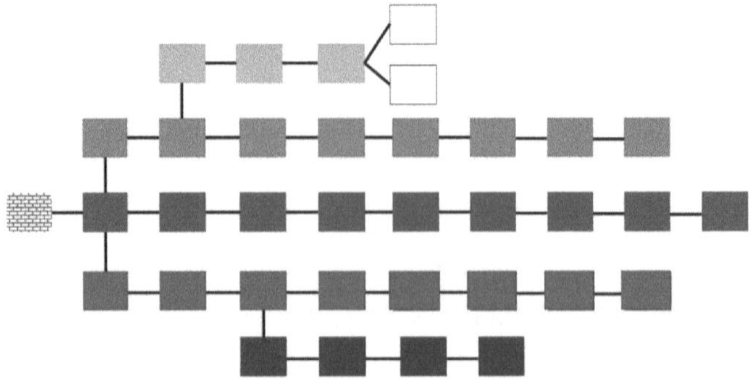

FIGURE 16.1 Blockchain formation.

From Fig. 16.1 it is evident that the main block is the left-most corner block which is called the origin block, technically called genesis. All the remaining blocks that are connected to the genesis are formed by the DLT that exchange data in a peer-to-peer network, wherein each node in the network mimics and further preserves a similar copy of the documents and updates itself individually. As it is evident that there is no monitoring done by the administrator, each node creates a fresh transaction making all the nodes smartly have the copy. In such a transaction, a general tendency would be with security. Here, in this case of having updated the ledger is carried out using cryptographic algorithms (Taylor, Dargahi, Dehghantanha, Parizi, & Choo, 2020). Now, it is obvious that once the data are recorded it cannot be altered without having an alteration in the subsequent blocks, again which requires authentication. One of the applications of blockchain technology is in trade. One of them is Bitcoin (Chaudhary et al., 2019; Nian, Chuen, & Lee Kuo Chuen, 2015), that performs ordering of the transactions for a limited sized structure, which is termed as a block that will be sharing its timestamp. The miners that are the nodes in the network perform the linking of the other blocks in sequential order with each block comprising of the previous block hash code to obtain a new blockchain. Such a framework provides better security for all the records in the transaction. However, it is argued that the traditional business process has been hindered by the introduction of the blockchain. The traditional way of having a business is to have a centralized administration or trusted third party to verify the transactions and having the same level of certainty in a decentralized manner on the other side. Many of the operators who tend to trust only with the traditional way of making transactions will have a technical glitch in converting or moving toward blockchain. As in the case of digital transactions or having payments services online at any retail store may cause panic to the shopkeeper, if money is not transferred in time to his account, maybe due to network delay or due to bank. This creates a loss of trust over digital transactions (Chen, Jiang, & Wang, n.d.). But, still, as time goes on, the situations may change consequently creating awareness over such breaches, making system nonvulnerable to failures as well as misconceptions.

Blockchain's underlying infrastructure and key strategies make it have transparency, robustness, accountability alongside providing security. Blockchain can be observed to be a distributed database that consists of a well-organized list of ordered blocks, which are immutable. Another application of blockchain is in the banking sector where all the banks can communicate or share their transactions under the same blockchain paving the way to have a better customer relationship. For the past decade, blockchain is thought of as a hype thinking that it is not possible to implement it. But organizations are investing in this infrastructure as they observe, predict an opportunity to decentralize their system, and reduce the cost in transactions or compliance costs making them serve in a better way, thus, having cheaper, simpler, and often quick transactions.

The concept of blockchain tells that it is a distributed, time-stamped, append-only data structure (Fran, Dasaklis, & Constantinos, 2019), thus making the connected miners to efficiently communicate with each other without having any centralized intermediary. Fig. 16.2 shows the architecture of the blockchain. The lowest layer of the blockchain architecture is formed by the basic transactions carried out by the different users. This may be physical and/or digital transactions that are carried out at the user end. Considering an example for this, if A gives cash in return to the goods purchased by him to B. This is a physical asset that is given to B from A. Now B deposits the same in his bank account and buys shares of a particular company. Now, B is having a digital asset with him. This forms the base layer. All these transactions are maintained as a record or ledger which is carried out on a network or through computer interface which is represented as a network in Fig. 16.2. A node is an object that is connected to the blockchain, which is called as a miner. All the miners need to obey the rules of the blockchain and they are called full miners. These miners combine the transactions into blocks and they must decide whether the process is legitimate and should be held in the blockchain or not. Hence, miners negotiate a deal to ensure that there are no corrupt branches and divergences exist. This task is accomplished by the next layer, i.e., consensus layer. Several algorithms are deployed in the consensus layer viz., Proof of Work (PoW), Proof of Stake (PoS), Delegated Proof of Stake (DPoS), Practical Byzantine Fault Tolerance (PBFT) (A review on consensus algorithm of blockchain, n.d.). To ensure authentication and verification PoW involves solving a complicated mathematical process that identifies hashes with unique patterns. Rather than having separate blocks among the miners in a proportional fashion based on their hash rates, PoS breaks the blocks in a proportional way to the resources held with miners. This precludes the network from being dominated by the top participant. Based on the substantial reduction in power and increased scalability most of the blockchains are slowly migrating to PoS. The consensus algorithm that is intended for

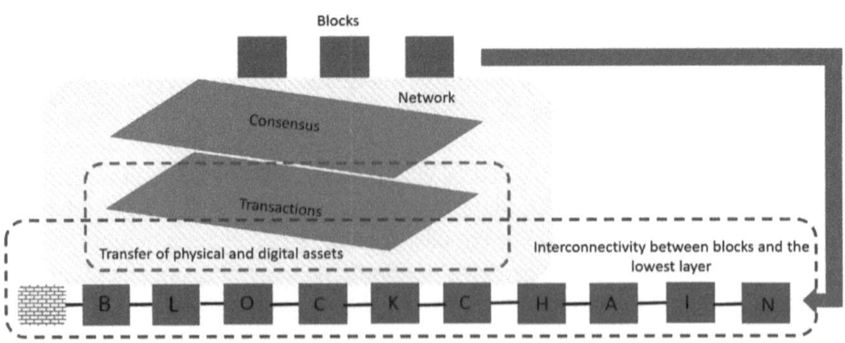

FIGURE 16.2 Architecture of blockchain.

usage depends on the type of blockchain. There are three types of blockchain: public, private, and permissioned (A review on consensus algorithm of blockchain, n.d.). A decentralized blockchain or public blockchain is the one that is open to everyone and can become miners and can further contribute to the benefits according to the norms. Furthermore, it is to be noted that no trusted relationships exist between miners. The transactions that are carried out can never be altered. Examples of public blockchain consensus algorithms are PoW, PoS, and DPoS.

In contrast to the public blockchain (Lai, Chuen, & Deng, 2018; Shi et al., 2020; Xiwei et al., 2019), in a private blockchain, the miners have limited power to change the information. Almost, the control is held only with the owner. On the side, it has an easy modification at low transaction costs. These are generally found in the intranet, which is a closed network. Permissioned blockchain is the network consisting of several entities and the key miners are well before specified by the parties wherein the representatives do not completely trust others. Each member identifies their consensus miner in compliance with the rules. The majority of these miners must accept these transactions. The amount of transparency and centralization lies in between the other two blockchains. This type of blockchain is ideal for a semiclosed network developed for various businesses and PBFT is used for this purpose. Hence, it is evident that the consensus algorithm is the main core component of the blockchain technology but is still in the primitive stage. It is the need of the hour to develop a common platform for the improvement of performance in a particular scenario. The existing algorithms are designed, developed, and deployed to address a rare and specific situation.

Another layer is the network layer or computer interface, which is shown in Fig. 16.2, which provides additional functionality that stores the transactions done by the users. However, for more implementations, complex states need to be maintained that are automatically modified using distributed computation, for example, when different conditions are met. This prerequisite has led to SCs using blockchain nodes to satisfy contract terms. There is an interconnectivity between blocks and the lowest layer that will be governing the communication with the physical world. The protocols that exist in the blockchain technology are well established but are affected by the inputs from heterogeneous groups of users who implement new techniques and algorithms. Hence, blockchain compacts on how these heterogeneous users arrive on a decision for changing the inputs and also to produce and maintain the blockchain.

The latest literature classifies blockchain networks in different forms. Such categories comprise public, private, and federated network management authorizations. Table 16.1 shows the comparison of different types of blockchain networks. Everyone can join as a miner in the public blockchain that is permission-free. Moreover, operations such as purchases or contracts can be carried out by any individual. A whitelist of approved users with unique

TABLE 16.1 Comparison of blockchain networks (Fran et al., 2019).

Parameter	Public	Private	Federated
Approval time for a transaction	Minutes	Milliseconds	Milliseconds
Efficiency	Low	High	High
Energy consumption	High	Low	Low

network features and permissions is specified in private blockchains, which, along with the federated ones, typically belongs to the permitted blockchain group. The hybrid mix of public and private blockchains is a federated blockchain. Although the degree of scalability and privacy with private blockchain is close, the key differences are that some nodes called lead nodes are chosen to validate transaction processes instead of a single business. This allows a partially decentralized architecture in which leading nodes will allow others.

Based on this discussion, it is evident and clear that what type of blockchain is to be selected for a particular type of application. Now, let us examine the applications related to smart cities. By exploring the possibilities of blockchain, the term "Smart" can be termed as the foundation to the areas connected with it. It provides a proactive strategy, which does not contain any public administration and thus never erases data. Extensive focus has been applied by several researchers to provide solutions for a broad variety of applications viz., providing security to the transactions, like protection by maintaining privacy and security among heterogeneous IoT gadgets that are increasing in the current digitized community, which is a Smart City. Blockchain offers a secure, open, scalable, and decentralized platform for real-life implementations. It also offers support to the smart cities that are having an overwhelming response to the upcoming technological development, hence, making the way to improve the value of life, management procedures, and for sustainable development, over the head providing services like decentralization, stability, anonymity, along with integrity and validity. To integrate and achieve physical, social, and commercial facilities, Smart City incorporates information technology, to offer its occupants better services while ensuring the best and best use of their existing properties. To develop an application using blockchain technology one has to concentrate or focus on the areas where blockchain is necessary, carry out a necessary survey to identify the hurdles to implement the solutions for a smart city (Ahmed et al., 2020). Combining IoT with blockchain is a great deal as most of the gadgets have no correlations for interoperability to the facilities that are required for the deployment in line with blockchain. The next section clearly outlines the applications of blockchain technology in smart cities.

2. Application of blockchain technology in smart cities

2.1 Unique identity

With the advent of the industrial revolution that has led to the development of Industry 4.0, the world is moving toward the intelligent as well as an autonomous system that is programmed to perform certain tasks. So, this can be considered as a cyber-physical system, which is a computer system that is monitored, controlled by algorithms that are deployed on it. Such systems can operate at various scales, exhibiting heterogeneous modalities such that they can interact with each other. Examples of such systems are autonomous vehicles, automated industries, robots, etc. In the virtual world, data protection and digital identity have become increasingly relevant as technology becomes combined with cyber-physical structures. In this digital world, where customers are increasingly using the Internet to carry out electronic transactions, both for the purchase of items and services and for bank transactions, the idea of having a unique identity that stores all of the data concerned to the user is quite an imperative requirement that includes all personal data accessible online for the user. All may include not only an e-mail or a physical address but also photographs, bank account details, shopping preferences, and also the information regarding the physical identity. Hence, the identity that is designed, developed, and deployed for authentication must be unique, any change in its part at any stage will create hackers to create other identities that are similar, same, and accurate in every aspect thus making way to have benami transactions. This flaw needs to be taken care of everywhere. Hence, a unique identity is required for each transaction. So, a unique identity is created and then it has to be authenticated digitally which has made it a digital identity. Digital identity is a combination of online and traditional identities where the latter are verified manually by observing its copy that is stored physically, digitally, or using an embedded system—based application that is an IoT-based system.

Most of the deployed systems that provide authentication based on identity have seen chaos in their systems wherein the security breaches need to be improved, thus making the system nonvulnerable. Hence, it is the need of the hour to improve security over the digital identity. Systems that employ digital identification will be using cryptographic techniques thus making them stronger even when the hackers try to break. Deploying such systems as a part of a smart city that includes blockchain needs to be verified for security and privacy issues making it a theft-proof. The application discussed in this chapter discusses the significance of a unique identity and how blockchain will contribute to protecting such confidential information.

Most of the applications of the unique identity have their share on providing identity for the government services that are provided to the public. Fig. 16.3 shows various sectors that use a unique identity as a part of

FIGURE 16.3 Usage of unique identity in various sectors of society.

authentication. First and prime most application is providing welfare measures that are provided to the public as a part of the policy. To verify the identity of the person who is taking services from the government a unique identity is necessary for authenticating the user, so that the same user may not get the same benefit at another place. Another sector that has the applicability of unique identity is secured transactions. In implementing citizen social welfare initiatives and examples, one context in which digital identity is used to fundamentally change socially and economically societies is significant. The utilization of the unique identity for such initiatives has made an upliftment in the status of some poor people. In India, more than 90% of the population have an Aadhaar ID which is a digital identity that links up with all the social schemes and has changed how subsidies are paid to lower-income sections of the society. The effective utilization of online identity does not only require authentication but it can also be captured in a sophisticated form that several systems can safely transmit and use on a demand-based requirement.

Secured transactions on the one hand have increased the authenticity of a person who is doing the transactions on the fly. Previously there are many lapses on the transactions that are especially financial transactions. Unauthorized persons and persons who do not have proper authentication sued to carry out the transactions when the transacting agent or the officer is known to him. But, with the advent of the technology, this had made them layman, making the transactions that are carried out on the fly also to be authentically proving the authenticity of a person by using the unique identity. The unique identifier may be any of the biometrics issued by the user at the time of registration along with the unique number provided to the user. If both get matched, then only the transaction will be carried out. Does a question arise in this context that Is this system secure enough? Yes, of course as the blockchain technology uses cryptography for securing the transactions, certainly there is high-level security maintaining privacy only to the intended user maintained. In India, the government has taken a digital initiative and has introduced UPI transactions that are carried out with such a unique identity only. Another sector is property registration. Such properties may be immovable or movable. Immovable properties are properties such as land, house, farms, etc.

These properties are registered with different names of the persons previously. For example, A.B.C. D is the name of the person. The initials A.B.C with the name of the person as D. The name can be written as D A.B.C. Hence, making a confusion state at the time of registration. Now, the name of the person is linked to the number like Aadhaar that is further linked to the biometrics of the person. Hence, making the user's name and identity to be unique based on the number assigned to the user. This process makes the user have a registered property on the user's name based on the unique identification number that is submitted at the time of registration. Now, the government can track the user's property based on the unique identification number so that he can be classified based on his economic status, further making the government ease in providing welfare measures based on the economic status.

Another sector that has the application area is the workplace. A user who is working in an organization needs to provide attendance both at the beginning of the day and end of the day. Impersonification leads to the lesser workforce and creates chaos as a part of the work at the workplace. Hence, a proper workforce needs to be appointed who does not impersonate themselves by providing wrong identities. In the case of the traditional way of having an attendance register, then there is a chance of having false signatures or forgery. But, authenticating a user at the workplace eliminates such false transactions, making the management to identify the in and out times as well. Fig. 16.4 shows the four pillars of the blockchain technology that are used in particular to unique identity.

Another aspect of a unique identity is taxation. It is a well-known fact that most of the users try to evade taxes by hiding the property that they hold. But with the unique identity and digital initiatives that are launched, the properties that are held with a user and the financial transactions that are carried out by the user are recorded under the unique identity. This makes the taxation to be easy for the authorities to monitor, control, as well as further fine the users if they do not pay the taxes. The last aspect of the unique identifier is the security enforcement in the border area. A person who authenticates himself can be

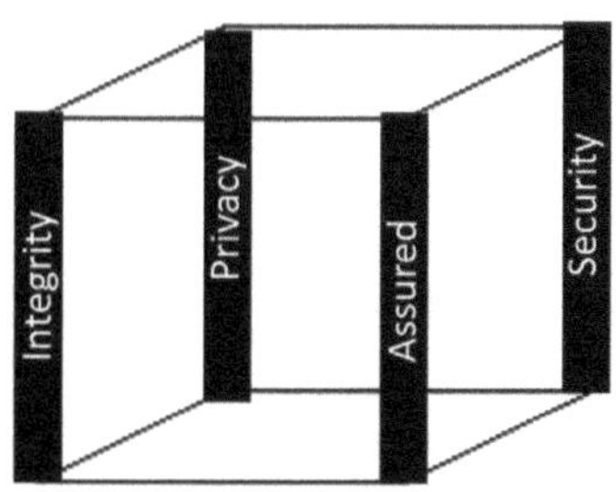

FIGURE 16.4 Four pillars of blockchain technology—based unique identity.

only allowed to cross the border, provided the user scans his identity at the entry point. This ensures that only authentic persons are given entry. Unauthorized persons need to have a proper and valid visa to enter the country thus making the country secure. The technology from blockchain has been developed to keep the data encrypted and unchanged and to ensure that the ID is protected and followed via encryption. Blockchain-based systems also eliminate the password authentication limitation. The DLT used for blockchain technology uses metadata for communication and the data validity is confirmed by the use of consensus mechanism through various miners. This type of identity system is beneficial over conventional systems as it can manage every identity across the miners who are connected over the network. The data are constantly harmonized and kept current even as it is spread through peer-to-peer networks. Furthermore, the blockchain network has no single point of failure, which makes the credibility of the dataset difficult for attackers to crack. As the number of transactions carried out worldwide on a large scale is high, hence there is a need for the protection of such sensitive data used for authentication. The digital signature which is nothing but the biometric identity or any other unique identity provided by the user that is paired with the blockchain encryption mechanism offers privacy, thus, making all the transactions foolproof. Hence, in this context blockchain can be viewed as a structure that provides a framework with its pros and cons for having a unique identity for all the transactions. Taking full advantage of the transparency and confidence that blockchain systems offer, several organizations and countries join forces to guarantee cross-border interoperability.

2.2 Smart properties

The main concern of every individual in this physical world is the property. This can be of two types:

1. Physical
2. Nonphysical

The former is the one that is also considered as the fixed/movable asset. Examples include cars, land, etc. The latter are shares which are stored in an account in an intangible form. The application of the blockchain toward the development of a smart city has expanded its claws into the smart property that has eliminated intermediary, making less intrusion of the others over the discussion of the property by the seller to the buyer. In the past, the falsification of properties and contracts had significant consequences and was mostly overlooked and ignored. Hence, blockchain needs to be applied to develop a proper management system that attempts to clear the flaws of the traditional registration system that registers the property on the owner's name.

Commercial businesses and real estate experts understand the significant impact of blockchain innovations in maximizing both retail and commercial real estate transactions, streamlining payments, and enhancing access and growth potential for assets. The digitization of records is nothing but tokenization. These can be configured to provide trusteeship, transaction history, and protection to ensure a secure transaction. Fig. 16.5 shows the smart property architecture using blockchain technology. The properties shown are physical as well as nonphysical forms. These include vacant land sites in rural as well as the urban areas, farmlands, shares, movable vehicles, etc. When any transaction is completed, it creates a block that will be notified in the ledger which is nothing but the distributed ledger in this case. This creates a chain that is governed by the computer interface which will be recording all the transactions carried out under one particular unique identity. Such a smart property registration eliminates all types of violations. This also makes the government easy to monitor all the transactions that are been carried out at a particular place of interest. For example, if new capital is announced, obviously the land rates at that particular place will increase and there is a chance of fake registrations. If a particular property is registered with the government or by private people, then there is a chance of notification of particular property at the time of registration as well. If these are not maintained as records, conquering of the property by filthy, faulty, and fraudulent activities takes place which almost everyone is aware of. The inherent trust mechanism of blockchain makes it the perfect smart property. Blockchain smart contracts and leading capability are being used by real estate companies throughout the globe to make leasing, purchasing, investment, and even lending transparent and effective. Since the daily rate for property transactions is so high, it is more than ever important to create a common repository of leasing and sales. This can be supplied by a blockchain. The improvements to a blockchain platform will build a much simpler header framework that would enable brokers and agents to access the full background of an estate's transaction.

FIGURE 16.5 Smart property architecture using blockchain.

It is necessary to always have reliable documentation for any form of property, wherein the current proprietor needs to be identified appropriately and provide evidence that the same person indeed is the proprietor. Such documents can help to secure proprietary rights against robbery, forgery, and settle conflicts, ensuring proper transfer of land ownership to a new owner after-sales avoids sales fraud. It is therefore critical that this information stays complete and accurately such that illegal transactions are prevented. From the viewpoint of a buyer who generally relies upon a third party who inquires about the property. The government has to have a track of the transactions and the transfer of ownership that are carried out. If so, these transactions are properly maintained and made trackable, then it creates a well-distributed ledger so that it will avoid all types of fraudulent activities. Does a general question arise that Can property ownership be monitored by a sort of distributed framework that does not have confidence? Yes, it can be monitored. The consensus layer that is existing in the blockchain technology takes care of this task of holding the past transactions and keeps evidence that makes it expensive to change the records by breaking the protocols. There is no requirement for any third party who in turn takes 1% or more for the service provided if the transactions are secured by the blockchain. A blockchain-based property rights framework allows verification of cryptographer semantics that are generated by the miners preventing most possible vulnerabilities and attacks. This will minimize third-party dependency, reduce costs, and provide actual, secure transactions. This technological advancement and interconnectivity between digital assets and connected networks will grow business opportunities for secondary companies and also value.

2.3 Smart governance

For many business and applications, blockchain technology is considered as a vital revolutionary innovation. This builds trust between the payments that are being carried out and the stakeholders. It also supports zero downtime, including immutability for nonduplication and unable to hack, enforce cryptography-based safe confidence between the miners using the consensus algorithms. The smart governance system that is based on the blockchain technology will have the following advantages:

Data integrity—secured data storage of citizens, government, and businesses carried out.

A decrease in the processing time that is being carried out by the traditional systems.

Decrease in fraud and improves trust.

Combining blockchain technology with governance benefits in various ways. The data that are stored, that is the citizens' details that are confidential, need to be authenticated whenever a transaction is carried out. This can be achieved by the cryptographic algorithms that are followed in the blockchain, thus, making the system highly secure on the fly as well. Such a system has an advantage of the reduction in the processing time. In a traditional system, manual verification of the data requires much time, whereas with such a digital system the authentication can be provided in seconds and also at a remote place. This reduces the fraudulent transactions that are carried out by giving some bribe and thus improves faith in the governance. Pilot projects are launched by all most all of the governments worldwide to have the implementation of blockchain. The applications are complex in nature that include unique identity, maintaining a database of the judiciary, financial transactions, economical and marital status, permissions viz., visas, passports, crime analysis records, taxation, etc. The blockchain technology allows several parties to engage in a transaction. A noteworthy example of this is the granting of permits to organizers of public activities such as concerts and protests that include local authorities, police, the fire department, and health service providers. Another instance is that the services provided for any public institution and the modification or transfer of authority and the capacity only if the nodes that are listed as strongest in the hierarchy agree on the transaction. Blockchain is thus a technology that eliminates individual repositories with a decentralized directory of shared data, leading to improved facilitating access.

Fig. 16.6 shows the traditional system that shares the repository of transactions that are carried out by the public and all are recorded in the repository. The drawback of such a technique is that transactions that are carried out are not shared among heterogeneous applications. In contrast to this, Fig. 16.7 shows the blockchain-based transactions that are carried out. This ensures that all the transactions are stored in a distributed ledger, thus forming a blockchain. The advantage of this system is that it will have a protected scenario of transactions and keeps a track record. This further ensures that only certain

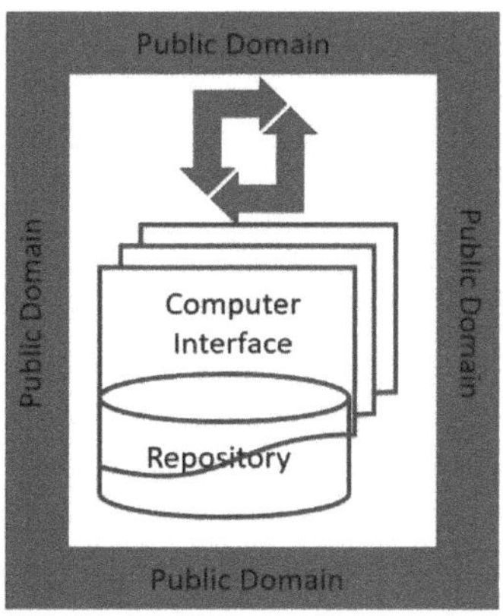

FIGURE 16.6 The traditional repository.

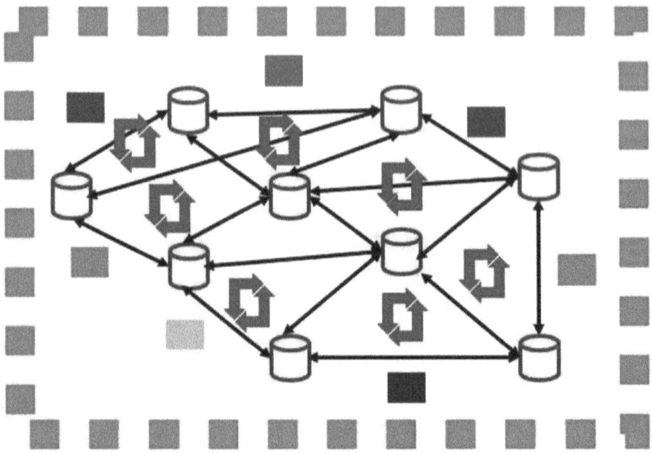

FIGURE 16.7 blockchain-based governance.

people with certain parameters will be selected for a particular criterion sponsored by the government. Such a search technique makes governance smart. Such a scenario is especially helpful when the ownership transfer is carried out as a part of the sale of the property. Under smart governance, the change in the title that is the land title is difficult to identify. This has the advantage of avoiding the data theft and loss. A point needs to be noted here that, blockchain will not contribute toward the reliability of the titles, but still aims at the validity of the title. Hence, blockchain is used as a tool for combating corruption that is still being carried out in the registration of the land. This makes governance easier. These examples illustrate the way of workflow that is structured which can dramatically impact blockchain applications. Such an implementation of the blockchain technology in smart governance needs frameworks that are well planned, implemented, maintained, and adapted. This determines how the entire framework functions and let the users operate with it and in it. Such an implementation can be done only if governments come forward ensuring the public interests and social requirements (Svein, Jolien, & Marijn, 2017).

2.4 Energy and resource management

Energy systems have evolved rapidly to meet the growing amounts of integrated renewable sources viz., wind, solar. The recent past has experienced significant development enabling them to improve their capability of development financially, industrially, and thus leading globally. This made private players in conjunction with public partnerships have made this sector so strong financially promoting incentives and initiatives for further development. These resources that are renewable sources are complex in nature, hard to anticipate, and entirely dependent on the environment, thus posing high-level problems for energy generation, distribution, and serving that require more flexibility for safer operation. These systems require large investments to meet aggressive targets that are intended to. Public enterprises that deploy such renewable energy–based systems are now capable of generating up to 30%–40% of the total generation. In other words, the total power generated out of renewable energy sources is limited to 40% of the total power generated. There is a rapid demand over the energy requirements for the upcoming scenarios like electric vehicles. Hence, this makes private players invest more in this sector. The projects that are developed for this purpose are highly demanding as energy markets are more involved, decentralized, dynamic, and at multiple levels with a growing number of players, so, potential steps must be taken to regulate appropriate investment management and control. Sophisticated data exchange among the various components of the system has become more prominent. To adapt these developments decentralization and digitalization with a local distribution system that can effectively control and manage such techniques are needed. The main aim is to make distributed transactions simpler by eliminating central control employing blockchains that maintain the blocks by the way of the distributed ledger. Managements from the energy sector have thought of introducing blockchain that would potentially be the solution to the challenges of the current industry. This will enhance the productivity, speed up the growth of IoT networks and digital applications that are used in the energy sector. The current energy production systems are undergoing a shift of transition that is induced by the latest technologies. The main concern is the upcoming decentralization and application of digitization. Hence, blockchain is thought of as a promising solution that could monitor and manage complex microgrids. The integration of these renewable energy sources with the current market trend is more acquisitive. As a part of billing that is carried out for the usage of the power by the customers, transactions are recorded. Active customer transactions must be secured enough, transparent, and must be tamperproof. The ability to offer these digital payment services will provide pricing information, cost per unit that is utilized by the customers, while at the same time making them adapt to their requirements and also to smartly manage. One of the key advantages of this strategy is to have a reduction in transmission and distribution losses and have a reduced time in improving the infrastructure. The infrastructure needs to be improved only when there is a requirement. Still, the supply is maintained only through the physical grid which requires continuous monitoring. Hence, blockchain provides a safe way to record the energy consumed by the consumer. If this task of billing is made a prepaid, then the consumer will be charging as per the requirement. This can be done using an intelligent and smart recharge in a smart city. Several application scenarios are related to energy with blockchain technology. Current literature determines the future implementations and related business models and how these are affected.

The first use case is the billing. Smart metering, distributed utilities my render automatic billing using blockchains. Pay-by-go solutions for the metering will be benefiting. Sales and marketing are some of the areas where these entirely depend on the energy consumption of the consumer. These utilize advanced technologies such as artificial intelligence and machine learning that precisely calculate the energy consumed by the consumer. It also monitors and identifies the patterns that are further tailored to create products based on demand. Fig. 16.8 shows the blockchain architecture—based power management. In the traditional way of power generation to the utilization, many layers are not visible. The intermediaries will have more control over the system thereby creating flaws, and fraudulent methods are adopted in every aspect from generation to utilization and in the latter case for billing as well. So, the latest blockchain-based energy management system will create blocks that are the records that are shared through the computer interface which is nothing but the consensus algorithm. This eliminates the requirement of intermediaries which delay the entire process. Decentralized payment systems that are allowed by the blockchain technology will disrupt the existing trading system and maintain risk management. As the systems are automated with control and monitoring will boost control over the microgrids and decentralized power systems. As most of the transactions are carried out online, the use of blockchain in energy management will have an edge on the utilization of smart apps to upload, store data. Smart grid products like smart meters, intelligent systems designed using smart sensors, energy storage, and control systems, and surveillance systems are used. It is a well-known fact that blockchain utilizes cryptography, hence, making all the transactions secure. Resource sharing is one of the areas where blockchain technology combines with energy management. The best example of resource management is electric vehicle charging stations. Stations are provided at a particular distance from each other like that of petrol bunks thus making each electric vehicle charge on the fly. This helps the government to monitor how many people are using electric vehicles and the usage of power as well (Merlinda et al., 2019).

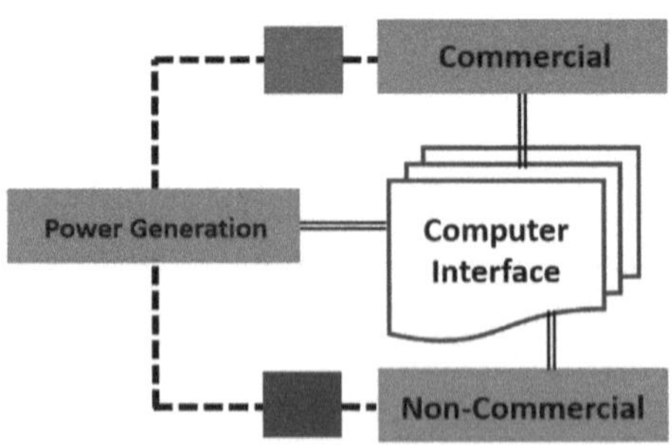

FIGURE 16.8 Blockchain architecture for power management.

2.5 E-voting

Securing the election is of more prominence in any democratic country. Most of the countries are thinking of having possibilities of moving toward online elections to reduce the costs but are in vain. Hence, electronic voting machines are deployed everywhere that are attached to further security devices maintaining high-level security and privacy. This prevents fraudulent techniques that are carried out like stealing ballot boxes and making further modifications or changing the box. This also eliminates the use of a pen—paper mechanism that is carried out in the traditional voting system. This also ensures security over a large group of people alongside the monitoring of physical devices. The administrators in the government create an election notification which is further expedited by editing or verifying the electoral rolls. This also enhances the existing electoral roll thus notifying the number of people living in a particular area. Fig. 16.9 shows the traditional way of having voting done by the public. In this traditional system, the vote is cast by the users using pen and paper. This has many lapses of having fraudulent mechanisms, hence, making the system vulnerable. This is eliminated by the use of blockchain technology in a voting system that uses consensus algorithms for voting which is shown in Fig. 16.10. Based on the e-voting mechanism that is developed using

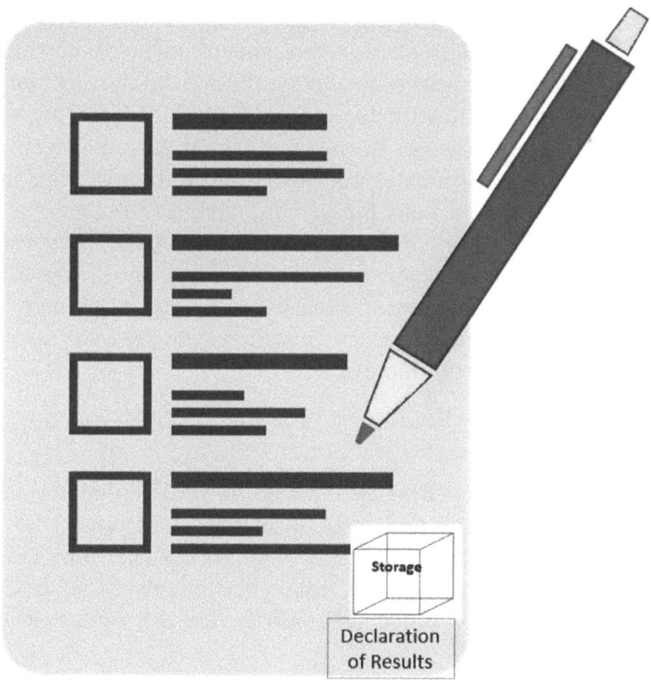

FIGURE 16.9 Traditional voting system.

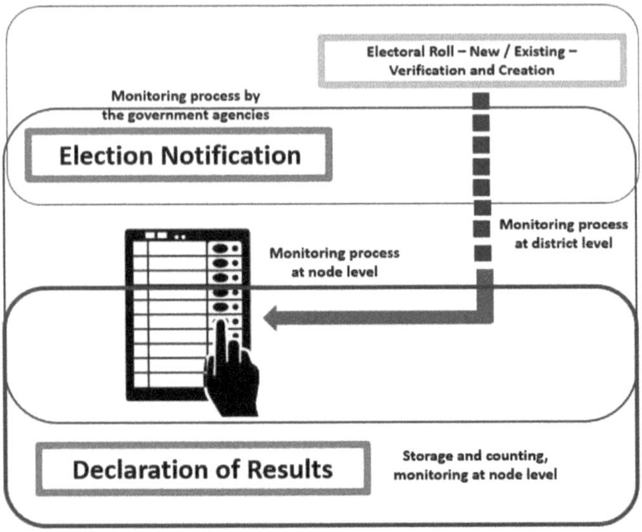

FIGURE 16.10 blockchain-based e-voting system.

blockchain technology, higher level analysis can be performed, which further improves the system. Election can be defined as a smart contract. Such smart contracts are observed to be listed as a part of the election at various places. For each of the user who is a voter corresponds to a location defines the process that includes registration to voting. Modern society is convinced to adopt automated voting infrastructure to ensure public elections affordable, quicker, and easier. It legitimizes the election system for the voters to make it inexpensive and simple, eliminates any power barrier between the elector and the public politician, and puts any pressure on the elect. It opens the door to a more direct democratic structure that allows voters to convey their will on their policies and bills. The blockchain technology presents a new opportunity for democratic countries while growing security measures under today's scheme and gives new options for transparency, to switch from pen and paper elections to expensive and time-consuming elections.

3. Conclusion

With the recent decision on the development of smart cities, the technology has crawled into developing applications for the smart cities that intend to acquire data on the fly, process, analyze, and take a decision. This made a requirement to transfer data at a much faster pace than the existing to make the system smarter. The devices that are used in the day-to-day life are also becoming smarter enough such that they can communicate with humans based on artificial intelligence that is programmed in them for an intentional purpose. Such an intelligence provided to a machine creates hindrances to the human activity until unless it is monitored for any flaws based on the inputs that are provided to it. Hence, it is quite obvious to make a connected record that helps the next level programmers or product manufactures to have modifications that will exploit the problems in the previous levels. Such an analysis provides a time-stamped record which is a series of records that are maintained by a user or by an algorithm. This time-stamped record is also called a blockchain which is distributed between computers and even algorithms for better analysis. This blockchain is also called as a free chain system which is not controlled by anyone. In other words, this is an open blockchain, when placed in a network creates a ledger or have a time-series data of the networked systems, called an open blockchain network. This is an immutable time-series data that are open to everyone for further access and research. Blockchain entirely depends on three parameters which are considered as pillars of blockchain technology. They are transparency, decentralization, and immutability. Applications of blockchain technology in smart cities try to provide solutions to the development of smart cities considering problems at various levels. The solutions that are provided to the consumers of smart cities viz., universal identification cards for citizens, management of land, property, and housing, urban planning, improving public transport, management of energy, water and pollution, rewarding citizens based on their decisions, interoperability of devices, security for internet-connected devices, universal platforms, transparency and interoperability platforms for various departments, keyless signature interface, development of E-commerce interface. All of these are being developed for the applications of blockchain technology as a part of smart city development. Development of identification (Id) cards for citizens using blockchain technology eradicates the issuance of duplicate Id cards based on the parameters issued by the citizens. Management of energy, water, pollution, housing, land, property, and public transport systems collects the data from various sensors that are placed in the various parts of the city. Such systems data will be heterogeneous in nature and need to develop a novel platform for sharing data. This also benefits the interoperability of devices and also provides a way for eCommerce platforms that can increase their productivity and also advertise based on the interest of the citizens. The devices that operate on such heterogeneous data will try to acquire data that are required based on the application deployed for service. As the consensus algorithm that is the main component of the blockchain technology takes care of the data that are being recorded and transferred, while cryptography which is a part of the blockchain takes care of the security.

References

Ahmed, S., Shah, M. A., & Wakil, K. (2020). Blockchain as a trust builder in the smart city domain: A systematic literature review. *IEEE Access, 8*, 92977–92985. https://doi.org/10.1109/ACCESS.2020.2993724.

(n.d.). *Application of blockchain technology in smart city infrastructure*. https://doi.org/10.1109/SmartIoT.2018.00056.

Bhushan, B., Khamparia, A., Sagayam, K. M., Sharma, S. K., Ahad, M. A., & Debnath, N. C. (2020). Blockchain for smart cities: A review of architectures, integration trends and future research directions. *Sustainable Cities and Society, 61*, 102360. https://doi.org/10.1016/j.scs.2020.102360.

Chaudhary, R., Jindal, A., Aujla, G. S., Aggarwal, S., Kumar, N., & Choo, K.-K. R. (2019). Best: Blockchain-based secure energy trading in SDN-enabled intelligent transportation system. *Computers and Security, 85*, 288–299. https://doi.org/10.1016/j.cose.2019.05.006.

Chen, P., Jiang, B., & Wang, C. (n.d.). Blockchain-based payment collection supervision system using pervasive bitcoin digital wallet. In: *2017 IEEE 13th international conference on wireless and mobile computing, networking and communications* (*WiMob*), 139–146. https://doi.org/10.1109/WiMOB.2017.8115844.

Fran, C., Dasaklis, T. K., & Constantinos, P. (2019). A systematic literature review of blockchain-based applications: Current status, classification and open issues. *Telematics and Informatics*, 55–81. https://doi.org/10.1016/j.tele.2018.11.006.

Lai, R., Chuen, D. L. K., & Deng, R. (2018). *Chapter 7 - blockchain – from public to private* (pp. 145–177). Academic Press. https://doi.org/10.1016/B978-0-12-812282-2.00007-3.

Licheng, W., Xiaoying, S., Jing, L., Jun, S., & Yixian, Y. (2019). Cryptographic primitives in blockchains. *Journal of Network and Computer Applications*, 43–58. https://doi.org/10.1016/j.jnca.2018.11.003.

Merlinda, A., Valentin, R., David, F., Simone, A., Dale, G., David, J., et al. (2019). Blockchain technology in the energy sector: A systematic review of challenges and opportunities. *Renewable and Sustainable Energy Reviews*, 143–174. https://doi.org/10.1016/j.rser.2018.10.014.

Nian, L. P., Chuen, D. L. K., & Lee Kuo Chuen, D. (2015). *Chapter 1 - introduction to Bitcoin* (pp. 5–30). Academic Press. https://doi.org/10.1016/B978-0-12-802117-0.00001-1

Roeck, D., Sternberg, H., & Hofmann, E. (2020). Distributed ledger technology in supply chains: A transaction cost perspective. *Null, 58*(7), 2124–2141. https://doi.org/10.1080/00207543.2019.1657247.

Satoshi. (2008). Bitcoin: A peer-to-peer electronic cash system. In *Consulted* (Vol. 1, p. 2012).

Shi, S., He, D., Li, L., Kumar, N., Khan, M. K., & Choo, K.-K. R. (2020). Applications of blockchain in ensuring the security and privacy of electronic health record systems: A survey. *Computers and Security, 97*. https://doi.org/10.1016/j.cose.2020.101966, 101966–101966.

SMS. (n.d.). *A review on consensus algorithm of blockchain*. https://doi.org/10.1109/SMC.2017.8123011.

Svein, Ø., Jolien, U., & Marijn, J. (2017). Blockchain in government: Benefits and implications of distributed ledger technology for information sharing. *Government Information Quarterly*, 355–364. https://doi.org/10.1016/j.giq.2017.09.007.

Taylor, P. J., Dargahi, T., Dehghantanha, A., Parizi, R. M., & Choo, K.-K. R. (2020). A systematic literature review of blockchain cyber security. *Digital Communications and Networks, 6*(2), 147–156. https://doi.org/10.1016/j.dcan.2019.01.005.

Theodorou, S., Sklavos, N., Rawat, D. B., & Ghafoor, K. Z. (2019). *Chapter 3 - blockchain-based security and privacy in smart cities* (pp. 21–37). Elsevier. https://doi.org/10.1016/B978-0-12-815032-0.00003-2.

Xiwei, X., Qinghua, L., Yue, L., Liming, Z., Haonan, Y., & Vasilakos, A. V. (2019). Designing blockchain-based applications a case study for imported product traceability. *Future Generation Computer Systems*, 399–406. https://doi.org/10.1016/j.future.2018.10.010.

Chapter 17

Standards and Protocols of Blockchain

C.M. Naga Sudha[1], K. Gokulakrishnan[2], J. Jesu Vedha Nayahi[2]
[1]*Anna University-MIT Campus, Chrompet, Tamil Nadu, India;* [2]*Anna University-Regional Campus, Tirunelveli, Tamil Nadu, India*

Chapter outline

1. Introduction

Distributed Ledger Technologies (DLTs) form distributed ledger networks with a larger group of nodes without any central authority to record information and reach an agreement. Blockchain is the most remarkable implementation of DLT. A distributed ledger is a type of ledger which can be replicated, synchronized, and shared in a distributed environment. Though

Blockchain for Smart Cities. https://doi.org/10.1016/B978-0-12-824446-3.00001-6

there are several kinds of distributed ledgers in practice, there are certain common characteristics namely direct communication of ledger nodes, node mechanism of the network involved in the addition of transactions to the existing block, computer programs and in managing processes; information about the consensus mechanism in which the existing network validates the agreed-upon instructions on the new block. (Sporny, 2017) An important feature that makes blockchain more prominent in the DLT solutions is the data storage which is provided as blocks. These blocks are linked to the previous blocks cryptographically and form a chain of data that will be ever-growing. The only difference is that without the central authority on data storage, the ledger is shared or distributed among the nodes which always keep a copy of the information on every node. Nodes added on the network must agree on the existing chain of blocks, and also with the new valid blocks when added (Alessandro, 2013; Androulaki et al., 2018; Technical Specification, 2019).

The main concept behind blockchain is "Trust None." With these magical words, security threats can be eliminated to a greater extent. Thus, an initiative to integrate blockchain with smart cities has started in countries such as United States and Singapore. Blockchain network emerged from the financial sectors in a decentralized form which will help in transactional data ordering. Blockchain is defined as a distributed ledger or shared records in which multiple transactions are shared between multiple parties in an efficient and verifiable way (Aristidou & Marcou, 2019). When new technological development is progressed into the implementation phase, there is a possibility of conflict among various vendors in view of similar functionalities. Therefore, there is a need of international standards in every sector. Hence, International Standards Development Organizations has been formed to maintain blockchain standards. In this chapter, categories of standards are to be explained in detail. In general, technically developed standards are promoted into international standards based on the usage among the vendors.

DLT components comprised of the user, user groups, DLT service provider, and DLT node. All these components are dedicated to separate organizations. Components of DLT are described in Fig. 17.1 (Ashiq, Manu, & Alan, 2017). Every node is considered as an individual system in a distributed ledger. Some nodes are defined as "full nodes" which stores ledger data and passes to other nodes to ensure that new blocks are valid. Service providers are another important component that provides DLT-based service to other parties using the service interfaces. A user is considered as one of the main components who can use the services or utilize the service output from other components. These components can act as both the provider and consumer of other services. The user group can be defined as a group of people and organizations which is a subset of DLT system users. The distributed ledger consists of digital information which is validated through consensus mechanisms and stored in different nodes through replication (Ashiq et al., 2017).

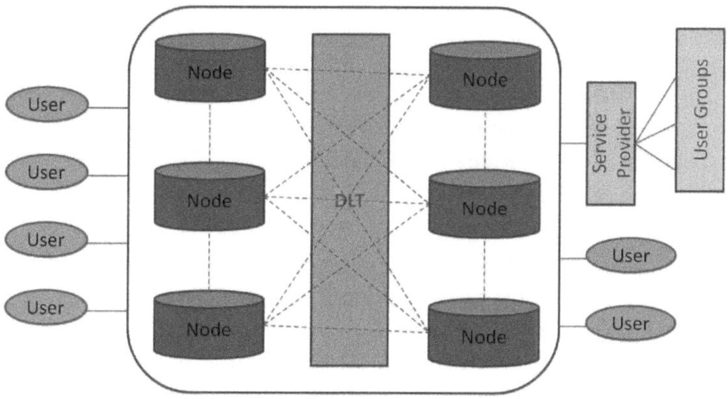

FIGURE 17.1 Components of distributed ledger technology.

1.1 Key Characteristics of DLT

Key characteristics of DLT are as follows: Shared—Multiple nodes in the network have the entire state of ledger being shared either partially or entirely. It increases the efficiency and transparency among all the nodes of the DLT network.

Distributed—As nodes are distributed, scalability of nodes can be performed easily. With the increase in number of nodes, possibility of bad actor interrupt in the consensus protocol is reduced. Therefore, DLT has made more resilient to bad actors attacks.

Append only—An append only ledger helps in providing full transaction history. The advantage of this technique is that it cannot be overwritten as like the traditional databases.

Immutable—Tamper-proof and attestable data are maintained in the network, as it ensures that the ledgers have immutability property.

There are three types of DLT namely permissionless distributed ledger systems, permissioned distributed ledger system, and hybrid distributed ledger system (Brilliantova & Thurner, 2019).

Permissionless distributed ledger allows everyone who is validating blocks. It does not require permission from authority. Permissioned distributed ledger system needs permission through which the validating blocks are authorized. These systems can be developed using open-source software which can be downloaded by anyone.

Permissioned distributed ledger requires permission where the users who are validating the blocks must be authorized. As the authorized users can only have access to the blocks, restrictions on read access, and issue transactions can be given to the users.

Hybrid distributed ledger systems help in maintaining privacy and security with the help of permissioned distributed ledger system and shares transparency property from permissionless distributed ledger system. It has the significant flexibility option by which data can be made public or private.

1.2 Ecosystem of DLT

Ecosystem is defined as collection of stakeholders which includes users and organizations in combination with other entities of different roles. Hardware aspect, protocol development aspect, business aspect, and software development aspect are the four kinds of DLT ecosystems.

1.2.1 Hardware Aspects

Hardware aspects of the DLT ecosystem have numerous nodes which include a server, computer, or storage device. Node modalities are categorized according to the functionalities of nodes such as block producing validating nodes, nonblock producing full validating nodes, and partial or light nodes. A block that produces full validating node participants in consensus protocol can contain a complete replication of distributed ledger which includes each and every transaction executed from the inception stage. A nonblock which produces full validating nodes does not participate in the consensus process. But, it has the entire replication of distributed ledger from the inception stage. A partial or light node consists of partial transaction which is connected to a full network in some way, in order to validate whether the data are useful and accurate (Smarter, 2018).

Permissionless DLT allows anyone to create a new node with adequate storage capacity and processing power. If there is more number of nodes available in the network, distribution capacity among various stakeholders is likely to be more. It will result in lower risk of fraud, system failure, or error. Proper working of DLT system requires reliable, stable, and sufficient supply of electricity. Energy consumption on DLT networks uses PoW consensus mechanisms which are numerically high and must consider energy sourcing. Also, DLT users store public and private keys along with their passwords in a physical wallet which also becomes a part on hardware ecosystem (Flood & McCullagh, 2020).

1.2.2 Business Aspects

Business aspects of DLT ecosystem comprises of users, investors, corporates, developers, and block producers. DLT users are considered as entities who are engaged in DLT using an application, service, or product for accomplishing a particular task such as transferring of asset. Organizations or people who provide capital for creating the DLT are considered as investors. These people are encouraged through the profit which is value-based and mission-based that helps in solving economic and social issues. Full validating nodes or block producers are actively participating in DLT network consensus mechanism. For Proof-of-Work (PoW) based DLT systems, block producers are considered as miners who are intended to get profit on validating transactions. In the Bitcoin network, initially miners try to solve computationally intensive puzzles which are highly complex. The reward will be provided in the form of bit coins (Gonzalez, Garca-Daz, Nunez-Valdez, & Crespo, 2020).

Corporates can utilize DLT for their business activities through which new technologies are pushed to end users or large number of customers. They are able to create a new space where end-users can interact with stakeholders related to money and can be able to make their time more productive. Corporates can be benefitted from data integrity and security on the backend. However, DLT is used in compliance which is applicable with regulations and legislation. Developers are those who build products, services, or applications which utilize DLT in networks and protocols. They help in developing distributed applications and assist in technical manner (Gramoli & Staples, 2018).

1.2.3 Software Aspects

DLT supports various software applications which are termed as software ecosystem. It is defined as a group of entities which can interact with shared market to access the software and services. These applications are developed in various languages namely Python, Java, Rust, Solidity, and JavaScript and many more. Languages are compared with the strengths and weakness and are chosen based on the applications that are to be developed. In basic, DLT should be language neutral which can fully compliant with requirements as well as software specifications. Applications are broadly categorized into three aspects namely nonfinancial, financial, semifinancial, and financial. Financial applications are related to more money-related businesses. Semifinancial applications also depend on money but the major focus will be on completion of tasks or smart contract execution. Nonfinancial applications are open-ended which involves ballot system, ID authentication, and data record storage (Huh & Kim, 2019).

1.2.4 Protocol Development Aspect

Protocol development aspect includes the developers and academicians. Developers are more involved in designing the DLT protocols which can model the networks. The protocol layer focuses mainly on the cryptographic keys which can interact with the networks more easily. Two kinds of protocols are categorized as open-ended and closed-source. Literally, as the name suggests, protocols can be downloaded by anyone and it is auditable through open-source communities. Closed-source networks can be accessed only by private entities and specific class operations. Developers can be able to encode and alter private information before providing the access to the aggregator. Academicians are involved in educating others in the implementation of DLT systems along with the limitations. In combination with the private companies and open-source communities, new DLT software is released (Hyland-Wood and Khatchadourian, 2018; Lacity, 2018).

1.3 Consensus Mechanism

Consensus mechanisms are defined as procedures for validation of nodes on a distributed ledger. A main criterion to be considered in DLT is to determine which user can validate the upcoming block. It can be achieved through the implementation of one of the possible consensus mechanisms. It provides the convergence ability toward a single, immutable version of the ledger. This allows the actors within the network to agree with the recorded information available in the distributed ledgers, with the consideration that some actors might be malicious. As there is no authority control over permissionless distributed ledger networks, many validating nodes participate in process of validation of next block. This process is carried out to generate new crypto-currency or network transaction fees. Therefore, validating process comprises of users who are mutually distributed and know each other only by public addresses. Three kinds of consensus mechanisms are namely Proof of Stake, Proof of Work, and Byzantine Fault Tolerant–based mechanism as shown in Fig. 17.2 (Lima, 2018).

Proof of Work (PoW)—A node which gets authority to validate the next block with their work on solving the puzzle which is computationally intensive. The solution of the puzzle act as the "proof" for the work done. Validation processes are highly in need of instantaneous power allotted for completion of the particular task. On completion of task, reward will be given to the node or miner in form of transaction fees or crypto assets.

Proof of Stake (PoS)—It is defined as a consensus process in which amount of stored value known as stake is held within the distributed ledger system. It is used to achieve consensus without any energy-intensive

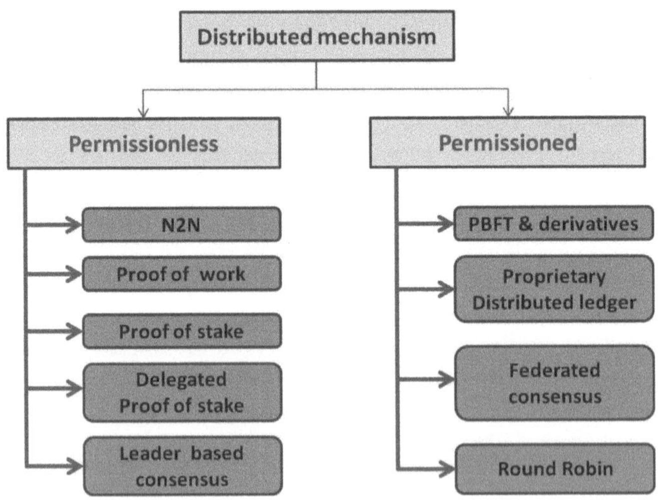

FIGURE 17.2 Explanation on distributed mechanism.

computations. Moreover, it is based on the idea that instead of investing directly into the system, users want the system to succeed rather than subvert the system. For instance, stake is referred to as crypto asset in which DLT has invested. Nodes which are involved in PoS consensus mechanisms are provided with the transaction fees collected from each block and they initiate the successful validation process. A group of nodes are chosen selectively which can function as a block to produce numerous validating nodes on the network. This approach is named as Delegated Proof of Stake (DPoS) which is an alternate approach to PoS.

Byzantine Fault Tolerant (BFT)—In DLT systems, faults can occur in the network due to the abnormal behavior of the nodes. These BFT algorithms are designed in such a way that the distributed ledger systems must ensure that the functions are normal even when abnormal nodes are involved in network. According to BFT-based consensus, each node participates in consensus process which is involved in voting and communication. These processes are performed on the block to reach consensus. It has the advantage that it is compatible with the small systems which has limited resources on nodes. In addition, BFT requires all participants to agree on the list of the participants. Therefore, this kind of protocol is used only under permissioned distributed ledger system.

Consensus algorithms have made DLT to be applied for decentralization and automation processes in various sectors. Attributes of DLT allow various nodes or entities which might be collaborator or competitors which process consensus mechanism on information and supports immutability. Use cases of DLT are in vast manner which includes industries automotive, banking, e-governance, education, healthcare, law, art, music, travel, and real estate. Comparison of main consensus protocols are shown in Table 17.1 (Monrat et al., 2019).

Smart contract is defined as the computer program which is applied on transactions in distributed ledger network which are cryptographically signed. Hyperledger Fabric's chain code, Ethereum smart contract are the examples of distributed ledger network. Smart contracts are executed by nodes in

TABLE 17.1 Comparison of consensus protocols.

Property	PoW	PoS	DPoS	BFT
Application	Public	Public	Public	Permissioned
Category	Probabilistic	Probabilistic	Probabilistic	Absolute
Power consumption	High	Low	Low	Not applicable
Fault tolerance	Medium	Medium	Medium	Low
Scalability	Excellent	Excellent	Excellent	Poor

distributed ledger system. These execution results are validated and recorded through consensus algorithm in the distributed ledger. Automation through smart contract reduces risks of errors, reduces cost, and mitigates fraud risks which can streamline business process potentially. A programmable blockchain and associated standard tool known as smart contract is needed for managing, debugging, monitoring, and developing smart contract systems. Comparison on performance of blockchain based on their types such as Parity Ethereum, and Hyperledger is carried out by National University of Singapore. As a result, a publicly available tool called Block Bench, which can measure performance like throughput, latency, scalability, and fault tolerance at different layers of blockchain stack is proposed. Hyperledger model which describes the entities and relationships is shown in Fig. 17.3 (Merlinda et al., 2019).

2. Standards of Blockchain

As blockchain applications are growing rapidly, there is an urgent need for the standardization of technology for optimization of interoperability and usefulness. Development of standards depends on simplifying the field by overlapping multiple suppliers in particular which promotes the markets for

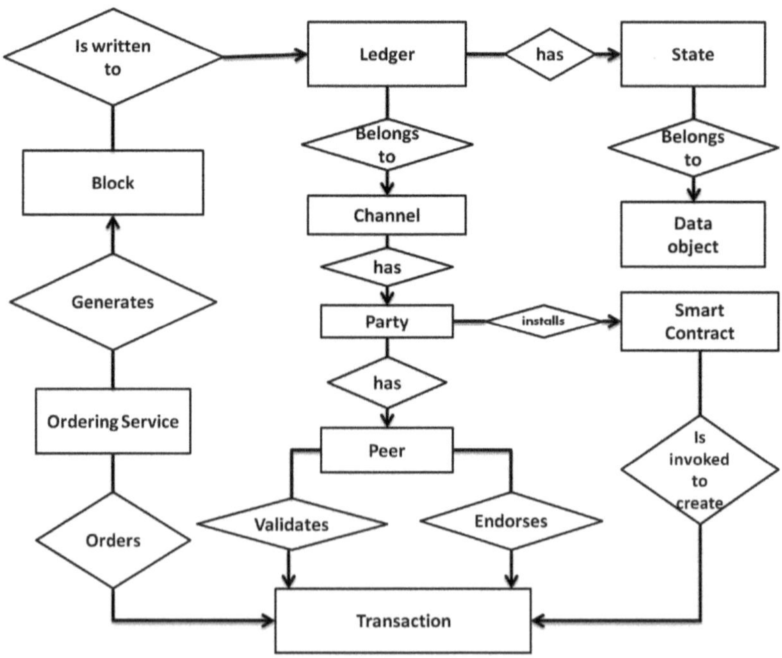

FIGURE 17.3 Hyper ledger model.

blockchain. A crucial effort is needed for the two-way feedback and communication between user and organizations. Targets which are considered for further standardization include consensus algorithms such as PoW, hash graph, and stellar consensus. Basic data models for Blockchain includes events, state machines, and blocks. Storage algorithms such as linked lists, merkle trees, signature algorithms such as chain point, hierarchically deterministic keys, Javascript Oriented Signing and Encryption (JOSE), linked data signatures (Aristidou and Marcou, 2019).

The World Wide Web Consortium (W3C) organized a workshop on June 2016 for examining the blockchain aspects which relates the web technologies and helps in identifying specific technologies which are considered for standardization. A new group was formed to address issues which arise on Credentials, Digital Verification, Blockchain Community, Verifiable Claims, and Interledger Community. W3C has a Web Ledger Protocol initiated as a work in incubation at W3C. The International Organization for Standardization (ISO) has organized a Technical Committee (TC) 307 for blockchain and DLT along with several ISO committees in combination with standard developing organizations. It was initially proposed by Standards Australia, which has framed an outline for blockchain standards on March 2017.

Newer version of TC 307 has formed study groups (SG) on taxonomy, reference architecture, ontology, security and privacy, smart contracts, identity, and use cases. These have deferred the challenging work in establishing the standards for auditing, governance of such technology impacts. Standardization sector of International Telecommunications Unionist-T (ITU-T) has formed focus group on DLT which will focus more on developing the outline of interoperable DLT-based services. Other standard organizations are actively participating for the promotion of the DLT standards which includes IEEE as a blockchain member interest group for the coordination along with the dissemination of information in various activities. Society for World Wide Interbank Financial Telecommunication (SWIFT) has also stepped into the area for collaborating with several industry-related organizations (Nakamoto, 2008).

2.1 International Telecommunication Union—Telecommunication Standardization Sector

ITU is a United Nations–specialized agency which have the main contribution in International Communication and Information Technologies (ICTs). ITU-T role is more important in analyzing technical, tariff, and operating areas and to make recommendations with a standard view on worldwide basis. ITU-T has classified the work items into three categories namely, Work for security for DLT, Work for security by DLT, and Security management.

2.2 ISO

ISO TC 307, Blockchain and DLT were established on September 2016. It has comprised of six working groups (WG) such as WG 1, WG 2, WG 3, WG 4, WG 5, WG 6, and one study group (SG) named as SG 7 which focuses on foundations, security and privacy, applications of smart contracts, blockchain and IT security techniques, governance, use cases, and interoperability of DLT, respectively.

2.3 IEEE standards association

DLT-related work items which are under development of IEEE working groups are tabulated as shown in Table 17.2.

2.4 W3C

Working groups and community groups have published draft documents namely, Verifiable Claims Working Group, Blockchain Community Group, and Credentials Community Group.

TABLE 17.2 IEEE working groups.

DLT-WG	Name of work item
CEWG	Crypto currency Exchange Working Group
BACWG	Blockchain Against Corruption Working Group
EIBCTWG	E-Invoice Business Using Blockchain Technology Working Group
CPWG	Crypto currency Payment Working Group
TIDMWG	Trusted IoT Data Management Working Group
DTLA	DLT in Agriculture
DLTCAV	DLT in Connected and Autonomous Vehicles
BDLTH WG	Blockchain and DLT in Health
BSCF_WG	Blockchain in Supply Chain Finance Working Group
BGAWG	Blockchain for Government Affairs Working Group
CBSTWG	Crypto currency Based Security Tokens Working Group
DAWG	Digital Asset Working Group

Technical specification FG DLT D1.1 focus group on application of distributed ledger technology, International Telecommunication Union August, 2019.

2.5 United Nations Center for Trade Facilitation and Electronic Business (UN/CEFACT)

UN/CEFACT which is working under the UN Economic Commission for Europe (UNECE) has initiated the regulatory and policy frameworks and international standards for the emerging technologies, such as blockchain. UN/CEFACT has published white papers on two criteria namely the introduction of blockchain technology particularly for trade policymakers, and research gaps of technical aspects of blockchain in relation to UN/CEFACT deliverables. Also, an individual paper which exclusively focuses on use cases specific to the challenges faced in government services, finance, and healthcare which are in the initial stage and has to be explored more to reach Sustainable Development Goals (Nakamoto, 2008).

2.6 Community Standards

DLT standards have been initiated more actively by community and industry organizations. Implemented work is different from that of formal standards developing organizations which have published specifications along with repository of open-source implementation code. One of the examples of community-driven organization which supports a range of cross-industry blockchain applications is Linux Foundation. Hyper ledger, which is hosted by Linux Foundation has about 200 companies which are focusing on the development as well as the implementation of enterprise blockchain. One of the works of Hyperledger is considered as a defacto standard for enterprise blockchain platforms. Ethereum platform is continually maintained by Ethereum community through the process known as Ethereum Improvement Proposals (EIP). It includes protocol specifications, interface/client APIs specifications, and application-level standards which are known as Ethereum Request for Comment (ERC) standards. Most important standard of ERC-20 is used in smart contracts on Ethereum blockchain which is accessed for token implementation. As there is cross-industry applicability of DLT standards, there is a need for industry-specific standards development. Blockchain in Transport Alliance (BiTA) was established in August 2017, which is working to satisfy such needs. It has focused the standardization efforts with respect to the transportation industry (Nyaletey, Parizi, Zhang, & Choo, 2019).

2.7 Other Standards

European Telecommunication Standard Institute (ETSI) has initiated an Industry Specification Group (ISG) on Permissioned Distributed Ledger (PDL) which comprises of three work items namely landscape of standards and technologies, applicability, and compliance to data processing requirements for connected machines and application scenarios. CEN/CENELEC is a

European organization which has formed a Focus Group on Blockchain and DLT in 2017. Focus Group has identified the research gaps in blockchain and DLT standardization in Europe and mapped accordingly to the ongoing work items. National Institute on Science and Technology (NIST) has been initiated under the US Department of Commerce. It has set up the Blockchain for Industrial Application Community of Interest (BIA COI) in 2017. The main goal is to provide guidelines which will enhance the synergy among all the stakeholders in blockchain ecosystem. It will reduce the complexity, delay, and cost of adoption on blockchain technologies. German Institute of Standardization, is contributing to DLT-related standardization which is a participating member of ISO/TC 307. Spanish Association for Standardization, UNE has established the work item named "Modelo Decentralized deIdentidad" which is a DLT-based model for decentralized identities(DIDs). It is used by various pilots who travel within European Self-Sovereign Identity Framework (ESSIF) (Ashiq, Manu, & Alan, 2017; Raikwar, Gligoroski, & Kralevska, 2019).

Current DLT work items include security architecture, security capabilities and threats, privacy and security considerations, identity management, security assurance, and security services. In addition to the new standards, effective work on creation and development of cloud standards has to be accomplished through the industry, community, and trade organizations. These are distinguished from the formal standards organizations which are accompanied through the open-source code. Many organizations have emerged for pursuing the goals on DLT which are independent from Bitcoin community. Cloud Standards Customer Council has summarized the existing requirement from the business view and has offered reference architecture for further standardization.

2.8 Blockchain Standards for Compliance and Trust

When RFID tags are used to enable tracking of items that are produced at large volumes, no one can use the same registry tag which can be assigned to another part. The introduction of other labels can be made intentionally for duplication of information. A debate is also in progress where the distributed ledgers could replace labels and physical tags, but there is an intersection on both technologies. Machine-readable labels are definitely needed in some situations. In addition to the digital signature, the incorporation of signature in the blockchain is easily checked for its uniqueness and authentication. Duplications are eliminated with this method. It can be applied to any kind of record management for storing tamper-proof business records which needs authentication. Such transactions are very small which are not suitable for affixing human or machine-readable labels like bar codes and RFID. These labels can themselves permit a record of their own data like temperature, shock protection during the record handling on shipment (Sayeed & Marco-Gisbert, 2019).

Blockchain applications are readily available for verification of associated records with shipment parts or food at low costs. Advertising verification methods can add trust and compliance for the business models which can ensure the added value. Blockchain methods are applied directly on advertising impressions, in addition to billing verification. Thus, spoofed domains, fake views, and other mechanisms are avoided. The registry file named adChain explains the decentrally owned domain which has been launched in collaboration with industry groups. Ethereum blockchains smart contract helps in storing the domain names which are accredited as fraudulentless mechanism.

2.9 Blockchain Standards and Government

A distributed ledger that is encrypted using different keys, immutable and democratic are the four principles upon which the government steps into the investment on blockchain. With these principles, blockchain helps to provide less expensive solutions to solve the problem. Traditional databases will assist in fulfilling the requirement, if tamper evidence is required. A unique elegant solution will be provided through blockchain technology for trust problems. Consensus requirement is important in the context of government usage on blockchain-based systems, which is unlikely to be acceptable to the United States government, whereas none of the entities in sovereign government using the blockchain with regard to their peers with any degree of trust. A basic element of the social contract is that the government should be trusted in its conduct of any transaction. The Temporary Worker Visa Program in the United States is one of the proposed initiatives of the use of blockchain technologies. Under the H2 temporary visa program, thousands of workers come to the United States from various fields. In order to process this technique, interested employers should submit the temporary labor certification application to the Department of Labor (DOL). Followed by that, the employer submits the US Citizenship & Immigration Services (USCIS) after certified by DOL. The worker must apply for a visa at a US consultate after approval by USCIS. One of the blockchain programs is a pilot that explores so many ways and making it more efficient for employers. To track and manage the physical assets, a pilot prototype will be launched by the Department of Treasury, Bureau of the Fiscal Service's Office of Financial Innovation and Transformation (FIT). The inventory of the agency's physical assets can be continuously monitored and reconciled by using the pilot project. With the aim to build real-time applications in 2018, the Centers for Disease Control (CDC), Center for Surveillance, Epidemiology, and Laboratory Services are pursuing several blockchain-based Proof-of-Concept (PoC) activities. The establishment of better health surveillance will be the most focusing work of PoC, which results in efficient use of blockchain technology in managing data during a public health crisis or related incident. General Services

Administration (GSA) is currently working with machine learning and artificial intelligence, which is also looking for blockchain to reduce "the amount of human interaction required to review new proposal documents to improve offer or experience." To understand the technology along with its features and to identify areas of potential interest for the Postal Service, a consulting firm with interest and expertise in blockchain technology in the US Postal Service and Office of Inspector General (OIG) has contracted with Swiss Economics.

2.9.1 Future research in blockchain

To deliver specific and tangible benefits, DLT appears in many areas of government. To encompass the different ways of doing business, good candidates with thoughtful and holistic consideration is required and the need to amend legislation or rethink public policies should be included which is actually made possible by a blockchain-enabled solution. For a tamperproof, redundant, and transparent ledger, any type of government records will have an obvious use. To recognize the digital record which was tied to the blockchain, there might be some difficulties exist in including legally binding document and to enable the use case legislative changes. Security depends upon the strength of the cryptographic algorithm. Cryptosystems become weakened since computers become powerful. There could be a chance of leaks or rise of modifications to underlying data if the selected DLT is cryptographically vulnerable. To conduct organizations' activities, each received a budget and is charged with the mission within the Federal government. At every step of the process, a permissioned blockchain should enable the budget to be allocated. From allocation through final spending, for a total audit of fund transfer, a system would be included. All transfers of funds in an organization should be recorded. Only authorized users should see complete details of transactions because sensitive expenditures could have been encrypted by certain masking aspects. The time required to obtain the necessary approvals for some activities is a perennial complaint. To streamline some of the processes, capabilities (called "smart contracts") have been built on DLT. Applications which exist as an integral part of the distributed ledger is known as smart contracts. With the help of some stimulus, a smart contract has worked on some set of automated functions. To enforce conditions or actions within a set of defined parameters, smart contracts might be used in the potential "streamlining" use case. Timeliness, cost comparison, and so many other factors are those defined parameters.

2.9.2 Recommendations

In order to track the development of the international systems and to increase the technical understanding of blockchain within government, there is a need for familiarity in developing decentralized and distributed paradigms of DLT, blockchain, and Directed Acyclic Graph (DAG). Internal USG blockchain expert workforce is necessary to build pilot projects. Collaborations with the institutions such as World Economic Forum's Center for Industry 4.0, Hyperledger Project, Blockchain Research Institute, Enterprise Ethereum Alliance, Digital Commerce Chamber, MIT Media Lab, and Hyperledger Project are recommended. From foreign-built cryptocurrencies like Petrocoin and theoretical Crypto-Ruble, the government awareness of malign cryptofinancial activity has increased knowledge on blockchain-based national security threats. Permissionless DLT Consortium Research can be encouraged to have partnership with private industry. The privacy and legality implications regarding the "right to be forgotten" and accidental private key destruction should be considered and studied. For many applications there is some other way to accomplish the same task in view with skepticism claimed as "successful uses" of blockchain.

2.9.3 Future Scope

Currently, blockchain methods have limitations and gaps in many areas related to scalability, governance, and flexibility. Architectural choices have been made for products that need data integrity and security than scalability and flexibility. Bitcoin-based systems can only initiate process transactions lesser than seven. This drawback has led to the specialized platforms which have increased to 70 so far which can address specific sector problems and application domains. Therefore, standardization activity can enable technologies which can make the property of interoperability to work efficiently. Improvements that can address identity, confidentiality, and collaboration among the network participants are expected to be implemented in the future.

3. Protocols of Blockchain

Blockchain protocols are defined as a pair of algorithms (Π, C) in which Π denotes stateful algorithm. It can receive the security factor κ as inputs and maintains in local state. Algorithm C(κ, state) has outputs in series chain known as "records" or "batches." It is parameterized with a validity predicated which has encapsulated semantic properties in which blockchain application can achieve. Protocols are defined as set of rules by which the blockchain works for processing transactions. Thus, blockchain exhibit several properties as follows (Zhang & Lee, 2020)

- Consistency
- g-chain growth
- Future self-consistence
- μ-chain quality

Nakamoto performed an initial analysis for the consistency which is assumed that the third party can mount a particular attack strategy that an attacker tries for chain generation faster than that of honest players. For instance, such analysis cannot be considered for sophisticated attack strategies where the third party might attempt to "split the players" and include them in different chains. When synchronous network is considered, no matter how strong it is, which is possibly unrealistic assumption, Nakamoto protocol is designed to work in network with message delays and executed in such a network.

Proof of Network Delay consensus works even in the setting of partial synchrony where there is an apriori bound Δ on the network latency, which continues as the mining hardness parameter ρ exceeds in calculation of $1/\rho * n * \Delta$ where ρ is defined as the fraction of computational power consumed by the third party and n is defined as the number of players. Decker and Wattenhofer have experimentally defined that by increasing the network delays in Nakamoto protocol results in increased forks, by which it is concluded that attacker use these delays and thereby consistency is violated with an attack which requires less than 50% of mining power. Robustness is achieved through the majority of the honest participants in which major computing power is provided by them. Consistency properties are achieved for a financial transaction through Bitcoin digital currency which needs stronger properties for fraud prevention and for preventing double-spending attacks.

3.1 Layers of Blockchain Protocols

Four different layers of blockchain protocols are described for ensuring the transactions which is shown in Fig. 17.4 (Plasma, 2017).

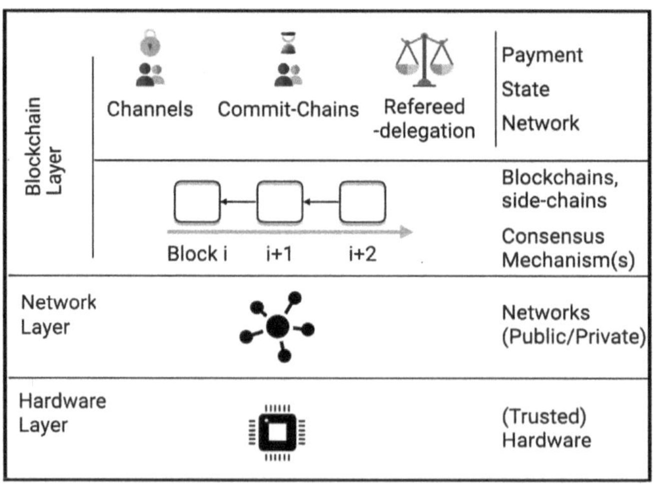

FIGURE 17.4 Layers of blockchain protocols.

- **Hardware Layer:** Trusted Execution Environments (TEE) has substituted the blockchain clock with the alternate option on trusted hardware assumption, which can enable the protocols efficiently in other such off chain layers which helps in removing the backward compatibility and dispute processes. TEE can execute security-critical applications or sensitive code, tamper-resistant codes provided through operating system or any other licensed software.
- **Network Layer**: Network layer or layer-zero is typically a peer-to-peer (P2P) layer in which asynchronous exchange of information is carried out in blockchain nodes. Network layer is very important for the scalability, security, and privacy of the blockchain. An efficient layer-zero can enable higher level transaction which can achieve throughput and resilience. Blockchain miners are those who write the blockchain transactions which are connected via dedicated miner P2P networks. It consists of a complete network stack on traditional network architecture. It is very much essential for the communication between two participants of blockchain.
- **Blockchain Layer:** Blockchain layer is defined as the layer-one hosts which plays an immutable append-only chain of blocks which helps in accumulation of transactions from network parties. It is much useful in public verifiability. Each transaction can be able to encode the updation of the blockchain state. It can exchange the digital assets between the parties and invoke the application such as smart contracts. Integrity of blockchain is ensured with the consensus algorithm which is executed by the participants.

3.2 Application of Blockchain Protocols in Email

Nowadays, email has become the major communication channel for the organizations. But, spam and phishing attacks are being increased day-by-day. To overcome this issue, email services have built their own email filters to protect the users from the attacks. These attacks are possibly occurred because of the lack of authentication. Thus, a blockchain protocol using smart contract has been implemented over Ethernet protocol. It is based on Proof of Concept mechanism as shown in Fig. 17.5 (González, García-Díaz, & Núñez-Valdez, 2020).

Functionalities of smart contract in email protocol are as follows:

- "Send" functions approve the sender for the initialization of transaction, otherwise it gets reverted. Sender of the message is stored as message sender.
- "Get Index Sent" allows checking the identifiers of email which are sent to yourself.

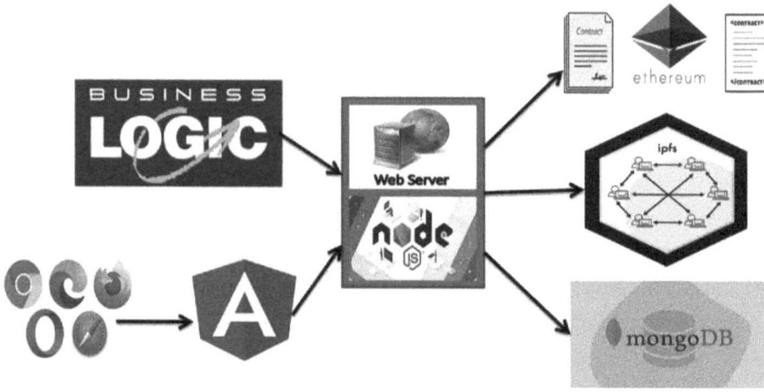

FIGURE 17.5 Blockchain mechanism in email protocols.

- "Get Message" helps in reading the message and to retrieve the message from blockchain.
- "Approve Sender" helps in approving the user to send new messages. If user has already approved, the transaction gets roll back.
- "Revoke Sender" helps in revoking the user to send messages. Only the previous users are able to revoke or else the execution will get rollback.

4. Conclusion

Nowadays, blockchain is the most interesting technology which is more focused. Initially, blockchain applications are more popular in the financial sectors. But, in recent days, it has stepped into various sectors such as healthcare, supply chain management, vehicle tracking, and many more. Blockchain technology has the main contribution toward the DLT. Therefore, basic components of DLT, key characteristics of DLT, and types of DLT are explained in Section 1. Also, ecosystem under which DLT is divided are discussed in Section 1.4. Consensus mechanism which is the basic working principle of blockchain is explained in Section 2 and a comparison of the main consensus mechanisms is tabulated. Various standards which are categorized based on the use cases are explained in Section 3. Other standards which are not particularly categorized are explained in Sections 2.6 and 2.7. As the main attribute of blockchain is trust and compliance, it is explained in detail in Section 2.8. Finally, the future scope of the blockchain standards is depicted in Section 2.9. Blockchain protocols which are applied layer wise are explained with an email application in Section 3.1 and Section 3.2. Blockchain has started its own revolution worldwide with prime advantages on edge of security and privacy.

References

Alessandro, M. (2013). The EU proposal for a general data protection regulation and the roots of the 'right to be forgotten. *Computer Law and Security Report*, 229–235. https://doi.org/10.1016/j.clsr.2013.03.010.

Androulaki, E., et al. (2018). Hyper ledger fabric: A distributed operating system for permissioned blockchain. In *Proceedings of the Thirteenth EuroSys conference* (p. 30).

Aristidou, C., & Marcou, E. (2019). Blockchain standards and government applications. *Journal of ICT Standardization, 7*(3), 287–312. https://doi.org/10.13052/jicts2245-800X.736.

Ashiq, A., Manu, S., & Alan, S. (2017). Blockchain standards for compliance and trust. *IEEE Cloud Computing*, 84–90. https://doi.org/10.1109/MCC.2017.3791019.

Brilliantova, V., & Thurner, T. W. (2019). Blockchain and the future of energy. *Technology in Society, 57*, 38–45. https://doi.org/10.1016/j.techsoc.2018.11.001.

Flood, & McCullagh, A. (2020). Blockchain's future: Can the decentralized blockchain community succeed in creating standards? *The Knowledge Engineering Review, 35*. https://doi.org/10.1017/S0269888920000016.

Get Smarter. (June 15, 2018). *What stakeholders are involved in the blockchain strategy system?*. https://www.getsmarter.com/blog/career-advice/what-stakeholders-are-involved-in-the-blockchain-strategy-system/.

Gonzalez, J. C., Garca-Daz, V., Nunez-Valdez, E. R., & Crespo, A. G. R. G. (2020). Replacing email protocols with blockchain-based smart contracts. *In Cluster Computing*. https://doi.org/10.1007/s10586-020-031289.

Gramoli, V., & Staples, M. (2018). Blockchain standard: Can we reach consensus? *IEEE Communications Standards Magazine, 2*(3), 16–21. https://doi.org/10.1109/MCOMSTD.2018.1800022.

Huh, J.-H., & Kim, S.-K. (2019). The blockchain consensus algorithm for viable management of new and renewable energies. *Sustainability, 11*, 3184.

Hyland-Wood, D., & Khatchadourian, S. (July 2018). A future history of international blockchain standards. *The JBBA, 1*(1).

Lacity, M. C. (2018). Addressing key challenges to making enterprise blockchain applications a reality. *MIS Quarterly Executive, 17*(3), 201–222. http://misqe.org/ojs2/index.php/misqe/article/viewFile/853/498.

Lima, C. (2018). Developing open and interoperable DLT/blockchain standards [standards]. *Computer, 51*(11), 106–111. https://doi.org/10.1109/MC.2018.2876184.

Merlinda, A., Valentin, R., David, F., Simone, A., Dale, G., David, J., et al. (2019). Blockchain technology in the energy sector: A systematic review of challenges and opportunities. *Renewable and Sustainable Energy Reviews*, 143–174. https://doi.org/10.1016/j.rser.2018.10.014.

Monrat, A. A., Schelén, O., & Andersson, K. (2019). A survey of blockchain from the perspectives of applications, challenges, and opportunities. *IEEE Access, 7*, 117134–117151. https://doi.org/10.1109/ACCESS.2019.2936094.

Nakamoto, S. (2008). *Bit coin: A peer-to-peer electronic cash system*. http://www.bitocin.org.

Nyaletey, E., Parizi, R. M., Zhang, Q., & Choo, K.-K. R. C. (2019). Block IPFS - blockchain-enabled interplanetary file system for forensic and trusted data traceability. In *Proceedings of 2nd IEEE international conference on blockchain, IEEE blockchain-2019*.

Plasma. (2017). *In plasma: Scalable autonomous smart contracts* (pp. 1–47).

Raikwar, M., Gligoroski, D., & Kralevska, K. (2019). SoK of used Cryptography in blockchain. *IEEE Access, 7*, 148550–148575. https://doi.org/10.1109/ACCESS.2019.2946983.

Sayeed, S., & Marco-Gisbert, H. (2019). Assessing blockchain consensus and security mechanisms against the 51% attack. *Applied Sciences, 9*(9). https://doi.org/10.3390/app9091788.

Sporny, M. (Apr. 3–7 April, 2017). Building better blockchain: Linked data in distributed ledgers. *Proc. 26th Int'l Conf. World Wide Web Companion*, 1429.

Technical Specification. (August, 2019). *Technical specification FG DLT D1.1 focus group on application of distributed ledger technology.* International Telecommunication Union.

Zhang, S., & Lee, J. H. (2020). Analysis of the main consensus protocols of blockchain. *ICT Express, 6*(2), 93–97. https://doi.org/10.1016/j.icte.2019.08.001.

Index

Note: 'Page numbers followed by "f" indicate figures and "t" indicate tables.'

C

D

J

K

L

M

N

O

P

Q

R

S

Z